LONDON MATHEMATICAL SOCIETY ST

Managing editor: Professor E.B. Davies, Depa...
King's College, Strand, London WC2R 2LS

1 Introduction to combinators and λ-calculus, J.R. HINDLEY & J.P. SELDIN
2 Building models by games, WILFRID HODGES
3 Local fields, J.W.S. CASSELS
4 An introduction to twistor theory, S.A. HUGGETT & K.P. TOD
5 Introduction to general relativity, L. HUGHSTON & K.P. TOD
6 Lectures on stochastic analysis: diffusion theory, DANIEL W. STROOCK
7 The theory of evolution and dynamical systems, J. HOFBAUER & K. SIGMUND
8 Summing and nuclear norms in Banach space theory, G.J.O. JAMESON
9 Automorphisms of surfaces after Nielsen and Thurston, A.CASSON & S. BLEILER
10 Nonstandard analysis and its applications, N.CUTLAND (ed)
11 Spacetime and singularities, G. NABER
12 Undergraduate algebraic geometry, MILES REID
13 An introduction to Hankel operators, J.R. PARTINGTON
14 Combinatorial group theory: a topological approach, DANIEL E. COHEN
15 Presentations of groups, D.L. JOHNSON
16 An introduction to noncommutative noetherian rings, K.R. GOODEARL & R.B. WARFIELD,JR.
17 Aspects of quantum field theory in curved spacetime, S.A. FULLING
18 Braids and coverings, V.L. HANSEN

London Mathematical Society Student Texts. 1

Introduction to Combinators and λ-calculus

J. ROGER HINDLEY
Department of Mathematics and Computer Science,
University College, Swansea

JONATHAN P. SELDIN
Department of Mathematics
Concordia University, Montreal

The right of the University of Cambridge to print and sell all manner of books was granted by Henry VIII in 1534. The University has printed and published continuously since 1584.

CAMBRIDGE UNIVERSITY PRESS
Cambridge
New York Port Chester
Melbourne Sydney

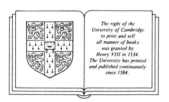

Published by the Press Syndicate of the University of Cambridge
The Pitt Building, Trumpington Street, Cambridge CB2 1RP
40 West 20th Street, New York, NY 10011, USA
10, Stamford Road, Oakleigh, Melbourne 3166, Australia

First published 1986
Reprinted 1988, 1990

Printed in Great Britain at the University Press, Cambridge

Library of Congress cataloguing in publication data:
Hindley, J. Roger
An introduction to combinators and the (lambda)-calculus
(London Mathematical Society student texts ; 1)
Bibliography: p.
Includes index.
1. Combinatory logic. 2. Lambda calculus
I. Seldin, J.P. II. Title III. Series.
QA 9.5.H56 1986 511'.6. 85-29908

British Library cataloguing in publication data:
Hindley, J. R.
An introduction to combinators and the lambda-calculus
(London Mathematical Society student texts ; v. 1)
1. Combinatorial analysis. 2. Algorithms
I. Title. II. Seldin, J.P. III. Series.
511'.6. QA 164

ISBN 0 521 26896 6 Hardcover
ISBN 0 521 31839 4 Paperback

For
Carol,
Julie,
Raymonde,
and to the memory of
Haskell B. Curry

TABLE OF CONTENTS

PREFACE

The theories of combinators and λ-calculus both have the same purpose; to describe some of the most primitive and general properties of operators and combinations of operators. In effect they are abstract programming languages. They contain the concept of computation in full generality and strength, but in a pure 'distilled' form with syntactic complications kept to a minimum.

The λ-calculus was invented in the 1930's by the logician Alonzo Church as part of a system of higher-order logic and function theory. In fact the language of λ-calculus, or some notation essentially equivalent to it, is a key part of any non-trivial higher-order language, whether for logic or for programming. And it was in terms of λ-calculus, not idealized computing machines, that the first uncomputability results were discovered and expressed.

Combinatory logic is a way of doing logic or computations without using variables. For first-order logic this is just a curiosity, but for higher orders it is a technique with considerable advantages. The basic idea is due to two logicians; Moses Schönfinkel who first thought of it in the early 1920's, and Haskell Curry who re-discovered it a few years later and turned it into a workable technique.

This book has been written to introduce the reader to the basic techniques and results in both fields. Its aim is to give the reader just the flavour of the subjects, without drowning him (or her) in too many details. Most long proofs are omitted and replaced by references to the literature.

The reader is assumed to have no previous knowledge of combinators or λ-calculus, but to have attended a beginning course on predicate calculus and recursive functions.

Simple exercises are included in the earlier chapters; some are starred, and answers to these are at the end of the book.

Some chapters on special topics have been included after the initial basic chapters. These have been chosen purely because the authors are currently interested in them, not because they believe them to be the most important. In fact λ-calculus and combinatory logic are surprisingly rich fields, and many interesting topics have had to be omitted. These are well covered in other publications, and references for further reading are given at the end of each chapter.

There is no chapter on programming-language theory, despite the fact that the main applications are currently in this field. There are two reasons for this. First, the applications are rather diverse, and what would interest one reader would be irrelevant to another, even within computer science. Second, and mainly, the applications are of ideas and proof-techniques rather than of specific theorems, and these methods only appear in programming-theory in a considerably modified form. So the core ideas are presented in this book in a 'pure' form, and the applications are left to each individual reader.

This book has been developed from 'Introduction to Combinatory Logic' by J.R. Hindley, B. Lercher and J.P. Seldin, London Mathematical Society Lecture Notes Series No. 7, Cambridge University Press 1972. The emphasis in the new book is on λ-calculus as much as combinators, and much new material has been added, including four chapters on semantics. Also all the old material has been thoroughly re-written, updated, and, where necessary, corrected.

ACKNOWLEDGEMENTS

The authors are very grateful to all those who have helped in producing this book, especially the following:

Hendrik Boom, Martin Bunder, Mario Coppo, Mariangiola Dezani-Ciancaglini, Peter Grogono, Joachim Lambek, Giuseppe Longo, Albert Meyer, Garrel Pottinger, Manfred Szabo, and the late Michael Moss, for reading draft extracts from the manuscript and offering detailed comments, which have saved quite a few mistakes from getting into print (no doubt some have escaped, though, and the authors alone are responsible for these);

Myrtle Prowse, for the more than ordinary care and skill she has put into the typing of the camera-ready copy;

the staff of the Cambridge University Press, especially David Tranah and Rufus Neal, for their help and advice (and patience during many delays in the writing!);

the Quebec Ministry of Education, for financial support from grants nos. EQ1648 and CE110 of the programme F.C.A.C.;

and, last but not least, Raymonde Gagnon and Carol Hindley for continuing interest and encouragement, the latter also for help with the proof-reading and for permission to include Appendix 3.

On the negative side, the authors have greatly missed the advice and enthusiasm of Haskell Curry, who died in 1982 at the age of 81. He was the moving force behind combinatory logic from the 1920's to the 1970's, and it was a privilege to know him and be associated with him.

NOTES ON THE TEXT

The popular book *Mendelson* 1964 will be used as a basic reference for any predicate logic background needed. And Mendelson's (fairly standard) mathematical notation will be used for functions, sets, etc. In particular, *ordered-pair* notation will be '< , >' and a *binary relation* will be regarded as a set of ordered pairs. (Cf. Mendelson 1964, pp. 5-9.)

A *function* will be regarded as a binary relation such that no two of its pairs have the same first member. The words *function*, *mapping* and *map* will be used interchangeably, as usual in mathematics. But the word *operator* will be reserved for a slightly different concept of function. This will be explained in Chapter 3, though in earlier chapters the reader should think of 'operator' and 'function' as meaning the same.

Reference to *Henk Barendregt*'s comprehensive book *'The Lambda Calculus'* will be frequent. A reference 'Barendregt 1981(I)' will mean the first edition, and '1981(II)' the second (actually published in 1984). A simple '1981' will refer to both editions.

Finally, a note about 'we': in this book, 'we' will almost always mean the reader and the authors together, not the authors alone.

J.R.H., J.P.S., 25 October 1985

CHAPTER ONE

λ-CALCULUS

1A INTRODUCTION

λ-calculus is a collection of several formal systems, based on a function-notation invented by Alonzo Church in the 1930's. They are designed to capture the most basic aspects of the ways that operators or functions can be combined to form other operators.

In practice, each λ-system has a slightly different grammatical structure, depending on its intended application. Some have extra constant-symbols, and most have built-in syntactic restrictions, for example type-restrictions. But to begin with, it is best to avoid these complications; hence the system presented in this chapter will be the 'pure' one, which is syntactically the simplest.

To motivate the λ-notation, consider the mathematical expression '$x-y$'. This can be thought of as defining either a function f of x or a function g of y;

$$f: x \mapsto x-y, \qquad g: y \mapsto x-y.$$

And there is a need for a notation that gives f and g different names in some systematic way. In practice, mathematicians usually avoid this need by various 'ad-hoc' special notations, but these can get very clumsy when higher-order functions are involved (functions which act on other functions).

Church's notation is a systematic way of constructing, for each expression involving 'x', a notation for the corresponding function of x (and similarly for 'y', etc.). Church chose 'λ' as an auxiliary symbol and wrote

$$f = \lambda x.x-y, \qquad g = \lambda y.x-y.$$

For example, consider the equations

$$f(0) = 0 - y, \qquad f(1) = 1 - y.$$

In the λ-notation these become

$$(\lambda x.x - y)(0) = 0 - y, \qquad (\lambda x.x - y)(1) = 1 - y.$$

These equations are clumsier than the originals, but do not be put off by this; the notation is principally intended for denoting higher-order functions, not first-order, and for this it is no worse than others.

The λ-notation can be extended to functions of more than one variable. For example, the expression 'x - y' determines two functions h and k of two variables, defined by

$$h(x,y) = x - y, \qquad k(y,x) = x - y.$$

These can be denoted by

$$h = \lambda xy.x - y, \qquad k = \lambda yx.x - y.$$

However, we can avoid the need for a special notation for functions of several variables by using functions whose values are not numbers but other functions. For example, instead of the two-place function h above, consider the one-place function h^* defined by

$$h^* = \lambda x.(\lambda y.x - y).$$

For each number a, we have

$$h^*(a) = \lambda y.a - y;$$

hence for each pair of numbers a,b,

$$\begin{aligned}(h^*(a))(b) &= (\lambda y.a - y)(b)\\ &= a - b\\ &= h(a,b).\end{aligned}$$

Thus h^* can be viewed as 'representing' h. For this reason, we shall largely ignore functions of more than one variable in this book. And from now on, 'function' will mean 'function of one variable' unless explicitly stated otherwise.

Having looked at λ-notation in an informal context, let us now construct a formal system of λ-calculus.

DEFINITION 1.1 (λ-<u>terms</u>). Assume that there is given an infinite sequence of distinct symbols called <u>variables</u>, and a finite, infinite or empty sequence of distinct symbols called <u>constants</u>. (When the sequence of constants is empty, the system will be called *pure*, otherwise *applied*.) The set of expressions called *λ-terms* is defined inductively as follows:

(a) All variables and constants are λ-terms (called *atoms*).

(b) If M and N are any λ-terms, then (MN) is a λ-term (called an *application*).

(c) If M is any λ-term and x is any variable, then $(\lambda x.M)$ is a λ-term (called an *abstraction*).

Examples of λ-terms.

$$(\lambda x.(xy)), \qquad ((\lambda y.y)(\lambda x.(xy))),$$
$$(x(\lambda x.(\lambda x.x))), \qquad (\lambda x.(yz)).$$

NOTATION 1.2. *Capital Roman letters* will denote arbitrary λ-terms in this chapter.

Letters 'x', 'y', 'z', 'u', 'v', 'w' will denote variables throughout the book, and distinct letters will denote distinct variables unless stated otherwise.

Parentheses will be omitted in such a way that, for example, 'MNPQ' denotes $(((MN)P)Q)$. (This convention is called *association to the left*.) Other abbreviations will be

$$\lambda x.PQ \quad \text{for} \quad (\lambda x.(PQ)),$$
$$\lambda x_1 \ldots x_n.M \quad \text{for} \quad (\lambda x_1.(\lambda x_2.(\ldots(\lambda x_n.M)\ldots))).$$

Syntactic identity of terms will be denoted by '≡'; for example

$$M \equiv N$$

will mean that M is exactly the same term as N. (The symbol '=' will be used in formal theories of equality, and for identity of objects that are not terms, such as numbers.) It will be assumed of course that if $MN \equiv PQ$ then $M \equiv P$ and $N \equiv Q$, and if $\lambda x.M \equiv \lambda y.P$ then $x \equiv y$ and $M \equiv P$. It will also be assumed that the four classes of terms (variables, constants, applications, abstractions) have no members in common. Such assumptions are always made when sets of expressions are defined, and will be left unstated in future.

The cases $k = 0$, $n = 0$ in statements like '$P \equiv MN_1 \ldots N_k$ $(k \geq 0)$' or 'M has form $\lambda x_1 \ldots x_n.PQ$ $(n \geq 0)$' etc. will mean '$P \equiv M$', 'M has form PQ', etc.

'λ' will often be used carelessly to mean 'λ-calculus in general'.

'Iff' will be used for 'if and only if'.

INFORMAL INTERPRETATION 1.3. Definition 1.1 allows more terms to be formed than we need in every application of the system. Some terms are usually left uninterpreted, as we shall see later.

But in general, if M has been interpreted as a function or operator, then (MN) is interpreted as the result of applying M to argument N, provided this result is defined. (The more usual notation for function-application is M(N), but historically (MN) has become standard in λ-calculus and combinatory logic. This book will use the (MN) notation when writing formal terms (following Definition 1.1(b)), but will revert to the common notation, e.g. $\phi(a)$, in informal discussions on functions.)

A term (λx.M) represents the operator whose value at an argument N is calculated by substituting N for x in M.

For example, λx.x(xy) represents the operation of applying a function twice to an object denoted by y; and the equation

$$(\lambda x.x(xy))N = N(Ny)$$

holds for all terms N, in the sense that both sides have the same interpretation.

For a second example, λx.y represents the constant-function that takes the value y for all arguments, and the equation

$$(\lambda x.y)N = y$$

holds in the same sense as before.

This is enough on interpretation for the moment; but the theme will be taken up again in Chapter 3.

EXERCISE 1.4.* (Answers to all starred exercises are given at the end of the book.)

(a) Express the following in an informal λ-notation, using 'D' for the differential operator. Notice how the distinction in (iii) below is made clearer by this notation.

(i) The derivative of x^2 is $2x$.

(ii) The derivative of x^2 at the point 3 is 6.

(iii) Let a function f be given, and let g be defined by saying $gx = f(x^2)$; the derivative of g at x is distinct from the derivative of f at x^2.

(b) Insert the full amount of parentheses in the following abbreviated λ-terms.

(i) ux(yz)(λv.vy).

(ii) (λxyz.xz(yz))uvw.

(iii) w(λxyz.xz(yz))uv.

1B TERM-STRUCTURE AND SUBSTITUTION

The main topic of the chapter will be a formal procedure for calculating with terms, that will closely follow their informal meaning. But to define it we shall need to know how to substitute terms for variables, and this is not entirely straightforward. The present section will cover the technicalities involved. The definitions are important, but the proofs of the lemmas are boring and best omitted. (And the reader only interested in main themes can omit the lemmas too!)

By the way, combinatory logic will avoid most of these technicalities, although at some cost.

DEFINITION 1.5. The <u>length</u> of a term M (called <u>lgh</u>(M)) is the total number of occurrences of atoms in M. More precisely, define

(a) lgh(a) = 1 for atoms a;

(b) lgh(MN) = lgh(M) + lgh(N);

(c) lgh(λx.M) = 1 + lgh(M).

The phrase '<u>induction on</u> M' will mean 'induction on the length of M'.

DEFINITION 1.6. The relation P <u>occurs in</u> Q (or P <u>is a subterm of</u> Q, or Q <u>contains</u> P) is defined by induction on Q, thus:

(a) P occurs in P;

(b) if P occurs in M or in N, then P occurs in (MN);

(c) if P occurs in M or $P \equiv x$, then P occurs in $(\lambda x.M)$.

The meaning of '<u>an occurrence of</u> P <u>in</u> Q' is assumed to be intuitively clear. For example, there are two occurrences of (xy) in $((xy)(\lambda x.(xy)))$, and three occurrences of x. The reader who wants more precision can define an occurrence of P in Q to be a pair $\langle P,p \rangle$ where p is some indicator of the position at which P occurs in Q; one suitable definition of p is in Hindley 1974, p.5. But it is best to avoid such details for as long as possible.

EXERCISE 1.7.*

(a) Does the term x(yz) occur in ux(yz)?

(b) Look at the terms in Exercise 1.4(b)(ii) and (iii); in which of these does $(\lambda xyz.xz(yz))u$ occur?

DEFINITION 1.8 (<u>Scope</u>). For a particular occurrence of $\lambda x.M$ in a term P, the occurrence of M is called the <u>scope</u> of the occurrence of λ on the left.

EXAMPLE 1.9. Let

$$P \equiv (\lambda y.yx(\lambda x.y(\lambda y.z)x))vw.$$

The scope of the left-most λ is $yx(\lambda x.y(\lambda y.z)x)$, that of the next λ is $y(\lambda y.z)x$, and that of the third is z.

DEFINITION 1.10 (<u>Free and bound variables</u>). An occurrence of a variable x in a term P is <u>bound</u> iff it is in a part of P with the form $\lambda x.M$; otherwise it is <u>free</u>. If x has at least one free occurrence in P it is called a <u>free variable of</u> P; the set of all such variables is called

$$FV(P).$$

A <u>closed term</u> is a term without any free variables.

<u>Examples</u>. In the term $xv(\lambda y.yv)w$, both y's are bound, both v's are free, and x and w are free. In the term P in 1.9, all four y's

are bound, the left-hand x is free and the other two are bound, and z, v, w are free; hence

$$FV(P) = \{x,z,v,w\}.$$

DEFINITION 1.11 (Substitution). For any M, N, x, define $[N/x]M$ to be the result of substituting N for every free occurrence of x in M, and changing bound variables to avoid clashes. The precise definition is by induction on M, as follows (after Curry and Feys 1958, p.94).

(a) $[N/x]x \equiv N;$

(b) $[N/x]a \equiv a$ for all atoms $a \not\equiv x$;

(c) $[N/x](PQ) \equiv ([N/x]P\ \ [N/x]Q)$;

(d) $[N/x](\lambda x.P) \equiv \lambda x.P$;

(e) $[N/x](\lambda y.P) \equiv \lambda y.[N/x]P$ if $y \not\equiv x$, and $y \notin FV(N)$ or $x \notin FV(P)$;

(f) $[N/x](\lambda y.P) \equiv \lambda z.[N/x][z/y]P$ if $y \not\equiv x$ and $y \in FV(N)$ and $x \in FV(P)$.

In (f), z is chosen to be the first variable $\notin FV(NP)$.

REMARK 1.12. Clause (f) prevents the intuitive meaning of $[N/x](\lambda y.P)$ from depending on the bound variable y. For example, take three distinct variables w, x, y and look at

$$[w/x](\lambda y.x).$$

The term $\lambda y.x$ represents the constant-function whose value is always x, so we should intuitively expect $[w/x](\lambda y.x)$ to represent the constant-function whose value is always w. And this is what we get; by (e) we have

$$[w/x](\lambda y.x) \equiv \lambda y.w.$$

But now suppose we change y to w. Since $w \not\equiv x$, the term $\lambda w.x$ still represents the constant-function whose value is x, just as $\lambda y.x$ did. So we should hope that $[w/x](\lambda w.x)$ represents the same constant-function that $[w/x](\lambda y.x)$ did.

But if $[w/x](\lambda w.x)$ was evaluated by (e), our hope would be denied; we would have

$$[w/x](\lambda w.x) \equiv \lambda w.w,$$

which represents the identity-function, not a constant-function.

Clause (f) rescues our hope. By (f) with $N \equiv y \equiv w$, we have

$$\begin{aligned}[w/x](\lambda w.x) &\equiv \lambda z.[w/x][z/w]x \\ &\equiv \lambda z.[w/x]x \quad \text{by (b)} \\ &\equiv \lambda z.w,\end{aligned}$$

which represents the same constant-function as $\lambda y.w$.

Incidentally, Church's formulation of λ-calculus simply said that N could not be substituted for x in case (f); but it seems tidier to define substitution for all cases, even though the definition is complicated.

EXERCISE 1.13.* Evaluate

(a) $[(\lambda y.xy)/x](\lambda y.x(\lambda x.x))$,

(b) $[(\lambda y.vy)/x](y(\lambda v.xv))$.

LEMMA 1.14.

(a) $[x/x]M \equiv M$;

(b) $x \notin FV(M) \Rightarrow [N/x]M \equiv M$;

(c) $x \in FV(M) \Rightarrow FV([N/x]M) = FV(N) \cup (FV(M) - \{x\})$;

(d) $lgh([y/x]M) = lgh(M)$.

Proof. Trivial. □

LEMMA 1.15. *Let* x, y, v *be distinct (the usual notation convention), and let no variable bound in* M *be free in* vPQ. *Then*

(a) $v \notin FV(M) \Rightarrow [P/v][v/x]M \equiv [P/x]M$;

(b) $v \notin FV(M) \Rightarrow [x/v][v/x]M \equiv M$;

(c) $y \notin FV(P) \Rightarrow [P/x][Q/y]M \equiv [([P/x]Q)/y][P/x]M$;

(d) $y \notin FV(P),\ x \notin FV(Q) \Rightarrow [P/x][Q/y]M \equiv [Q/y][P/x]M$;

(e) $[P/x][Q/x]M \equiv [([P/x]Q)/x]M$.

Proof. Straightforward induction on M. (The restriction on variables bound in M ensures that Definition 1.11(f) is never used.) Part (b) follows from (a) and 1.14(a), and (d) from (c) and 1.14(b). □

DEFINITION 1.16 (*Change of bound variables*, *congruence*). Let a term P contain an occurrence of $\lambda x.M$, and let $y \notin FV(M)$. The act of replacing this $\lambda x.M$ by

$$\lambda y.[y/x]M$$

is called a *change of bound variable* in P. We shall say P *is congruent to* Q, or P α-*converts to* Q, or

$$P \equiv_{\alpha} Q,$$

iff Q has been obtained from P by a finite (perhaps empty) series of changes of bound variables.

Example.

$$\begin{aligned} \lambda xy.x(xy) &\equiv \lambda x.(\lambda y.x(xy)) \\ &\equiv_{\alpha} \lambda x.(\lambda v.x(xv)) \\ &\equiv_{\alpha} \lambda u.(\lambda v.u(uv)) \\ &\equiv \lambda uv.u(uv). \end{aligned}$$

Comment. Congruent terms have identical interpretations and play identical roles in any application of λ-calculus. In fact, most authors identify '$P \equiv_{\alpha} Q$' with '$P \equiv Q$', and this book will do the same in its later chapters.

Definition 1.16 comes from Curry and Feys 1958, p.91. The name 'α-converts' comes from the same authors, as do other Greek-letter names used later; they are now more or less standard notation.

LEMMA 1.17.

(a) *If* $P \equiv_{\alpha} Q$ *then* $FV(P) = FV(Q)$;

(b) *For any* P *and any* $x_1,\ldots,x_n$, *there exists* P′ *such that* $P \equiv_{\alpha} P'$ *and none of* $x_1,\ldots,x_n$ *is bound in* P′;

(c) *The relation* $\equiv_{\alpha}$ *is transitive, reflexive and symmetric.*

Proof.

(a) Use 1.14(c).

(b) Easy induction on P.

(c) Transitivity and reflexivity are obvious. But symmetry is not; we must prove that if $y \notin FV(M)$ (and $y \not\equiv x$), then

$$\lambda y.[y/x]M \equiv_\alpha \lambda x.M.$$

Now $x \notin FV([y/x]M)$ by 1.14(c), so

$$\lambda y.[y/x]M \equiv_\alpha \lambda x.[x/y][y/x]M.$$

To complete the proof we must show that

$$[x/y][y/x]M \equiv_\alpha M.$$

This is proved by a straightforward induction on M, cf. 1.15(b). □

LEMMA 1.18. *If we remove the condition on variables bound in* M *from Lemma* 1.15 *and replace* '≡' *by* '$\equiv_\alpha$', *that lemma stays true.*

Proof. Induction on M, or Curry and Feys 1958, §3E Thm. 2(c). □

LEMMA 1.19. $M \equiv_\alpha M'$, $N \equiv_\alpha N' \Rightarrow [N/x]M \equiv_\alpha [N'/x]M'$.

Proof. Induction on M, using 1.18. □

EXERCISE 1.20.* The above technicalities are rather dull. But their very dullness has made them a trap for even the most careful authors. For example, Curry and Feys give perhaps the most thorough analysis published of these things, in their 1958 pp.91ff., but even they slip up at the bottom of p.91, where they claim that the one-step change-of-bound-variable relation is symmetric, i.e. that the step backwards from $\lambda y.[y/x]M$ to $\lambda x.M$ can always be done by a single change of bound variable. Show that this is false. (Cf. the proof of 1.17(c).)

REMARK 1.21. By the way, there is a way of changing Definition 1.16 that makes the one-step change-of-bound-variable relation symmetric; insert the restriction that neither x nor y occurs bound in M. Then symmetry follows by 1.15(b). It is not hard to prove that this restriction leaves the $\equiv_\alpha$ relation unchanged (though the proof is tedious). The restriction makes the details of some technical proofs easier, e.g. Lemma A1.6 in Appendix 1.

1C β-REDUCTION

A term of form $(\lambda x.M)N$ represents an operator $\lambda x.M$ applied to an argument N. In the informal interpretation of $\lambda x.M$, its value at N is calculated by substituting N for x in M, so $(\lambda x.M)N$ can be 'simplified' to $[N/x]M$. This simplification-process is captured in the following definition.

DEFINITION 1.22 (β-<u>contraction</u>, β-<u>reduction</u>). Any term of form

$$(\lambda x.M)N$$

is called a β-<u>redex</u>, and the corresponding term

$$[N/x]M$$

is called its <u>contractum</u>. If a term P contains an occurrence of $(\lambda x.M)N$ and we replace that occurrence by $[N/x]M$, and the result is P', we say we have <u>contracted</u> the redex-occurrence in P, or P β-<u>contracts to</u> P' or

$$P \rhd_{1\beta} P'.$$

We say P β-<u>reduces to</u> Q, or

$$P \rhd_{\beta} Q,$$

iff Q is obtained from P by a finite (perhaps empty) series of β-contractions and changes of bound variables.

When there is no danger of confusion, 'β' may be omitted.

EXAMPLES 1.23.

(a) $(\lambda x.x(xy))N \quad \rhd_{1\beta} \quad N(Ny).$

(b) $(\lambda x.y)N \quad \rhd_{1\beta} \quad y.$

(c) $(\lambda x.(\lambda y.yx)z)v \quad \rhd_{1\beta} \quad [v/x]((\lambda y.yx)z) \equiv (\lambda y.yv)z$

$\rhd_{1\beta} \quad [z/y](yv) \equiv zv.$

(d) $(\lambda x.xxy)(\lambda x.xxy) \quad \rhd_{1\beta} \quad (\lambda x.xxy)(\lambda x.xxy)y$

$\rhd_{1\beta} \quad (\lambda x.xxy)(\lambda x.xxy)yy$

etc.

The last example shows that the simplification-process need not really simplify, and need not terminate either. In fact, it terminates iff it reaches a term containing no redexes.

DEFINITION 1.24. A term Q which contains no β-redexes is called a β-normal form (or a term in β-normal form). The class of all β-normal forms is called β-nf or λβ-nf. If a term P β-reduces to a Q in β-nf, then Q is called a β-normal form of P.

The 'β' may be omitted when this causes no confusion.

EXAMPLES 1.25.

(a) In 1.23(c), zv is a β-normal form of (λx.(λy.yx)z)v.

(b) Let L ≡ (λx.xxy)(λx.xxy). By 1.23(d) we have

$$L \rhd_{1\beta} Ly \rhd_{1\beta} Lyy \rhd_{1\beta} \dots \ .$$

(c) Let P ≡ (λu.v)L for the above L. Then P can be reduced in two different ways (at least), thus:

(i) $P \equiv (\lambda u.v)L \ \rhd_{1\beta} \ [L/u]v$

$\equiv v.$

(ii) $P \rhd_{1\beta} (\lambda u.v)(Ly)$ by contracting L

$\rhd_{1\beta} (\lambda u.v)(Lyy)$ by contracting L

etc.

So P has a normal form v, but also has an infinite reduction.

(d) The reduction shown for L in (b) above is the only possible one, so L has no normal form.

REMARK 1.26. We have seen that some terms can be reduced in more than one way. For another example, let P ≡ (λx.(λy.yx)z)v, the term in Example 1.23(c). It has two reductions:

$P \rhd_{1\beta} (\lambda y.yv)z$ by contracting P

$\rhd_{1\beta} zv$ by contracting (λy.yv)z;

$P \rhd_{1\beta} (\lambda x.zx)v$ by contracting (λy.yx)z

$\rhd_{1\beta} zv.$

In this case, both reductions reach the same normal form. Is this always true? Certainly, for any system claiming to represent computation the end-result should be independent of the path. So if this property failed for β-reduction, any claim by λ-calculus to be like a programming language would fail from the start.

The aim of the following theorem is to show that the normal form of a term is indeed unique, provided we ignore the difference between congruent terms. But it has many other applications too, and is probably the most often quoted theorem on λ-calculus.

Before the theorem, here are two routine lemmas.

LEMMA 1.27. *If* $P \equiv_\alpha P'$, $Q \equiv_\alpha Q'$ *and* $P \rhd_\beta Q$, *then* $P' \rhd_\beta Q'$.

Proof. The definition of $\rhd_\beta$ includes $\equiv_\alpha$, and $\equiv_\alpha$ is symmetric. □

LEMMA 1.28 (Substitution lemma for β-reduction).

(a) $P \rhd_\beta Q \implies (v \notin FV(P) \implies v \notin FV(Q))$;

(b) $P \rhd_\beta Q \implies [P/x]M \rhd_\beta [Q/x]M$;

(c) $P \rhd_\beta Q \implies [N/x]P \rhd_\beta [N/x]Q$.

Proof.

(a) By 1.14(b) and (c), if $v \notin FV((\lambda x.M)N)$ then $v \notin FV([N/x]M)$. (Roughly speaking, (a) says that nothing new can be introduced during a reduction.)

(b) Any contractions in P can also be made in [P/x]M.

(c) It is enough to prove (c) for just one contraction. By 1.27 and 1.19 we can assume no variable bound in P is free in N. Let $R \equiv (\lambda y.H)J$ be the redex contracted, and let $y \notin FV(N)$. Then

$$\begin{aligned} [N/x]R &\equiv (\lambda y.[N/x]H)[N/x]J && \text{by 1.11(c), (e)} \\ &\rhd_{1\beta} [([N/x]J)/y][N/x]H \\ &\equiv_\alpha [N/x][J/y]H && \text{by 1.15(c) and 1.18.} \end{aligned}$$

□

THEOREM 1.29 (Church-Rosser theorem for β-reduction). *If*

$$P \rhd_\beta M \quad \textit{and} \quad P \rhd_\beta N$$

(see Figure 1.1), then there exists T *such that*

$$M \rhd_\beta T \quad \textit{and} \quad N \rhd_\beta T.$$

Figure 1.1

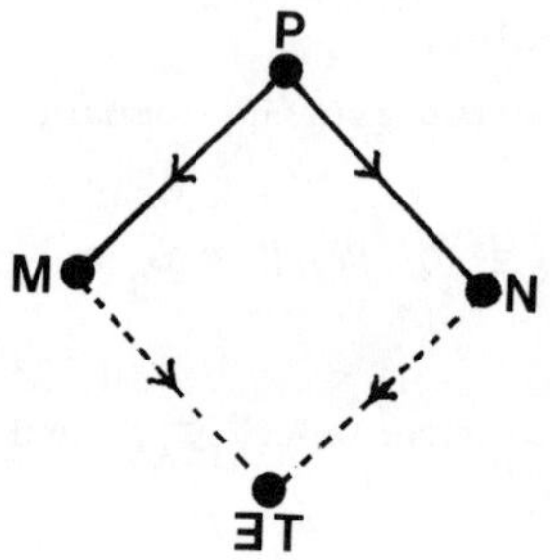

Proof. See Appendix 1. □

COROLLARY 1.29.1. *If* P *has β-normal forms* M *and* N, *then* M *is congruent to* N.

Proof. We have $P \rhd_\beta M$ and $P \rhd_\beta N$, so the theorem gives a T such that M and N reduce to T. But M and N contain no redexes, so they must both be congruent to T. □

The following is an alternative characterization of β-normal forms which will be used in a later chapter.

LEMMA 1.30. *The class β-nf is the smallest class such that*

(a) *all atoms are in β-nf;*

(b) *if* $M_1,\ldots,M_n$ *are in β-nf and* a *is any atom, then*

$$aM_1\ldots M_n \in \beta\text{-}nf;$$

(c) *if* $M \in \beta\text{-}nf$, *then* $\lambda x.M \in \beta\text{-}nf$.

Proof. M is in the class defined by (a) - (c) iff M contains no redexes. (Proof by induction on M.) □

EXERCISE 1.31.* A term Q is called <u>minimal with respect to</u> β-<u>reduction</u> iff

$$(\forall M)(Q \triangleright_\beta M \implies M \equiv_\alpha Q).$$

Of course, every normal form is minimal. But there are minimal terms that are not normal forms. Find one such. (In fact, if we demand that the term be also a β-redex, then there is only one such term, modulo congruence; Lercher 1976.)

1D β-EQUALITY

Reduction is non-symmetric, but it generates the following equivalence relation.

DEFINITION 1.32. P is β-<u>equal</u> or β-<u>convertible</u> to Q (notation $P =_\beta Q$) iff Q is obtained from P by a finite (perhaps empty) series of β-contractions and reversed β-contractions and changes of bound variables. That is, $P =_\beta Q$ iff there exist $P_0,\ldots,P_n$ $(n \geq 0)$ such that

$$P_0 \equiv P, \quad P_n \equiv Q,$$

$$(\forall i \leq n-1)(P_i \triangleright_{1\beta} P_{i+1} \text{ or } P_{i+1} \triangleright_{1\beta} P_i \text{ or } P_i \equiv_\alpha P_{i+1}).$$

LEMMA 1.33. *If* $P \equiv_\alpha P'$, $Q \equiv_\alpha Q'$ *and* $P =_\beta Q$, *then* $P' =_\beta Q'$.

LEMMA 1.34 (<u>Substitution lemma for</u> β-<u>equality</u>).

(a) $P =_\beta Q \implies [P/x]M =_\beta [Q/x]M$;

(b) $P =_\beta Q \implies [N/x]P =_\beta [N/x]Q$.

THEOREM 1.35 (Church-Rosser theorem for β-equality). *If* $P =_\beta Q$, *then there exists* T *such that*

$$P \rhd_\beta T \text{ and } Q \rhd_\beta T.$$

Proof. By induction on the number n in Definition 1.32. The basis, n = 0, is trivial. For the induction step, n to n+1, we assume:

$$P =_\beta P_n; \quad P_n \rhd_{1\beta} P_{n+1} \text{ or } P_{n+1} \rhd_{1\beta} P_n;$$

and the induction hypothesis gives a term T_n such that

$$P \rhd_\beta T_n, \quad P_n \rhd_\beta T_n.$$

(See Figure 1.2.) We want a T such that $P \rhd_\beta T$ and $P_{n+1} \rhd_\beta T$. If $P_{n+1} \rhd_{1\beta} P_n$, choose $T \equiv T_n$. If $P_n \rhd_{1\beta} P_{n+1}$, apply Theorem 1.29 to P_n, T_n, P_{n+1} as shown in Figure 1.2. □

Figure 1.2.

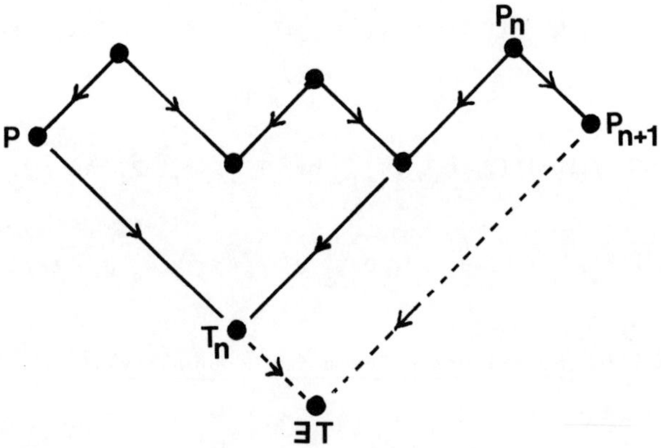

This theorem shows that two β-convertible terms both intuitively represent the same operator, since they can both be reduced to the same term. (This is why convertibility is called '='.)

COROLLARY 1.35.1. *If* $P =_\beta Q$ *and* Q *is in β-normal form, then* $P \rhd_\beta Q$.

Proof. By the Church-Rosser theorem, P and Q both reduce to some T. But Q contains no redexes, so $Q \equiv_\alpha T$. Hence $P \rhd_\beta Q$. □

COROLLARY 1.35.2. *If* $P =_\beta Q$, *then either* P *and* Q *do not have β-normal forms, or* P *and* Q *both have the same β-normal forms.*

COROLLARY 1.35.3. *Two β-equal terms in β-normal form must be congruent.*

By this corollary the relation $=_\beta$ is non-trivial, in the sense that there exist P and Q such that $P \neq_\beta Q$. For example, $\lambda xy.xy$ and $\lambda xy.yx$ are non-congruent β-normal forms, and hence are not β-equal.

COROLLARY 1.35.4. *A term can be β-equal to at most one β-normal form, modulo congruence.*

The following corollary will be needed in a later chapter.

COROLLARY 1.35.5. *If* $xM_1 \ldots M_m =_\beta yN_1 \ldots N_n$, *then* $x \equiv y$ *and* $m = n$ *and* $M_i =_\beta N_i$ *for all* i.

Proof. By the Church-Rosser theorem, both terms reduce to some T. But each β-redex in $xM_1 \ldots M_m$ must be in an M_i, so T must have form $xT_1 \ldots T_m$, where $M_i \rhd_\beta T_i$ for $i = 1,\ldots,m$. Similarly, $T \equiv yT'_1 \ldots T'_n$ where $N_j \rhd_\beta T'_j$ for $j = 1,\ldots,n$. Hence result. □

REMARK 1.36. (Church's original λ-terms). If P has a normal form, not every subterm of P need have a normal form. (For example see the P in 1.25(c).) When Church originally devised λ-calculus he excluded such cases by allowing $\lambda x.M$ to be a term only when x occurred free in M.

The reason was that Church regarded terms without normal form as meaningless, and he did not want meaningful terms to have meaningless subterms.

All the results in this chapter are valid for Church's original system (indeed, they were first formulated by Church), but this will not hold true for all future results. A modern account of Church's system is in Barendregt 1981, Ch.9.

Nowadays, Church's terms are called λI-terms, and the terms of the present book λK-terms.

EXERCISE 1.37.

(a)* (i) Prove that if the equation $\lambda xy.x =_\beta \lambda xy.y$ was added as an extra axiom to the definition of β-equality, then all terms would become β-equal.

(ii) Do the same for $\lambda x.x =_\beta \lambda xy.yx$.

(b) Find a closed term Y such that $Yx =_\beta x(Yx)$. Informally, such a Y represents an operator for constructing fixed-points of other operators; for every term M, if we let $P \equiv YM$, then $MP =_\beta P$. A suitable Y and further discussion will be given in §3B. (Hint: try $Y \equiv \lambda x.ZZ$ for a suitable Z.)

FURTHER READING 1.38. We shall leave λ-calculus now and return again in Chapter 3. In fact Chapter 3 and most later chapters will apply equally well to both λ-calculus and combinatory logic.

But before we move on, here are a few suggestions for deeper reading.

Barendregt 1981. An encyclopaedic, well-organized and up-to-date account of λ-calculus, including also some treatment of combinators. Presents the deeper ideas underlying the theory. Highly recommended for anyone reading further than the present book, and essential for the intending specialist. Second edition (1984) has an improved treatment of models.

Church 1941. Church originated λ-calculus, and this is the first expository account. Very readable and well-motivated. The early pages are worth reading for the basic ideas, even though the book as a whole is now out of date.

<u>Curry and Feys</u> 1958. Explains combinators and λ-calculus in parallel, and with details such as substitution lemmas which are skipped over in later books. Has interesting historical sections at the end of each chapter. But the emphasis is on combinators, not on λ. And on the technical side, much has happened since 1958.

<u>Hindley and Seldin</u> 1980. A collection of technical papers by some of the best authors in the field. A good indication of the range of research at the time it was published.

<u>Rezus</u> 1982. A bibliography of all the literature up to 1982. Very few omissions. Includes many unpublished manuscripts.

EXERCISE 1.39. Do not confuse <u>being</u> a β-nf with <u>having</u> a β-nf. Note that the second does not imply the first, and prove that

(a) $[N/x]M$ is a β-nf $\Rightarrow$ M is a β-nf,

(b) $[N/x]M$ has a β-nf $\not\Rightarrow$ M has a β-nf.

CHAPTER TWO

COMBINATORY LOGIC

2A INTRODUCTION

Systems of combinators are designed to perform the same tasks as systems of λ-calculus, but without using bound variables. In fact, the annoying technical complications concerned with substitution and congruence will be avoided completely. However, for this technical advantage we shall have to sacrifice the intuitive clarity of the λ-notation.

To motivate combinators, consider the commutative law of addition in arithmetic. It can be expressed as

$$(\forall x,y)\quad x+y \;=\; y+x\ .$$

But this law can also be expressed without using the bound variables 'x' and 'y'. To do this, we first define

$$A(x,y) \;=\; x+y \qquad\qquad (\text{for all}\ \ x,y),$$

and then introduce an operator C defined by

$$(Cf)(x,y) \;=\; f(y,x) \qquad\qquad (\text{for all}\ \ f,x,y).$$

Then the commutative law becomes simply

$$A \;=\; CA.$$

The operator C may be called a <u>combinator</u>; other examples of such operators are the following:

B, which composes two functions: $(B(f,g))x \;=\; f(gx)$;

I, the identity operator: $If \;=\; f$;

K, which forms constant-functions: $(Ka)x \;=\; a$;

S, a stronger composition operator: $(S(f,g))x \;=\; f(x,gx)$;

W, a 'diagonalizing' operator: $(Wf)x \;=\; f(x,x)$.

Instead of trying to define 'combinator' rigorously in this informal context, we shall build up a formal system of terms in which the above 'combinators' can be represented. Just as in the previous chapter, the system to be studied here will be the simplest possible one, with no syntactical complications or restrictions, but with the warning that systems used in practice are more complicated. The ideas introduced in the present chapter will be common to all systems, however.

DEFINITION 2.1 (Combinatory logic terms, or CL-terms). Assume that there is given an infinite sequence of symbols called variables, and a finite or infinite sequence of symbols called constants, including two called basic combinators; K, S. (If K and S are the only constants, the system will be called pure, otherwise applied.) The set of expressions called CL-terms is defined inductively as follows:

(a) All variables and constants, including K and S, are CL-terms.

(b) If X and Y are CL-terms, then so is (XY).

An atom is a variable or constant. A non-redex atom is an atom other than K, S. A non-redex constant is a constant other than K, S. A closed term is a term containing no variables. A combinator is a term whose only atoms are K, S. (In a pure system this is the same as a closed term.)

Examples of CL-terms (the one on the left is a combinator):

$$((S(KS))K), \qquad ((S(Kx))((SK)K)).$$

NOTATION 2.2. Capital Roman letters will denote CL-terms in this chapter, and 'term' will mean 'CL-term'.

'CL' will mean 'combinatory logic', i.e. the study of systems of CL-terms. In later chapters, particular systems will be called 'CLw', 'CLξ', etc., but never just 'CL'.)

The rest of the notation will be the same as in Chapter 1. For example 'x', 'y', 'z', 'u', 'v', 'w' will stand for variables (distinct unless otherwise stated), '≡' for syntactic identity of terms, and parentheses will be omitted following the convention of association to the left.

DEFINITION 2.3. The <u>length of</u> X (or <u>lgh</u>(X)) is the number of occurrences of atoms in X:

(a) lgh(a) = 1 for atoms a;

(b) lgh(UV) = lgh(U) + lgh(V).

DEFINITION 2.4. The relation X <u>occurs in</u> Y is defined thus:

(a) X occurs in X;

(b) if X occurs in U or in V, then X occurs in (UV).

The set of all variables occurring in Y is called FV(Y). (All occurrences of variables are free, because there is no λ to bind them.)

DEFINITION 2.5. (<u>Substitution</u>). [U/x]Y is defined to be the result of substituting U for every occurrence of x in Y:

(a) [U/x]x $\equiv$ x,

(b) [U/x]a $\equiv$ a for atoms a $\not\equiv$ x,

(c) [U/x](VW) $\equiv$ ([U/x]V [U/x]W).

For any $U_1,\ldots,U_n$ and any $x_1,\ldots,x_n$ (mutually distinct), define

$$[U_1/x_1,\ldots,U_n/x_n]Y$$

to be the result of simultaneously substituting U_1 for x_1, U_2 for $x_2,\ldots$, U_n for x_n, in Y.

EXERCISE 2.6.*

(a) Give a definition of simultaneous substitution by induction on lgh(Y). Give an example where

$$[U_1/x_1,\ldots,U_n/x_n]Y \equiv [U_1/x_1][U_2/x_2]\ldots[U_n/x_n]Y$$

fails. State a non-trivial condition sufficient to make it true.

(b) Define simultaneous substitution for λ-terms, and note how much more complicated it is than for CL-terms.

2B WEAK REDUCTION

In the next section, we shall see how K and S can be made to play a role equivalent to λ, and we shall need the following reducibility relation.

DEFINITION 2.7 (Weak reduction). Any term KXY or SXYZ is called a (weak) redex. Contracting an occurrence of a redex in a term U means replacing one occurrence of

$$\text{KXY} \quad \text{by} \quad \text{X},$$

$$\text{SXYZ} \quad \text{by} \quad \text{XZ(YZ)}.$$

If this changes U to U′, we say that U (weakly) contracts to U′, or

$$U \rhd_{1w} U'.$$

We say that U (weakly) reduces to V, or

$$U \rhd_{w} V,$$

iff V is obtained from U by a finite (perhaps empty) series of weak contractions. The '$_w$' will sometimes be omitted.

DEFINITION 2.8. A weak normal form (or term in weak normal form) is a term containing no weak redexes. If U weakly reduces to a weak normal form X, then X is called a weak normal form of U.

EXAMPLE 2.9. Define $B \equiv S(KS)K$.
Then $BXYZ \rhd_w X(YZ)$, because

$$\begin{aligned}
BXYZ &\equiv S(KS)KXYZ \\
&\rhd_{1w} KSX(KX)YZ && \text{by contracting } S(KS)KX \text{ to } KSX(KX), \\
&\rhd_{1w} S(KX)YZ && \text{by contracting } KSX \text{ to } S, \\
&\rhd_{1w} KXZ(YZ) && \text{by contracting } S(KX)YZ, \\
&\rhd_{1w} X(YZ) && \text{by contracting } KXZ.
\end{aligned}$$

EXAMPLE 2.10. Define $C \equiv S(BBS)(KK)$.

Then $CXYZ \rhd_w XZY$, because

$$\begin{aligned}
CXYZ &\equiv S(BBS)(KK)XYZ \\
&\rhd_{1w} BBSX(KKX)YZ && \text{by contracting } S(BBS)(KK)X, \\
&\rhd_{1w} BBSXKYZ && \text{by contracting } KKX, \\
&\rhd_w B(SX)KYZ && \text{by 2.9}, \\
&\rhd_w SX(KY)Z && \text{by 2.9}, \\
&\rhd_{1w} XZ(KYZ) && \text{by contracting } SX(KY)Z, \\
&\rhd_{1w} XZY && \text{by contracting } KYZ.
\end{aligned}$$

(Incidentally, in line 4 of this reduction, a redex KYZ seems to occur; but this is not really so, since $B(SX)KYZ$ is really $((((B(SX))K)Y)Z)$ when all its parentheses are put in.)

EXERCISE 2.11.* Find combinators I and W such that

$$IX \rhd_w X \qquad \text{(for all } X\text{)},$$

$$WXY \rhd_w XYY \qquad \text{(for all } X,Y\text{)}.$$

LEMMA 2.12 (<u>Substitution lemma for weak reduction</u>).

(a) $X \rhd_w Y \Rightarrow (v \notin FV(X) \Rightarrow v \notin FV(Y))$;

(b) $X \rhd_w Y \Rightarrow [X/v]Z \rhd_w [Y/v]Z$;

(c) $X \rhd_w Y \Rightarrow [U_1/x_1,\ldots,U_n/x_n]X \rhd_w [U_1/x_1,\ldots,U_n/x_n]Y$.

Proof.

(a) If $v \notin FV(KXY)$ then $v \notin FV(X)$; if $v \notin FV(SXYZ)$ then $v \notin FV(XZ(YZ))$.

(b) Any contractions in X can also be made in $[X/v]Z$.

(c) If R is a redex and contracts to T, then $[U_1/x_1,\ldots]R$ is also a redex and contracts to $[U_1/x_1,\ldots]T$. □

THEOREM 2.13 (Church-Rosser theorem for weak reduction). *If* $U \rhd_w X$ *and* $U \rhd_w Y$, *then there exists a* Z *such that*

$$X \rhd_w Z \quad \textit{and} \quad Y \rhd_w Z.$$

Proof. Appendix 1, Theorem A1.10. □

COROLLARY 2.13.1. *A* CL-*term can have at most one weak normal form.*

2C ABSTRACTION

In this section, we shall define a CL-term $\lambda^*x.M$ for each x and M, with the property that

$$(\lambda^*x.M)N \rhd_w [N/x]M.$$

In contrast to λ in the λ-calculus, λ^* is not part of the formal language of CL-terms; $\lambda^*x.M$ will be a combination of K's and S's and parts of M, built up as follows.

DEFINITION 2.14 (Abstraction). For each CL-term M and each x, a CL-term called $\lambda^*x.M$ is defined by induction on M, thus:

(a) $\lambda^*x.M \equiv KM$ if $x \notin FV(M)$;

(b) $\lambda^*x.x \equiv I$ where $I \equiv SKK$;

(c) $\lambda^*x.Ux \equiv U$ if $x \notin FV(U)$;

(f) $\lambda^*x.UV \equiv S(\lambda^*x.U)(\lambda^*x.V)$ if neither (a) nor (c) applies.

(The names '(a)' - '(f)' are from Curry and Feys 1958, §6A, and are now fairly standard. Curry and Feys called $\lambda^*x.M$ '[x].M'.)

Example.

$\lambda^*x.xy \equiv S(\lambda^*x.x)(\lambda^*x.y)$ by (f)

$\equiv SI(Ky)$ by (b) and (a).

THEOREM 2.15. $(\lambda^*x.M)N$ *behaves like a β-redex; that is*

$$(\lambda^*x.M)N \;\rhd_w\; [N/x]M.$$

Proof. By Lemma 2.12(c) it is enough to prove that

$$(\lambda^*x.M)x \;\rhd_w\; M.$$

This will be done by induction on the length of M.

Case 1: $M \equiv x$. Then Definition 2.14(b) applies, and

$$\begin{aligned}(\lambda^*x.x)x &\equiv Ix \equiv SKKx\\ &\rhd_w Kx(Kx)\\ &\rhd_w x.\end{aligned}$$

Case 2: M is an atom, $M \not\equiv x$. Then 2.14(a) applies, and

$$(\lambda^*x.M)x \equiv KMx \;\rhd_w\; M.$$

Case 3: $M \equiv UV$. By the induction hypothesis, we may assume

$$(\lambda^*x.U)x \;\rhd_w\; U, \qquad (\lambda^*x.V)x \;\rhd_w\; V.$$

Subcase (i): $x \notin FV(M)$. Like Case 2.

Subcase (ii): $x \notin FV(U)$ and $V \equiv x$. Then 2.14(c) applies, and

$$\begin{aligned}(\lambda^*x.M)x &\equiv (\lambda^*x.Ux)x\\ &\equiv Ux\\ &\equiv M.\end{aligned}$$

Subcase (iii): Neither of the above two subcases applies. Then

$$\begin{aligned}(\lambda^*x.M)x &\equiv S(\lambda^*x.U)(\lambda^*x.V)x && \text{by 2.14(f)}\\ &\rhd_w (\lambda^*x.U)x((\lambda^*x.V)x)\\ &\rhd_w UV && \text{by induction hypothesis}\\ &\equiv M.\end{aligned}$$

(Note how the redex-contractions for S and K in Definition 2.7 fit in with the cases in this proof; this is their purpose.) □

REMARK 2.16. There are several other possible definitions of $\lambda^* x.M$ besides the one above. For example, Barendregt 1981 Definition 7.1.5 omits (c); and one could restrict (a) to the case in which U is an atom. Both these changes would leave Theorem 2.15 still true, in fact they would shorten its proof. But restricting (a) would make part (c) of the useful substitution-and-abstraction lemma below fail, and omitting (c) would enormously increase the lengths of terms $\lambda^* x_1.(\ldots(\lambda^* x_n.M)\ldots)$ for most $x_1,\ldots,x_n,M$. Different definitions of λ^* will be compared in Chapter 9.

DEFINITION 2.17. For any $x_1,\ldots,x_n$ (not necessarily distinct),

$$\lambda^* x_1 \ldots x_n.M \equiv \lambda^* x_1.(\lambda^* x_2.(\ldots(\lambda^* x_n.M)\ldots)).$$

EXAMPLES 2.18.

(a) $\lambda^* xy.x \equiv \lambda^* x.(\lambda^* y.x)$

$\equiv \lambda^* x.(\mathsf{K}x)$ by 2.14(a)

$\equiv \mathsf{K}$ by 2.14(c).

(b) $\lambda^* xyz.xz(yz) \equiv \lambda^* x.(\lambda^* y.(\lambda^* z.xz(yz)))$

$\equiv \lambda^* x.(\lambda^* y.(\mathsf{S}(\lambda^* z.xz)(\lambda^* z.yz)))$ by 2.14(f)

$\equiv \lambda^* x.(\lambda^* y.\mathsf{S}xy)$ by 2.14(c)

$\equiv \lambda^* x.\mathsf{S}x$ by 2.14(c)

$\equiv \mathsf{S}$ by 2.14(c).

EXERCISE 2.19. Evaluate $\lambda^* xy.xyy$ and compare it with your answer to Exercise 2.11, combinator W.

THEOREM 2.20. *For any* $x_1,\ldots,x_n$ *(mutually distinct),*

$$(\lambda^* x_1 \ldots x_n.M)U_1 \ldots U_n \rhd_w [U_1/x_1,\ldots,U_n/x_n]M.$$

Proof. By Lemma 2.12(c) it is enough to prove the case in which $U_i \equiv x_i$ for all i. And this comes from Theorem 2.15 by an easy induction on n. □

LEMMA 2.21 (Substitution and abstraction).

(a) $FV(\lambda^* x.M) = FV(M) - \{x\}$;

(b) $\lambda^* y.[y/x]M \equiv \lambda^* x.M$ if $y \notin FV(M)$;

(c) $[N/x](\lambda^* y.M) \equiv \lambda^* y.[N/x]M$ if $y \notin FV(xN)$.

Proof. Induction on M. □

Comment. Part (b) shows that the analogue of the λ-relation $\equiv_\alpha$ is identity here. Part (c) is the analogue of Definition 1.11(e).

The last few results have shown that λ^* has similar properties to λ. But it must be emphasized again that, in contrast to λ, λ^* is not part of the formal system of terms; $\lambda^* x.M$ is defined in the metatheory by induction on M, and it is constructed from S, K, and parts of M.

2D WEAK EQUALITY

DEFINITION 2.22 (Weak equality or weak convertibility). We shall say X is weakly equal or weakly convertible to Y, or

$$X =_w Y,$$

iff Y can be obtained from X by a finite (perhaps empty) series of weak contractions and reversed weak contractions. That is, $X =_w Y$ iff there exist $X_0, \ldots, X_n$ $(n \geq 0)$ such that

$$X_0 \equiv X, \qquad X_n \equiv Y,$$

$$(\forall i \leq n-1)(X_i \rhd_{1w} X_{i+1} \text{ or } X_{i+1} \rhd_{1w} X_i).$$

LEMMA 2.23.

(a) $X =_w Y \Rightarrow [X/v]Z =_w [Y/v]Z$;

(b) $X =_w Y \Rightarrow [U_1/x_1, \ldots, U_n/x_n]X =_w [U_1/x_1, \ldots, U_n/x_n]Y$.

THEOREM 2.24 (Church-Rosser theorem for weak equality). *If* $X =_w Y$, *then there exists a* Z *such that*

$$X \rhd_w Z \quad \textit{and} \quad Y \rhd_w Z.$$

Proof. Like Theorem 1.35. □

COROLLARY 2.24.1. *If* $X =_w Y$ *and* Y *is in weak normal form, then* $X \rhd_w Y$.

COROLLARY 2.24.2. *If* $X =_w Y$, *then either* X *and* Y *have no weak normal forms, or they both have the same weak normal form.*

COROLLARY 2.24.3. *If* X *and* Y *are distinct weak normal forms, then* $X \neq_w Y$; *in particular* $S \neq_w K$. *Hence* $=_w$ *is non-trivial in the sense that not all terms are equal.*

COROLLARY 2.24.4. *A term can be weakly equal to at most one weak normal form.*

COROLLARY 2.24.5. *If* $xX_1 \ldots X_m =_w yY_1 \ldots Y_n$, *then* $x \equiv y$ *and* $m = n$ *and* $X_i =_w Y_i$ *for all* i.

REMARK 2.25 (Warning). By the above results, weak reduction and equality closely parallel $\lambda\beta$-reduction and equality. But the two systems do not correspond exactly. The main difference is that $\lambda\beta$-equality has the property which Curry and Feys 1958 calls (ξ), namely

$$(\xi) \qquad X =_\beta Y \implies \lambda x.X =_\beta \lambda x.Y.$$

((ξ) holds because any contraction or change of bound variable made in X can also be made in $\lambda x.X$.)

When translated into combinatory language, (ξ) becomes

$$X =_w Y \implies \lambda^* x.X =_w \lambda^* x.Y.$$

But for CL-terms λ^* is not part of the syntax, and (ξ) fails. For example, take

$$X \equiv Sxyz, \qquad Y \equiv xz(yz);$$

then $X =_w Y$, but

$$\lambda^*x.X \equiv S(SS(Ky))(Kz),$$

$$\lambda^*x.Y \equiv S(SI(Kz))(K(yz)).$$

These are normal forms and distinct, so by 2.24.3 they are not weakly equal.

For many purposes the lack of (ξ) is no problem and the simplicity of weak equality gives it an advantage over λ. This is especially true if all we want to do is define a set of functions in a formal theory, for example the recursive functions in Chapter 5, or the primitive recursive functionals of Gödel in Chapter 18.

But for some other purposes (ξ) turns out to be indispensable, and weak equality is too weak. We then either have to abandon combinators and use λ, or add new axioms to weak equality to make it stronger. Possible extra axioms will be discussed in Chapter 9.

EXERCISES 2.26.*

(a) Construct a pairing-combinator and its projections; i.e. construct D, D_1, D_2 such that

$$D_1(Dxy) \rhd_w x, \qquad D_2(Dxy) \rhd_w y.$$

(b) Show that there is no combinator that dinstinguishes between atoms and composite terms; i.e. show that there is no A such that

$$AX =_w S \text{ if } X \text{ is an atom,}$$

$$AX =_w K \text{ if } X \equiv UV \text{ for some } U, V.$$

(In general, syntactic operations on terms cannot be done by combinators.)

(c) Prove that a term X is in weak normal form iff X is minimal with respect to weak reduction, i.e. iff

$$X \rhd_w Y \Rightarrow Y \equiv X.$$

(Contrast λ-calculus, Exercise 1.31.) Show that this would be false if there were an atom W with an axiom-scheme

$$WXY \rhd_w XYY.$$

FURTHER READING 2.27.

Barendregt 1981. Contains only one chapter on combinators (Ch. 7). But most of the ideas in the book apply to combinatory logic as well as λ-calculus, and most theorems on weak reduction are proved by simplified versions of λ-proofs.

Curry and Feys 1958. The 'Bible' of combinatory logic for many years. Now very dated, but still valuable for, e.g.: its discussion of particular combinators and interdefinability questions (Ch. 5), alternative definitions of λ^* (§6A), strong equality and reduction (§§6D 6F), systems of type theory based on combinators (Chs. 7 - 9), and historical comments.

Curry et al 1972. Continues and updates Curry and Feys 1958. The main properties of weak reduction are proved in §11B, and definitions of λ^* are discussed in §11C. References for other topics will be given as they crop up later in the present book.

Rezus 1982. This bibliography includes all the literature on combinatory logic up to 1982, as well as λ-calculus.

Schönfinkel 1924. The first exposition of combinators, by the man who invented them. A very readable non-technical introduction.

Smullyan 1985. Contains a humorous and clever account of combinators and self-application.

Backus 1978. A plea for a functional style of programming, using combinators as an analogy. Has led to an upsurge of interest in combinators, and to several explicitly combinator-based programming languages. (Some precursors: Landin 1965, 1966, Böhm and Gross 1966, Turner 1976.)

CHAPTER THREE

THE POWER OF λ AND COMBINATORS

3A INTRODUCTION

The purpose of this chapter and the next two is to give some concrete evidence that λ and CL are non-trivial.

Chapter 4 will show that all partial recursive functions are definable in both systems, and Chapter 5 will deduce from this a general undecidability theorem.

The present chapter gives three interesting theorems which hold for both λ and combinators, and are used repeatedly in the published literature: the Fixed-point theorem, Böhm's theorem, and a version of the Standardization theorem due to Henk Barendregt.

After these results, §3E will touch on the history of λ-calculus and combinatory logic, and discuss the question of whether they have any meaning, or are just uninterpretable formal systems.

NOTATION 3.1. This chapter is written in a neutral notation, which may be interpreted in either λ or CL, as follows.

Notation	Meaning for λ	Meaning for CL
Term	λ-term	CL-term
$X \equiv Y$	X congruent to Y	X identical to Y
$X \rhd_{\beta,w} Y$	$X \rhd_{\beta} Y$	$X \rhd_{w} Y$
$X =_{\beta,w} Y$	$X =_{\beta} Y$	$X =_{w} Y$
λ	λ	λ^*

DEFINITION 3.2. A combinator is (in λ) a closed term containing no constant atoms, and (in CL) a term whose only atoms are S and K. In λ-calculus, three combinators are given special names:

$$S \equiv \lambda xyz.xz(yz), \qquad K \equiv \lambda xy.x, \qquad I \equiv \lambda x.x.$$

3B THE FIXED-POINT THEOREM

In the world of pure combinators and λ-calculus, every operator has a fixed-point. (Cf. Exercise 1.37(b).) That is, for each term X there is a term P such that

$$XP =_{\beta,w} P.$$

Furthermore, there is a combinator Y which finds these fixed-points, i.e. such that YX is a fixed-point of X for all X.

THEOREM 3.3 (Fixed-point theorem). *In both λ-calculus and combinatory logic, there is a combinator* Y *such that*

(a) $$Yx =_{\beta,w} x(Yx).$$

In fact, there is a Y *with the stronger property*

(b) $$Yx \rhd_{\beta,w} x(Yx).$$

Comment. Y is not unique; two definitions will be given below, and there are many others (Curry et al. 1972 §11F7, Barendregt 1981 §6.5). The first is due to Curry and is the simplest known, but only satisfies (a) above. The second is due to Alan M. Turing and satisfies (b) as well.

DEFINITION 3.4 (Two fixed-point combinators). Define

$$Y_{Curry} \equiv \lambda x.VV, \qquad V \equiv \lambda y.x(yy);$$

$$Y_{Turing} \equiv ZZ, \qquad Z \equiv \lambda zx.x(zzx).$$

Curry: Yx = λx. VV = λx(λy. x(yy)) λy. x(yy)
= λx. x

Proof of Theorem 3.3. Let Y be Y_{Turing}. Then Y satisfies (b), because

$$
\begin{array}{llll}
\mathsf{Y}x & \equiv & (\lambda z.(\lambda x.x(zzx)))\,Z\,x & \text{by definition of } Z \\
& \triangleright_{\beta,w} & ([Z/z](\lambda x.x(zzx)))\,x & \text{by Definition 1.22 or Theorem 2.15} \\
& \equiv & (\lambda x.x(ZZx))x & \text{by Definition 1.11 or Lemma 2.21(c)} \\
& & & \text{(Note that } FV(Z) \text{ is empty.)} \\
& \triangleright_{\beta,w} & x(ZZx) & \text{by Definition 1.22 or Theorem 2.15} \\
& \equiv & x(\mathsf{Y}x). &
\end{array}
$$

Note that the above reduction is correct for both λ and CL. For λ, each of the two steps above is a single contraction; for CL, it is a reduction given by Theorem 2.15. □

Exercise: Prove that Y_{Curry} satisfies (a).

COROLLARY 3.3.1. *In both* λ *and* CL; *for any* Z *and* $n \geq 0$, *the equation*

$$xy_1 \ldots y_n = Z$$

can be solved for x. *That is, there is a term* X *such that*

$$Xy_1 \ldots y_n =_{\beta,w} [X/x]Z.$$

Proof. Define $X \equiv \mathsf{Y}(\lambda xy_1 \ldots y_n.Z)$. □

Comment. Note that Z may contain any or none of $x, y_1, \ldots, y_n$. However, the most interesting cases occur when Z contains x.

This corollary will show some of the power of the fixed-point theorem. It can be used in representing the recursive functions by λ- and CL-terms (Chapter 4). If one is not careful, it can also be used to construct logical paradoxes (Chapter 17).

On a more trivial level, it provides the λ and combinatory world with a garbage disposer X_1 which swallows all arguments presented to it,

$$X_1 y =_{\beta,w} X_1,$$

and a bureaucrat X_2 which eternally permutes its arguments with no other effect,

$$X_2yz =_{\beta,w} X_2zy.$$

EXERCISE 3.5.* Extend Corollary 3.3.1 to show that every finite set of simultaneous equations of form

$$\begin{aligned} x_1y_1\ldots y_n &= Z_1 \\ &\vdots \\ x_ky_1\ldots y_k &= Z_k \end{aligned} \qquad (n \geq 0,\ k \geq 1)$$

is solvable for $x_1,\ldots,x_k$ in both λ and CL. ($Z_1,\ldots,Z_k$ may contain any or none of $x_1,\ldots,x_k,y_1,\ldots,y_n$.)

COROLLARY 3.3.2 (Double fixed-point theorem). *In both* λ *and* CL: *for any* X, Y *there exist* P, Q *such that*

$$XPQ =_{\beta,w} P, \qquad YPQ =_{\beta,w} Q.$$

Proof. By Exercise 3.5 with $n = k = 2$, there exist X_1,X_2 such that

$$X_iy_1y_2 =_{\beta,w} y_i(X_1y_1y_2)(X_2y_1y_2) \quad \text{for } i = 1,2.$$

Define $P \equiv X_1XY$ and $Q \equiv X_2XY$. (Barendregt 1981, §6.5.) □

For more on fixed-points, see Barendregt 1981 §§6.1, 6.5, 19.3.

3C BÖHM'S THEOREM

The next theorem is due to Corrado Böhm (Böhm 1968); it shows that the members of a significant class of normal forms can be distinguished from each other in a very powerful way, and it has applications in both the syntax and semantics of λ and CL.

To prepare for the theorem the relevant class of normal forms will now be defined, first in λ-calculus and then in combinatory logic. These two classes will gain further significance in later chapters, but for the moment they are simply aids to stating Böhm's theorem.

DEFINITION 3.6 ($\beta\eta$-normal forms). A λ-term X which contains no β-redexes and no parts of form

(i) $\lambda x.Ux, \quad x \notin FV(U),$

is called a $\beta\eta$-normal form (or a term in $\beta\eta$-normal form). The class of all such λ-terms is called $\beta\eta$-nf or $\lambda\beta\eta$-nf. (Terms of form (i) are called η-redexes, and we shall study them in Chapter 7.)

Example: The λ-term $\lambda ux.ux$, which is really $\lambda u.(\lambda x.ux)$, is in β-nf but not in $\beta\eta$-nf.

DEFINITION 3.7 (Strong normal forms). For CL-terms, the class strong nf is defined inductively as follows. Its members are called strong normal forms, or terms in strong normal form.

(a) All non-redex atoms are in strong nf;

(b) if $X_1,\ldots,X_n$ are in strong nf, and a is any non-redex atom, then $aX_1\ldots X_n$ is in strong nf;

(c) if X is in strong nf, then so is $\lambda^* x.X$.

EXERCISE 3.8.

(a) Notice that Definition 3.7 is like Lemma 1.30.

(b) Show that the class strong nf contains K, S, and all terms whose only atoms are non-redex atoms.

LEMMA 3.9. *Every* CL-*term in strong normal form is also in weak normal form.*

Proof. Induction on Definition 3.7. □

Remark. The converse to this lemma is false, by the way. For example, the fixed-point combinators in Definition 3.4 are easily seen to be both in weak nf; but Exercise 3.11(a) will show that they cannot be in strong nf.

THEOREM 3.10 (<u>Böhm's theorem</u>). *Let* M *and* N *be combinators, either in* βη*-normal form (in* λ*-calculus) or in strong normal form (in combinatory logic). If* $M \not\equiv N$, *then there exist* $n \geq 0$ *and combinators* $L_1, \ldots, L_n$ *such that*

$$ML_1 \ldots L_n xy \rhd_{\beta,w} x,$$

$$NL_1 \ldots L_n xy \rhd_{\beta,w} y.$$

<u>Comment</u>. This theorem says that by feeding M and N with suitable arguments, the same for both, they can be made to behave as different selectors.

The proof is a deep analysis of the structures of M and N, and is beyond this book. Böhm's original proof was for λ (Böhm 1968), and there is another proof for λ in Curry et al 1972 §11F8. A proof for λ is also contained in the thorough investigation of the theorem in Barendregt 1981 Ch. 10. (See especially Theorem 10.4.2(ii), of which the above theorem is the special case $P \equiv \lambda xy.x$, $Q \equiv \lambda xy.y$).

For combinatory logic, the theorem can be deduced from the λ-theorem (Hindley 1979). Alternatively, a careful check of the λ-proofs of Böhm and Curry shows that all the reductions in these proofs become correct weak reductions when translated into CL.

The theorem can be extended to more than two normal forms; Böhm et al 1979.

COROLLARY 3.10.1. *Let* M, N *be combinators, either in* βη*-normal form (in* λ*) or in strong normal form (in* CL*), and let* $M \not\equiv N$. *If we add the equation* $M = N$ *as a new axiom to the definition of* $=_\beta$ *or* $=_w$, *then all terms become equal.*

Proof. ('Add the equation $M = N$ as a new axiom' really means allowing any occurrence of M in a term to be replaced by N, and vice-versa.) For any X, Y, we have

$$X =_{\beta,w} ML_1 \ldots L_n XY \quad \text{by Bohm's theorem,}$$

$$=_{\beta,w} NL_1 \ldots L_n XY \quad \text{by the axiom } M = N,$$

$$=_{\beta,w} Y. \qquad \square$$

EXERCISES 3.11.

(a)* In combinatory logic, show that no term Y in strong normal form can satisfy the fixed-point equation

$$Yx =_w x(Yx).$$

Hence, to say that a CL-term is in strong nf is some restriction on the kinds of operator it can represent. In contrast, there is no such restriction on the weak normal forms. Show that if there is a combinator X such that

$$Xy_1 \dots y_n =_w [X/x]Z,$$

where Z is any given combination of $x, y_1, \dots, y_n, S, K$, then there is a weak normal form X' with exactly the same property, namely

$$X'y_1 \dots y_n =_w [X'/x]Z.$$

(Hint: why are the Y's of Definition 3.4 in weak nf?) Contrast 3.20.

(b)* Prove Böhm's theorem for λ or for CL, in the following two special cases:

(i) $M \equiv \lambda xyz.xz(yz)$, $N \equiv \lambda xyz.x(yz)$;

(ii) $M \equiv \lambda xy.x(yy)$, $N \equiv \lambda xy.x(yx)$.

(Hint for (ii): choose $n \geq 3$.) The repeated 'x' in N in (ii) shows the main difficulty in the proof of Böhm's theorem.

3D THE QUASI-LEFTMOST-REDUCTION THEOREM

DISCUSSION 3.12. The proofs of some of the deeper theorems in λ and CL depend on an underlying theory of reductions, most of which is well beyond the scope of this book. (See Barendregt 1981 Chs. 11-15.) The first major result of this theory is the Church-Rosser theorem (1.29, 2.13, and Appendix 1). It holds for all the kinds of reduction to be defined in this book, and for many others under certain broad conditions.

The second main result is the Standardization theorem, due to Curry. It says that if X reduces to Y, then there is a reduction from X to Y with a particularly simple form, called a standard reduction. One of its uses is in proving statements of form 'X does not reduce to Y'. But it has many others, for example it is needed in the proof of Bohm's theorem. It holds for both β and weak reduction, and for some but not all of the reductions to be defined later.

Unfortunately, even to define a standard reduction is a little beyond this book. However, there is a well-explained account of the theorem for λβ-reduction in Barendregt 1981 §11.4, based on a neat proof by Gerd Mitschke in his paper 1979. For combinatory weak reduction there is a proof in Curry et al. 1972 §11B3. For some more general kinds of reduction there is an abstract proof with discussion in Hindley 1978b, based on the original λ-proof in Curry et al. 1958 §4E1.

DISCUSSION 3.13. Instead of studying the theorem itself here, what we shall do is look at one of its more important consequences. It is a modification by Henk Barendregt of a theorem of Curry, and will be called here the Quasi-leftmost-reduction theorem. Its main use is to show that terms have no normal form, and it will be needed for this purpose in Chapter 4.

But first we need to say precisely what a contraction or a reduction is, as follows.

DEFINITION 3.14 (Contractions and reductions). In both λ-calculus and combinatory logic, a contraction is an ordered triple $\langle X,R,Y\rangle$ where X is a term, R is an occurrence of a redex in X, and Y is the result of contracting R in X. (For 'occurrence' see the end of Definition 1.6; for 'contracting' see 1.22 and 2.7.) Instead of '$\langle X,R,Y\rangle$' we may write

$$X \triangleright_R Y.$$

A reduction ρ is a finite or infinite series of contractions of the form

$$X_1 \triangleright_{R_1} X_2 \triangleright_{R_2} X_3 \triangleright_{R_3} \dots \;.$$

Its length is the number of its contractions (finite or ∞). If the length is finite, say n, then X_{n+1} is called the terminus of ρ.

DEFINITION 3.15. A reduction ρ has maximal length iff either ρ is infinite or its terminus contains no redexes (i.e. iff ρ continues as long as there are redexes to be contracted).

EXAMPLES 3.16.

(a) Consider the following combinatory weak reduction (the R_i contracted at each step is underlined):

$$
\begin{aligned}
S(Kx)Iy &\equiv \underline{S(Kx)(SKK)y} \\
&\rhd_{1w} Kxy(\underline{SKKy}) \\
&\rhd_{1w} Kxy(\underline{Ky(Ky)}) \\
&\rhd_{1w} Kxyy.
\end{aligned}
$$

The length of this reduction is 3, and it is not maximal. To make it maximal we must add the contraction $Kxyy \rhd_{1w} xy$.

(b) The following $\lambda\beta$-reduction has infinite length:

$$X_1 \equiv (\lambda x.xx)(\lambda x.xx), \quad R_1 \equiv X_1,$$

$$X_i \equiv X_1 \qquad \text{for all } i \geq 2.$$

DEFINITION 3.17. An occurrence R of a redex in X is called maximal iff it is not contained in any other redex-occurrence in X. It is leftmost maximal iff it is the leftmost of the maximal occurrences in X. (E.g. the maximal redex-occurrences in

$$SS(Kx(Kyz)(Kuv))$$

are $Kx(Kyz)$ and Kuv.) A reduction ρ such that for each i, R_i is the leftmost maximal redex-occurrence in X_i, and which has maximal length, is called the normal or leftmost reduction of X_0. (It is uniquely determined, given X_0.)

DEFINITION 3.18. A quasi-leftmost reduction of X_0 is a reduction ρ with maximal length and such that for each i, if X_i is not the terminus then there exists $j \geq i$ such that R_j is leftmost maximal.

An infinite ρ is quasi-leftmost iff an infinity of its R_i are leftmost maximal (and is leftmost iff they all are). In example 3.16(a), the second redex contracted is not leftmost maximal; hence the reduction

$$S(Kx)Iy \rhd_{1w} Kxy(Iy) \rhd_{1w} Kxy(Ky(Ky)) \rhd_{1w} Kxyy \rhd_{1w} xy$$

is quasi-leftmost but not leftmost. In contrast, the reduction

$$S(Kx)Iy \rhd_{1w} Kxy(Iy) \rhd_{1w} x(Iy) \rhd_{1w} x(Ky(Ky)) \rhd_{1w} xy$$

is leftmost maximal.

THEOREM 3.19 (Quasi-leftmost-reduction theorem). *For* λ-*terms and* $\rhd_\beta$, *or* CL-*terms and* $\rhd_w$: *if* X *has a normal form* X^*, *then every quasi-leftmost reduction of* X *is finite and terminates at* X^*.

COROLLARY 3.19.1. *For* λ-*terms and* $\rhd_\beta$, *or* CL-*terms and* $\rhd_w$: X *has no normal form iff some quasi-leftmost reduction of* X *is infinite.*

Comment. In the Corollary, 'has no normal form' means as usual, 'does not β-reduce to a β-normal form' (for λ), and 'does not weakly reduce to a weak normal form' (for CL).

Without this corollary, to show that a term X has no normal form we would have the impossibly complicated task of showing that all reductions starting at X could be continued for ever. The corollary reduces this to showing that just one reduction can be continued for ever.

Proof references. Theorem 3.19 depends on the standardization theorem. It was first formulated and proved by Barendregt, strengthening a similar result by Curry which said only that the normal reduction terminates at X^*. There is a proof for λ in Barendregt 1981 Theorem 13.2.6.

For CL, there is no explicit proof in the literature. But the above λ-proof applies also to CL. (The standardization theorem, on which it depends, is proved for weak reduction in Curry et al. 1972, §11B3.) A slightly different proof can be found by modifying the proof of a very similar result, also due to Barendregt, in Hindley 1978b, Theorem 8; just omit Lemma 5 and replace its use in Lemma 6 by the definition of 'quasi-leftmost'. □

EXERCISE 3.20. In λ-calculus, prove that the Y-combinators in 3.4 have no β-normal form. (In CL, in contrast, they not only possess weak normal forms, they are in weak nf; cf. Exercise 3.11(a). Cf. also 9.16(f).)

REMARK 3.21. Because the standardization theorem has had to be omitted from this book, two other important topics, whose proper understanding depends on knowing something of its proof, have had to be omitted also. These are unsolvable terms and Böhm trees. Unsolvable terms formalize the notion of 'undefined object'; see Barendregt 1981 §§2.2, 8.3, 9.4. The Böhm tree of a term encodes a procedure for searching for a normal form of the term; see Barendregt 1981, Chapter 10.

FURTHER READING 3.22. For the reduction-theory of λ and CL, there is a comprehensive treatment of all the main results in Barendregt 1981 Chs. 3 and 11-14, based on work by Lévy, Klop and Barendregt. This is written in terms of λ, but most of its results hold also for CL_w. For the theory of more general kinds of reductions, see for example Staples 1975, Rosen 1973, Huet 1980, Hindley 1969b, 1974, 1978a,b. (At the end of 1978b there is a table showing which main results hold for which kinds of reduction.) For reduction techniques in an important non-λ-CL-context, see for example Knuth and Bendix 1967, Huet 1981, and the surveys Huet and Oppen 1980, Book 1983 and the references they give.

But for the deeper properties of reductions, in λ and CL and in general, the essential reading is the thesis Klop 1980, which combines significant mathematical advances with a very clearly written overview of the field.

3E HISTORY AND INTERPRETATION

HISTORICAL COMMENT 3.23. Combinators were invented by Moses Schönfinkel in about 1920, and published in Schönfinkel 1924. They were re-invented independently a few years later by Haskell Curry, who was then responsible for the main line of development of combinatory logic until about 1970.

The λ-calculus was introduced by Alonzo Church in the early 1930's and developed by Church, Barkley Rosser and Stephen Kleene.

Both λ and CL were originally introduced as parts of systems of higher-order logic designed to provide a type-free foundation for mathematics.

The original logical systems of Church and Curry were proved inconsistent by Kleene and Rosser 1935, and everybody except Curry and F.B. Fitch lost interest in this application. Fitch's contribution was to develop a series of powerful but non-recursively-enumerable systems; see his 1936, 1963, 1967, 1980a,b, and for a general exposition, his book 1974. The work of Curry and his students will be described in Chapter 17. Its main drive was towards the detailed investigation of weak systems, and the last system of Curry, presented in his 1973, is still weaker than classical second-order logic, although it does have some interesting models (Scott 1975b, Aczel 1980). Somewhat later, Martin Bunder's 1983a-c introduced a new system in which second and higher-order logic can be interpreted.

Strong logical systems have also been devised by Solomon Feferman (see his 1975a,b and, for more recent work and references, his 1984), and interest in such systems is increasing again.

But interest in λ and CL has not been restricted to their use in logical theories. Right from the beginning, the pure systems also attracted attention, and major contributions to the pure theory were made by Church, Curry, Kleene, Rosser and others in the earlier days, and later by Dana Scott and Henk Barendregt, who, with their students and former students, are currently the leaders in the 'pure' field. The main recent advances in this field have been, first, the study of semantics, started by Scott's construction of a model of λ-calculus in set theory in 1969, and second, the insight into the deep structure of reductions gained by Jean-Jacques Lévy and Jan-Willem Klop. These are all described in Barendregt 1981.

On the 'applied' side, the main development has been the rapidly-growing interaction with programming-language theory. This began in the late 1950's when John McCarthy invented the language LISP, which was very close to λ-calculus. It was fuelled in the late 1960's when Scott's analysis of computability led simultaneously to a model of λ-calculus on the one hand, and the theory of the 'denotational semantics' of programming languages on the other. Currently, λ-calculus and combinatory logic are regarded as 'test-beds' in the study of higher-order programming languages: techniques are tried out on these two simple languages, developed, and then applied to other more 'practical' languages.

For further reading on the history of the subject, see Kleene 1981, Rosser 1984, Seldin 1980, and the historical sections at the ends of the chapters in Curry et al. 1958.

DISCUSSION 3.24. Up to now, λ-calculus and combinatory logic have been presented as uninterpreted formal systems, and the authors hope that even on this level they have been shown to have some technical interest.

However, these systems were originally developed to formalize primitive properties of functions or operators. In particular, I represents the identity operator, K an operator which forms constant-functions, S a substitution-and-composition operator, etc.

But just what kind of operators are these? Most mathematicians think of functions as being sets of ordered pairs in some standard set theory, for example Zermelo-Fraenkel set theory (ZF). To such a mathematician, I, K and S simply do not exist. In ZF each set $\mathscr{S}$ has an identity-function $I_{\mathscr{S}}$ with domain $\mathscr{S}$, but there is no 'universal' identity which can be applied to everything. (Similarly for K and S.)

In many practical applications of combinators this question does not arise. For example, the terms used to define Gödel's functionals in a later chapter will be limited by built-in type-restrictions, and instead of one 'universal' identity-term I there will be a term I_α for each type-symbol α. Type-symbols will denote sets, and I_α will denote the identity on the set denoted by α. But type-free systems also have their uses, and for these systems the question must still be faced: what kind of functions do the terms represent?

One possible answer is explained very clearly by Church in the introduction to his 1941. In the 1920's when λ and CL began, logicians did not automatically think of functions as sets of ordered pairs, with domain and range given, as they are trained to do today. Throughout mathematical history, right through to modern computer science, there has run another concept of function, less precise but strongly influential; that of a function as an operation-process (in some sense) which may be applied to certain objects to produce other objects. Such a process can be defined by giving a set of rules describing how it acts on an arbitrary input-object. (The rules need not produce an output for every input.) A simple example is the permutation-operation ϕ defined by

$$\phi(\langle x,y,z\rangle) = \langle y,z,x\rangle.$$

Nowadays one would think of a computer program, though the 'operation-process' concept was not intended to have the finiteness and effectiveveness limitations that are involved with computation.

From now on, let us reserve the word 'operator' to denote this imprecise function-as-operation-process concept, and 'function' and 'map' for the set-of-ordered-pairs concept.

Perhaps the most important difference between operators and functions is that an operator may be defined by describing its action without defining the set of inputs for which this action produces results, i.e. without defining its domain. (Compare for example the definition of a function by a not-everywhere-convergent power series in informal differential calculus.) In a sense, operators are 'partial functions'.

A second important difference is that some operators have no restriction on their domain; they accept any inputs, including themselves. The simplest example is I, which is defined by the operation of doing nothing at all. If this is accepted as a well-defined concept, then surely the operation of doing nothing can be applied to it. We simply get

$$II = I.$$

Other examples are K and S; in formal conbinatory logic we have

$$KKxyz =_w y, \qquad SSxyz =_w yz(xyz),$$

which suggest natural meanings for KK and SS. Of course, it is not claimed that every operator is self-applicable; this would lead to contradictions. But the self-applicability of at least such simple operators as K, S and I seems very reasonable.

The operator concept has never been successfully formalized, at least not in a theory that is both consistent and reasonably comprehensive. Church and others tried it in the 1930's but their systems turned out inconsistent. And later work by Curry, Seldin and Bunder has concentrated on weaker theories (Chapter 17).

But if a concept has not yet been formalized fully, we cannot infer that it does not exist. Otherwise, set-theory would never have progressed beyond the Russell paradox. Further insight and clarification are needed, but the operator concept is far too fertile a way of thinking to be simply abandoned.

An operator-style concept of function currently plays a role in theories of constructive mathematics, and to some extent in category theory. Also several recent attempts to formalize the foundations of mathematics have started from operators rather than sets. (E.g. Chauvin 1979, Feferman 1975b, 1977, Aczel and Feferman 1980, and the references they cite.) And of course it is fundamental in computer science; indeed, the reader with a computing background may be wondering why the authors feel the concept needs motivating at all.

However, it must be stressed that most of the rest of this book will be independent of any sympathy with the operator concept. (The only exceptions will be Chapters 14-17, and even these will have independent interest as formal studies.)

Also, despite the non-set-theoretic origins of λ and CL, there do exist models of both systems in ordinary ZF set-theory (see Chapter 12). Such a model cannot interpret terms as functions of course. In fact, the first non-trivial model interpreted them, roughly speaking, as infinite sequences of functions. (Dana Scott's model D_∞, see Chapter 12.) These models have been important contributions to the theory of computation.

Another way of interpreting combinators in set theory is to change the set theory. The simplest change is to drop the axiom of foundation which prevents functions from applying to themselves, and this allows a very natural model to be constructed (von Rimscha 1980).

CHAPTER FOUR

REPRESENTING THE RECURSIVE FUNCTIONS

In this chapter a sequence of terms will be picked out to represent the natural numbers. It is reasonable to expect, from the informal interpretation of terms as operators, that some of the other terms will then represent functions of the natural numbers in some sense. This sense will be defined precisely below, and it will be shown that the functions so representable are exactly the partial recursive ones.

This is further evidence that λ-calculus and combinatory logic are not the trivial notation-games that they might at first seem, for they are deep enough to give an alternative definition of the concept of computable function. Indeed, historically, the concepts of λ-definable function and combinatorially definable function grew up independently of that of recursive function, and the first undecidability result was obtained by Church in the λ-calculus, not in the theory of recursive functions. (See the next chapter.)

NOTATION 4.1. This chapter is written in the same neutral notation as Chapter 3, and its results will hold for both λ-calculus and combinatory logic unless explicitly stated otherwise.

Recall that 'X <u>has no normal form</u>' means 'X does not β-reduce to a β-normal form' in λ-calculus, and 'X does not weakly reduce to a weak normal form' in combinatory logic.

Recall also the definition of <u>combinator</u> in 3.2.

<u>Natural numbers</u> will be denoted by 'i', 'j', 'k', 'm', 'n', and the set of all natural numbers by $\mathbb{N}$. ($0 \in \mathbb{N}$.)

A function ϕ from a subset of $\mathbb{N}^k$ into $\mathbb{N}$ will be called <u>properly partial</u> iff $\phi(m_1,\ldots,m_k)$ is undefined for some $m_1,\ldots,m_k$, and <u>total</u> otherwise. As usual, a <u>partial</u> function may be either properly partial or total. The class of all <u>partial recursive functions</u> (partial functions that are recursive) will be defined below. A <u>recursive total function</u> will be a partial recursive function that is total.

For any terms X and Y, we shall use the abbreviation

$$X^nY \equiv \underbrace{X(X(\ldots(X}_{n\ X\text{-s}}Y)\ldots)) \quad \text{for } n \geq 1,$$

$$X^0Y \equiv Y.$$

DEFINITION 4.2 (<u>The Church numerals</u>). For each natural number n, the <u>Church numeral representing</u> n is (in λ-calculus) the term

$$\bar{n} \equiv \lambda xy.x^n y,$$

and (in CL) the term

$$\bar{n} \equiv (SB)^n(KI).$$

Sometimes $\bar{n}$ is called Z_n.

This representation comes from Church 1941, p.28; in both λ and CL it has the useful property that for all terms F, X,

(4.3) $$\bar{n}FX \triangleright_{\beta,w} F^nX.$$

(In particular, $\bar{1}$ represents the operation of applying a function to an argument, because $\bar{1}FX \triangleright_{\beta,w} FX$.)

Other representations exist in the literature, each with its own technical advantages; cf. Curry et al. 1972, §13A1, and Barendregt 1981, Def. 6.2.9 and §6.4.

DEFINITION 4.4. Let ϕ be an n-argument partial function of natural numbers. We say a λ-term X <u>λ-defines</u> ϕ, or a CL-term X <u>combinatorially defines</u> ϕ, iff for all $m_1,\ldots,m_n$,

$$X\bar{m}_1\ldots\bar{m}_n =_{\beta,w} \overline{\phi(m_1,\ldots,m_n)}$$

when $\phi(m_1,\ldots,m_n)$ is defined, and $X\bar{m}_1\ldots\bar{m}_n$ has no normal form otherwise.

The main result in this chapter will be that every partial recursive function can be both λ- and combinatorially defined.

Conversely, every λ- or combinatorially definable function can be shown to be partial recursive. The proof of this fact will not be given here; it comes from the fact that the definitions of $=_\beta$ and $=_w$ can be rewritten as recursively axiomatized formal theories (see Chapter 6), and it proceeds by tedious but fairly standard Gödel-numbering techniques (Kleene 1936). So, for both λ and CL, the definable functions are exactly the partial recursive ones.

LEMMA 4.5. *For λ-terms and $=_\beta$, or CL-terms and $=_w$: every primitive recursive function ϕ can be defined by a combinator $\bar{\phi}$.*

Proof and comments. The lemma is due in essence to Kleene. Incidentally, as a rather tedious exercise, the $\bar{\phi}$ constructed below can be shown to have a normal form.

The class of primitive recursive functions is defined by induction in Mendelson 1964, Chapter 3 §3. But for notational convenience we shall here include the natural numbers as 0-place functions and use the following definition from Kleene 1952, §44 Remark 1 Basis B:

(I) The successor function σ is primitive recursive;

(II) 0 is a 0-place primitive recursive function;

(III) for each $n \geq 1$ and $k \leq n$ the projection function

$$\phi(m_1,\ldots,m_n) = m_k \qquad (\forall m_1,\ldots,m_n)$$

is primitive recursive;

(IV) if ψ, $\chi_1,\ldots,\chi_p$ are primitive recursive, $n \geq 0$, and

$$\phi(m_1,\ldots,m_n) = \psi(\chi_1(m_1,\ldots,m_n),\ldots,\chi_p(m_1,\ldots,m_n)),$$

then ϕ is primitive recursive;

(V) If ψ and χ are primitive recursive, $n \geq 0$, and

$$\phi(0,m_1,\ldots,m_n) = \psi(m_1,\ldots,m_n),$$

$$\phi(k+1,m_1,\ldots,m_n) = \chi(k, \phi(k,m_1,\ldots,m_n),m_1,\ldots,m_n),$$

then ϕ is primitive recursive.

The term $\bar{\phi}$ is chosen by induction corresponding to these clauses, as follows.

(I) $\bar{\sigma} \equiv \lambda uxy.x(uxy)$ (in CL, this is SB).

(II) $\bar{0} \equiv \lambda xy.y$ (in CL, this is KI).

(III) $\bar{\phi} \equiv \lambda x_1\ldots x_n.x_k$.

(IV) Given $\bar{\psi}, \bar{\chi}_1,\ldots,\bar{\chi}_p$ defining $\psi, \chi_1,\ldots,\chi_p$ respectively, let

$$\bar{\phi} \equiv \lambda x_1\ldots x_n.(\bar{\psi}(\bar{\chi}_1 x_1\ldots x_n)\ldots(\bar{\chi}_p x_1\ldots x_n)).$$

(V) Given $\bar{\psi}$ and $\bar{\chi}$ defining ψ and χ respectively, let

$$\bar{\phi} \equiv \lambda u x_1\ldots x_n.R(\bar{\psi}x_1\ldots x_n)(\lambda uv.\bar{\chi}uvx_1\ldots x_n)u,$$

where R is a combinator to be constructed below, called a <u>recursion combinator</u>. R will have the property that for all X, Y, k,

$$(4.6)\qquad \begin{cases} RXY\bar{0} =_{\beta,w} X, \\ RXY(\overline{k+1}) =_{\beta,w} Y\bar{k}(RXY\bar{k}). \end{cases}$$

If such an R exists, then $\bar{\phi}$ will define ϕ; because

$$\bar{\phi}\bar{0}x_1\ldots x_n =_{\beta,w} R(\bar{\psi}x_1\ldots x_n)(\lambda uv.\bar{\chi}uvx_1\ldots x_n)\bar{0}$$

$$=_{\beta,w} \bar{\psi}x_1\ldots x_n \qquad \text{by (4.6);}$$

$$\bar{\phi}(\overline{k+1})x_1\ldots x_n$$

$$=_{\beta,w} R(\bar{\psi}x_1\ldots x_n)(\lambda uv.\bar{\chi}uvx_1\ldots x_n)(\overline{k+1})$$

$$=_{\beta,w} (\lambda uv.\bar{\chi}uvx_1\ldots x_n)\bar{k}(R(\bar{\psi}x_1\ldots x_n)(\lambda uv.\bar{\chi}uvx_1\ldots x_n)\bar{k}) \quad \text{by (4.6)}$$

$$=_{\beta,w} (\lambda uv.\bar{\chi}uvx_1\ldots x_n)\bar{k}(\bar{\phi}\bar{k}x_1\ldots x_n) \quad \text{by definition of } \bar{\phi}$$

$$=_{\beta,w} \bar{\chi}\bar{k}(\bar{\phi}\bar{k}x_1\ldots x_n)x_1\ldots x_n.$$

We shall now construct an R satisfying (4.6). There are many ways of doing this (Curry et al. 1972, §13A3); the one chosen here is due to P. Bernays, and is one of the easiest to motivate. It has a normal form, and is also stratified in the sense of Chapters 14 and 15.

To motivate Bernays' R, consider for example a primitive recursive function ϕ defined by

$$\phi(0) = m, \quad \phi(k+1) = \chi(k,\phi(k)).$$

To calculate $\phi(k)$ we may first write down the ordered pair $\langle 0,m\rangle$ and then iterate k times the operation f such that

$$f(\langle n,x\rangle) = \langle n+1,\chi(n,x)\rangle,$$

and finally take the second member of the last pair produced. Bernays' R

is the λ-analogue of this calculation procedure.

The first step in constructing R is to define an 'ordered-pair' combinator (cf. Exercise 2.26(a)):

(4.7) $$D \equiv \lambda xyz.z(\mathsf{K}y)x.$$

This D has the property that

(4.8) $$\begin{cases} \mathsf{D}XY\bar{0} =_{\beta,w} X, \\ \mathsf{D}XY(\overline{k+1}) =_{\beta,w} Y. \end{cases}$$

$\mathsf{D}\bar{m}\bar{n}$ can be thought of as a formal analogue of the ordered pair $\langle m,n \rangle$, since (4.8) gives a method of picking out the first or second element. Incidentally, D can also be viewed as a 'conditional' operator; we can define

(4.9) $$(\underline{\text{If}}\ \ Z = 0\ \ \underline{\text{then}}\ \ X\ \ \underline{\text{else}}\ \ Y) \equiv \mathsf{D}XYZ.$$

Now, using D we construct an analogue of the f above. Define

$$Q \equiv \lambda yv.\mathsf{D}(\bar{\sigma}(v\bar{0}))(y(v\bar{0})(v\bar{1})),$$

where $\bar{\sigma}$ was defined in (I) above. Then for any X, Y, n,

$$\begin{aligned} QY(\mathsf{D}\bar{n}X) &=_{\beta,w} \mathsf{D}(\bar{\sigma}(\mathsf{D}\bar{n}X\bar{0}))(Y(\mathsf{D}\bar{n}X\bar{0})(\mathsf{D}\bar{n}X\bar{1})) \\ &=_{\beta,w} \mathsf{D}(\bar{\sigma}\bar{n})(Y\bar{n}X) \quad \text{by (4.8)} \end{aligned}$$

(i) $$=_{\beta,w} \mathsf{D}(\overline{n+1})(Y\bar{n}X).$$

Thus Q imitates the operation f above, if Y is a term defining χ. Also, for all X, Y, k we have

(ii) $$(QY)^{k}(\mathsf{D}\bar{0}X) =_{\beta,w} \mathsf{D}\bar{k}X_{k}$$

for some term X_k. (The details of X_k will play no role below; but if Y defined χ and $X \equiv \bar{m}$, then X_k would correspond to the value of $\phi(k)$ above.)

Now define

(4.10) $$R_{\text{Bernays}} \equiv \lambda xyu.u(Qy)(\mathsf{D}\bar{0}x)\bar{1}.$$

Then if R is R_{Bernays}, we have

$$\begin{aligned}
\mathrm{RXY}\bar{k} &=_{\beta,w} \bar{k}(\mathrm{QY})(\mathrm{D}\bar{0}\mathrm{X})\bar{1} && \\
&=_{\beta,w} (\mathrm{QY})^{k}(\mathrm{D}\bar{0}\mathrm{X})\bar{1} && \text{by (4.3)}\\
&=_{\beta,w} \mathrm{D}\bar{k}\mathrm{X}_k\bar{1} && \text{by (ii) above}\\
\text{(iii)} \qquad &=_{\beta,w} \mathrm{X}_k && \text{by (4.8).}
\end{aligned}$$

From this the two parts of (4.6) follow, thus:

$$\begin{aligned}
\mathrm{RXY}\bar{0} &=_{\beta,w} (\mathrm{QY})^{0}(\mathrm{D}\bar{0}\mathrm{X})\bar{1} && \text{by (4.10), (4.3)}\\
&\equiv \mathrm{D}\bar{0}\mathrm{X}\bar{1} && \text{by definition of } (\mathrm{QY})^{0}\text{, see 4.1}\\
&=_{\beta,w} \mathrm{X} && \text{by (4.8)}\\
\mathrm{RXY}(\overline{k+1}) &=_{\beta,w} (\mathrm{QY})^{k+1}(\mathrm{D}\bar{0}\mathrm{X})\bar{1} && \text{by (4.10), (4.3)}\\
&=_{\beta,w} (\mathrm{QY})((\mathrm{QY})^{k}(\mathrm{D}\bar{0}\mathrm{X}))\bar{1} && \text{by definition of } (\mathrm{QY})^{k+1}\text{, see 4.1}\\
&=_{\beta,w} \mathrm{QY}(\mathrm{D}\bar{k}\mathrm{X}_k)\bar{1} && \text{by (ii) above}\\
&=_{\beta,w} \mathrm{D}(\overline{k+1})(\mathrm{Y}\bar{k}\mathrm{X}_k)\bar{1} && \text{by (i) above}\\
&=_{\beta,w} \mathrm{Y}\bar{k}\mathrm{X}_k && \text{by (4.8)}\\
&=_{\beta,w} \mathrm{Y}\bar{k}(\mathrm{RXY}\bar{k}) && \text{by (iii) above.} \qquad \Box
\end{aligned}$$

REMARK 4.11. The fact that R and $\bar{\phi}$ have normal forms in the λ-calculus has a certain philosophical interest. But if we did not care about normal forms, could R be simplified? Well, suppose we could construct a simple term $\bar{\pi}$ such that

(4.12) $$\bar{\pi}\bar{0} =_{\beta,w} \bar{0}, \quad \bar{\pi}(\bar{\sigma}\bar{n}) =_{\beta,w} \bar{n}.$$

(Such a $\bar{\pi}$ is called a _predecessor combinator_.) Then it would be enough to solve the equation

$$\mathrm{R}xyz = (\underline{\text{if}}\ \ z = 0\ \ \underline{\text{then}}\ \ x\ \ \underline{\text{else}}\ \ y(\bar{\pi}z)(\mathrm{R}xy(\bar{\pi}z))),$$

(cf. (4.9)), or equivalently the equation

$$\mathrm{R}xyz = \mathrm{D}x(y(\bar{\pi}z)(\mathrm{R}xy(\bar{\pi}z)))z.$$

By the proof of Corollary 3.3.1, one solution is

$$\text{R} \equiv \text{Y}(\lambda uxyz.\text{D}x(y(\bar{\pi}z))(uxy(\bar{\pi}z))z). \tag{4.13}$$

This has no normal form, but is simpler in concept than Bernays' R; or rather, it would be so if we could find a simple $\bar{\pi}$. No very simple $\bar{\pi}$ is known for the Church numerals, but Barendregt 1981, Lemma 6.2.10, gives a simple $\bar{\pi}$ for a different set of numerals.

REMARK 4.14 To check that all the conversions in the proof of Lemma 4.5 hold for $=_w$ as well as for $=_\beta$ seems a tedious task. But after Chapter 6 it will become clear that only one fact need be checked: never is a redex-occurrence contracted when it is in the scope of a λ. In fact, with one exception, all contractions in the proof of 4.5 have form

$$P_1 \ldots P_r((\lambda x.M)NQ_1 \ldots Q_s) \rhd P_1 \ldots P_r(([N/x]M)Q_1 \ldots Q_s)$$

$(r, s \geq 0)$, and these translate into CL as correct weak reductions. The same will be true in other combined proofs for $=_w$ and $=_\beta$ later. The one exception is the β-reduction $\bar{\sigma}\bar{n} \rhd \overline{n+1}$; but we do not need to translate this into CL because in that system $\bar{\sigma}\bar{n} \equiv \overline{n+1}$ by definition of $\bar{n}$. (This is why the definitions of $\bar{n}$ are different in each system.)

THEOREM 4.15 (Kleene). *For* λ*-terms and* $=_\beta$, *or* CL*-terms and* $=_w$: *every recursive total function* ϕ *can be defined by a combinator* $\bar{\phi}$ *with a normal form.*

Proof. By the Kleene Normal Form theorem for partial recursive functions (Kleene 1952, §58 or Mendelson 1964, Proposition 5.17), for every partial recursive ϕ there exist primitive recursive functions ψ and χ such that

$$\phi(m_1,\ldots,m_n) = \psi(\mu k[\chi(m_1,\ldots,m_n,k) = 0]),$$

where $\mu k[\chi(m_1,\ldots,m_n,k) = 0]$ is the least k, if any, for which $\chi(m_1,\ldots,m_n,k) = 0$. If k fails to exist for some $m_1,\ldots,m_n$, then $\phi(m_1,\ldots,m_n)$ is undefined for those $m_1,\ldots,m_n$. But the theorem has assumed ϕ is total, so we can assume k always exists.

One way of computing μk is to define a program $\theta(k)$ which prints k if $\chi(m_1,\ldots,m_n,k) = 0$, and moves on to $\theta(k+1)$ otherwise; when this program is started with $k = 0$, it will output the first k for which $\chi(m_1,\ldots,m_n,k) = 0$. The formal analogue, H, of this program is the solution of the equation

$$Hx_1\ldots x_n y = \underline{\text{if}}\ \bar{\chi}x_1\ldots x_n y = 0\ \underline{\text{then}}\ y\ \underline{\text{else}}\ Hx_1\ldots x_n(\bar{\sigma}y).$$

If one does not mind an H without normal form, this equation can be solved using the fixed-point combinator Y. But here is how to define an H with normal form.

Begin by defining

$$(4.16)\qquad \begin{cases} T \equiv \lambda x.D\bar{0}(\lambda uv.u(x(\bar{\sigma}v))u(\bar{\sigma}v)), \\ P \equiv \lambda xy.Tx(xy)(Tx)y. \end{cases}$$

P has the property

$$(4.17)\qquad \begin{cases} PXY =_{\beta,w} Y & \text{if } XY =_{\beta,w} \bar{0}, \\ PXY =_{\beta,w} PX(\bar{\sigma}Y) & \text{if } XY =_{\beta,w} \overline{m+1} \text{ for some } m. \end{cases}$$

Proof of (4.17). Let X, Y be any terms and let $u, v \notin FV(XY)$. Then

$$\begin{aligned} PXY &=_{\beta,w} TX(XY)(TX)Y \\ &=_{\beta,w} D\bar{0}(\lambda uv.u(X(\bar{\sigma}v))u(\bar{\sigma}v))(XY)(TX)Y. \end{aligned}$$

If $XY =_{\beta,w} \bar{0}$, then by (4.8),

$$\begin{aligned} PXY &=_{\beta,w} \bar{0}(TX)Y \\ &=_{\beta,w} Y \quad \text{because } \bar{0} \equiv \lambda xy.y. \end{aligned}$$

If $XY =_{\beta,w} \overline{m+1}$, then by (4.8),

$$\begin{aligned} PXY &=_{\beta,w} (\lambda uv.u(X(\bar{\sigma}v))u(\bar{\sigma}v))(TX)Y \\ &=_{\beta,w} TX(X(\bar{\sigma}Y))(TX)(\bar{\sigma}Y) \\ &=_{\beta,w} PX(\bar{\sigma}Y). \end{aligned}$$

This proves (4.17).

Now define

$$H \equiv \lambda x_1 \dots x_n y . P(\bar{\chi} x_1 \dots x_n) y .$$

Then for any $X_1, \dots, X_n, Y$, we have by (4.17)

$$HX_1 \dots X_n Y =_{\beta,w} P(\bar{\chi} X_1 \dots X_n) Y$$

$$=_{\beta,w} \begin{cases} Y & \text{if } \bar{\chi} X_1 \dots X_n Y =_{\beta,w} \bar{0} \\ HX_1 \dots X_n (\bar{\sigma} Y) & \text{if } \bar{\chi} X_1 \dots X_n Y =_{\beta,w} \overline{m+1} . \end{cases}$$

Finally, define

$$\bar{\phi} \equiv \lambda x_1 \dots x_n . \bar{\psi}(Hx_1 \dots x_n \bar{0}) .$$

This $\bar{\phi}$ defines ϕ. Also, $\bar{\phi}$ has a normal form, by a boring routine proof. □

THEOREM 4.18. *For* λ-*terms and* $=_\beta$, *or* CL-*terms and* $=_w$: *every partial recursive function* ϕ *can be defined by a combinator* $\bar{\phi}$ *with a normal form.*

Proof and comments. Just as in the proof of Theorem 4.15, ϕ can be expressed as

$$\phi(m_1, \dots, m_n) = \psi(\mu k[\chi(m_1, \dots, m_n, k) = 0]) ,$$

but now μk need not be defined for all $m_1, \dots, m_n$. We need to construct a $\bar{\phi}$ so that it is clear that $\bar{\phi}\bar{m}_1 \dots \bar{m}_n$ has no normal form when there is no k such that $\chi(m_1, \dots, m_n, k) = 0$. This is done by a device due to Bruce Lercher.

We first take the $\bar{\phi}$ from the proof of Theorem 4.15 and call it 'F':

$$F \equiv \lambda x_1 \dots x_n . \bar{\psi}(Hx_1 \dots x_n \bar{0}) .$$

When $\phi(m_1, \dots, m_n)$ is defined we have

$$F\bar{m}_1 \dots \bar{m}_n =_{\beta,w} \overline{\phi(m_1, \dots, m_n)} .$$

However, it is not clear that $F\bar{m}_1 \dots \bar{m}_n$ has no normal form when $\phi(m_1, \dots, m_n)$ is undefined.

Next, define

$$\phi \equiv \lambda x_1 \ldots x_n . P(\bar{\chi} x_1 \ldots x_n)\bar{0} I(F x_1 \ldots x_n),$$

where P is defined in (4.16).

Suppose $m_1, \ldots, m_n$ are such that there is a k such that

$$\chi(m_1, \ldots, m_n, k) = 0,$$

and let j be the least such k. Then

$$\begin{aligned}
\bar{\phi}\bar{m}_1 \ldots \bar{m}_n &=_{\beta,w} \bar{j} I(F\bar{m}_1 \ldots \bar{m}_n) && \text{by proof of 4.15} \\
&=_{\beta,w} I^j(F\bar{m}_1 \ldots \bar{m}_n) && \text{by (4.3)} \\
&=_{\beta,w} F\bar{m}_1 \ldots \bar{m}_n && \text{by definition of } I \\
&=_{\beta,w} \overline{\phi(m_1, \ldots, m_n)} && \text{by definition of } F.
\end{aligned}$$

Suppose now that $m_1, \ldots, m_n$ are such that there is no k such that

$$\chi(m_1, \ldots, m_n, k) = 0;$$

then for each k there is a p_k such that

$$\chi(m_1, \ldots, m_n, k) = p_k + 1.$$

(Note that χ is total, being primitive recursive.) Let

$$X \equiv \bar{\chi}\bar{m}_1 \ldots \bar{m}_n, \qquad G \equiv F\bar{m}_1 \ldots \bar{m}_n.$$

Then for each k, $X\bar{k} =_{\beta,w} \overline{p_k+1}$. Furthermore, by the Church-Rosser theorem we have

$$X\bar{k} \rhd_{\beta,w} \overline{p_k+1}$$

because the Church numerals are in normal form in both λ and CL.

We must show that $\bar{\phi}\bar{m}_1 \ldots \bar{m}_n$ has no normal form. By Corollary 3.19.1, it is enough to exhibit an infinite quasi-leftmost reduction of this term. Consider the following reduction (not every contraction will be shown):

$\bar{\phi}\bar{m}_1 \ldots \bar{m}_n \rhd_{\beta,w} PX\bar{0}IG$ — by definition of $\bar{\phi}$

$\rhd_{\beta,w} TX(X\bar{0})(TX)\bar{0}IG$ — by definition of P

$\rhd_{\beta,w} TX(\overline{p_0+1})(TX)\bar{0}IG$ — by reducing $X\bar{0}$

$\rhd_{\beta,w} (\lambda uv.u(X(\bar{\sigma}v))u(\bar{\sigma}v))(TX)\bar{0}IG$ — by definition of T, D

$\rhd_{\beta,w} TX(X(\bar{\sigma}\bar{0}))(TX)(\bar{\sigma}\bar{0})IG$

$\rhd_{\beta,w} TX(X\bar{1})(TX)\bar{1}IG$ — by definition of σ

$\rhd_{\beta,w} \ldots$

$\rhd_{\beta,w} TX(X\bar{2})(TX)\bar{2}IG$ — similarly

$\rhd_{\beta,w} \ldots$

Clearly, this reduction is infinite, and each part

$$TX(X\bar{i})(TX)\bar{i}IG \rhd_{\beta,w} TX(X(\overline{i+1}))(TX)(\overline{i+1})IG$$

contains at least one leftmost maximal contraction. □

DISCUSSION 4.19 (Abstract numerals). Instead of representing the numbers in λ-calculus or combinatory logic by terms of the pure system, it is possible to add two new constant atoms $\bar{0}$, $\bar{\sigma}$ to the definition of 'term', and to represent each number n by

$$\bar{n} \equiv \bar{\sigma}^n \bar{0}.$$

These are called the abstract numerals.

In this case an R with property (4.6) cannot be constructed. (Curry et al. 1972, §13A Thm. 2.) But R can be added as a third new atom, and the following new contractions added to the definition of reduction (1.22 or 2.7):

$$(4.20) \qquad \begin{cases} RXY\bar{0} \rhd_1 X, \\ RXY(\overline{k+1}) \rhd_1 Y\bar{k}(RXY\bar{k}). \end{cases}$$

If this is done, every recursive total function becomes definable. (The proof is essentially the same as those of Lemma 4.5 and Theorem 4.15; in fact, those proofs stay valid when we replace the Church numerals by abstract ones, provided we replace the D in (4.7) by

(4.21) $$D^* \equiv \lambda xy.Rx(K(Ky)).)$$

But for properly partial functions the non-normal-form part of the definability proof breaks down, because reduction with (4.20) fails to satisfy the Quasi-leftmost-reduction theorem (Hindley 1978b, §7).

Instead of adding R, it is better to add an atom Z, called an <u>iterator</u>, and add to the definition of reduction the following new contractions (one for each $n \geq 0$):

(4.22) $$Z\bar{n} \rhd_1 Z_n,$$

where Z_n is the Church numeral for n. For both combinatory weak and $\lambda\beta$-reduction, the reduction obtained by adding (4.22) can be shown to satisfy the Church-Rosser theorem, the Standardization theorem, and a theorem close enough to the Quasi-leftmost-reduction theorem to make the proof of the definability of the partial recursive functions work. (Hindley 1978b, §7 and Thm. 8.)

Given Z and (4.22), an R can be constructed as follows:

(4.23) $$R \equiv \lambda xyu.Zu(Qy)(D\bar{0}x)Z_1,$$

where D is the same as in (4.7), Z_n is the Church numeral for n, and

$$Q \equiv \lambda yv.D(\bar{\sigma}(vZ_o))(y(vZ_o)(vZ_1)).$$

This is very close to Bernays' R, and by following the proof of Lemma 4.5 it can be shown to satisfy (4.6).

Thus if we add the new constants $\bar{0}$, $\bar{\sigma}$, Z to λ or CL, and add (4.22) to the definition of β- or weak reduction, we can define all partial recursive functions in terms of the abstract numerals.

EXERCISES 4.24.

(a) Without using R, construct terms which λ-define addition, multiplication, exponentiation. (Answer: Church 1941, p. 10.) Do the same using R.

(b) In the system with abstract numerals (Remark 4.19), prove that if R and Z were absent, we could not even define the predecessor-function π: $\pi(0) = 0$, $\pi(n+1) = n$. (Answer: Curry et al. 1972, §13A3 Theorem 2.)

(c) In λ or CL with abstract numerals but neither R nor Z, add two new atoms D, $\bar{\pi}$ and add (4.8) and (4.12) as new contractions to the definition of reduction; construct a Z satisfying

$$Z\bar{n} \rhd Z_n.$$

(Answer: Curry et al. 1972, §13A3.)

CHAPTER FIVE

THE UNDECIDABILITY THEOREM

The aim of this chapter is to prove a general undecidability theorem which shows in particular that the β- and weak equality relations are recursively undecidable, and that there is no recursive way of deciding whether a term has a normal form or not. These results for Church's λ-system were the first undecidability results to be discovered in mathematics, and it was from them that Church deduced the undecidability of pure first-order predicate calculus in 1936.

But the general theorem was first proved by Scott in 1963 (unpublished notes), and rediscovered independently by Curry (1969b, 1972 §13B2).

NOTATION 5.1. The neutral notation of 3.1 will be used, which can be read in both λ-calculus and combinatory logic.

Numerals will be those of Church:

$$\bar{n} \equiv \lambda xy.x^n y.$$

We shall assume that every term X has been given a number n called the <u>Gödel number of</u> X, or <u>gd</u>(X). This can be done in many different ways, for example as in Mendelson 1964, Chapter 3, §4, but the details do not matter here. All we shall need to assume are the following:

(5.2) There is a recursive total function τ of natural numbers, such that for all terms X, Y,

$$\tau(gd(X), gd(Y)) = gd(XY).$$

(5.3) There is a recursive total function ν such that for all natural numbers n,

$$\nu(n) = gd(\bar{n}).$$

(The reason that numberings with these properties exist is that the operation of forming the term (XY) from terms X and Y is effectively computable, and so is the operation of forming $\bar{n}$ from n.)

For any X, the Church numeral corresponding to the Gödel number of X will be called $\ulcorner X \urcorner$:

$$\ulcorner X \urcorner \equiv \overline{gd(X)} .$$

DEFINITION 5.4. Two sets $\mathcal{A}$ and $\mathcal{B}$ of natural numbers are called <u>recursively separable</u> iff there is a recursive total function ϕ whose only values are 0 and 1, such that

$$n \in \mathcal{A} \implies \phi(n) = 1,$$

$$n \in \mathcal{B} \implies \phi(n) = 0.$$

Two sets of terms are <u>recursively separable</u> iff the corresponding sets of Gödel numbers are recursively separable. A set (of numbers or terms) is <u>recursive</u> iff it and its complement are recursively separable.

Informally, $\mathcal{A}$ and $\mathcal{B}$ are recursively separable iff they are non-overlapping and there is a recursive way of telling whether a number or term is in $\mathcal{A}$ or in $\mathcal{B}$.

DEFINITION 5.5. For λ-terms and β-equality, or CL-terms and weak equality: a set $\mathcal{A}$ of terms is <u>closed under equality</u> iff, for all terms X, Y,

$$X \in \mathcal{A} \text{ and } Y =_{\beta,w} X \implies Y \in \mathcal{A}.$$

THEOREM 5.6 (<u>Scott-Curry undecidability theorem</u>). *For λ-terms and β-equality, or CL-terms and weak equality: no pair of non-empty sets of terms which are closed under equality is recursively separable.*

Proof. Suppose ϕ separates $\mathcal{A}$ and $\mathcal{B}$, where $\mathcal{A}$ and $\mathcal{B}$ are disjoint sets of terms both of which are non-empty and closed under equality. Let F define ϕ; then

(i) $X \in \mathcal{A} \implies F\ulcorner X \urcorner =_{\beta,w} \bar{1}$,

(ii) $X \in \mathcal{B} \implies F\ulcorner X \urcorner =_{\beta,w} \bar{0}$.

Let T and N be closed terms which define the functions τ and ν of (5.2) and (5.3) respectively; then for all X, Y, n,

$$T\ulcorner X \urcorner \ulcorner Y \urcorner =_{\beta,w} \ulcorner XY \urcorner,$$

$$N\bar{n} =_{\beta,w} \ulcorner \bar{n} \urcorner.$$

Choose any terms A in $\mathcal{A}$ and B in $\mathcal{B}$. We shall construct a term J (depending on A and B), such that

(iii) $F\ulcorner J\urcorner =_{\beta,w} \bar{1} \Rightarrow J =_{\beta,w} B$,

(iv) $F\ulcorner J\urcorner =_{\beta,w} \bar{0} \Rightarrow J =_{\beta,w} A$.

This will cause a contradiction because, letting j = gd(J), we shall have

$$\begin{aligned}
\phi(j) = 1 &\Rightarrow F\ulcorner J\urcorner =_{\beta,w} \bar{1} &&\text{since } \ulcorner J\urcorner \equiv \bar{j}\\
&\Rightarrow J =_{\beta,w} B &&\text{by (iii)}\\
&\Rightarrow J \in \mathcal{B} &&\text{since } \mathcal{B} \text{ closed under } =_{\beta,w}\\
&\Rightarrow \phi(j) = 0;
\end{aligned}$$

and

$$\begin{aligned}
\phi(j) = 0 &\Rightarrow F\ulcorner J\urcorner =_{\beta,w} \bar{0} &&\text{since } \ulcorner J\urcorner \equiv \bar{j}\\
&\Rightarrow J =_{\beta,w} A &&\text{by (iv)}\\
&\Rightarrow J \in \mathcal{A} &&\text{since } \mathcal{A} \text{ closed under } =_{\beta,w}\\
&\Rightarrow \phi(j) = 1.
\end{aligned}$$

(ϕ is a total function whose only values are 0 and 1.)

To get (iii) and (iv) it will be enough to build J such that

(v) $J =_{\beta,w} DAB(F\ulcorner J\urcorner)$

where D is defined by (4.7), since D has the property that

$$DAB\bar{1} =_{\beta,w} B, \quad DAB\bar{0} =_{\beta,w} A.$$

Define

$$J \equiv H\ulcorner H\urcorner, \quad H \equiv \lambda y.DAB(F(Ty(Ny))).$$

This J satisfies (v), because

$$\begin{aligned}
J &=_{\beta,w} DAB(F(T\ulcorner H\urcorner(N\ulcorner H\urcorner))) &&\text{by definition of } H\\
&=_{\beta,w} DAB(F(T\ulcorner H\urcorner\ulcorner\ulcorner H\urcorner\urcorner)) &&\text{by definition of } N\\
&=_{\beta,w} DAB(F\ulcorner(H\ulcorner H\urcorner)\urcorner) &&\text{by definition of } T\\
&\equiv DAB(F\ulcorner J\urcorner) &&\text{by definition of } J.
\end{aligned}$$

□

COROLLARY 5.6.1. *For λ-terms and β-equality, or CL-terms and weak equality: if $\mathcal{A}$ is closed under equality and both $\mathcal{A}$ and its complement are non-empty, then $\mathcal{A}$ is not recursive.*

Proof. In the theorem let $\mathcal{B}$ be the complement of $\mathcal{A}$. □

COROLLARY 5.6.2. *For λ-terms and β-equality, or CL-terms and weak equality: the set of all terms which have normal forms is not recursive.*

COROLLARY 5.6.3. *The β- and weak equality relations are not recursive; that is, there is no recursive total ψ such that*

$$\psi(\mathrm{gd}(X),\ \mathrm{gd}(Y)) = \begin{cases} 1 & \textit{if}\ \ X =_{\beta,w} Y, \\ 0 & \textit{if}\ \ X \neq_{\beta,w} Y. \end{cases}$$

Proof. In Corollary 5.6.1, let $\mathcal{A}$ be the set of all terms equal to one particular term (I, for example.) □

REMARK 5.7. Church's original proof that pure classical first-order predicate calculus is undecidable can be summarized as follows. When λ-terms are Gödel-numbered, then β-equality corresponds to a relation between natural numbers. Natural numbers can be coded in a pure predicate calculus which has function-symbols, by choosing a variable z and a function-symbol f and letting

$$\begin{array}{ll} z & \text{represent } 0, \\ f(z) & \text{represent } 1, \\ f(f(z)) & \text{represent } 2, \text{ etc.} \end{array}$$

Let $\bar{\bar{n}}$ be the representative of n in this coding. The definition of $=_\beta$ can be rewritten as a formal theory with eight axiom-schemes and rules of inference, as we shall see in the next chapter (the theory λβ). These axiom-schemes and rules can be translated, via Gödel-numbering, into predicate-calculus formulas $F_1,\ldots,F_8$ containing a predicate-symbol E, such that the formula

$$(F_1 \wedge \ldots \wedge F_8) \supset E(\bar{\bar{m}},\bar{\bar{n}})$$

is provable in pure predicate calculus iff m, n are Gödel numbers of β-equal terms.

Hence, if we would decide all questions of provability in pure predicate calculus, we could decide whether arbitrary λ-terms are β-equal, contrary to Corollary 5.6.3. (The details of this undecidability proof are in Church 1936a, 1936b.)

REMARK 5.8. Church's proof of the undecidability of $=_\beta$ was a direct one; he did not deduce it from a general theorem as we have done here. In fact the proof of the Scott-Curry theorem used K, which is not allowed in Church's original system (Remark 1.36). And the theorem itself fails for that system (Exercise 5.9(b)).

EXERCISES 5.9.

(a)* In λ-calculus or combinatory logic, the range of a combinator X may be defined to be the set of all Z such that for some Y,

$$XY =_{\beta,w} Z.$$

Prove that the range of X is either infinite or a singleton, modulo $=_{\beta,w}$. (This was conjectured by Böhm and proved by Barendregt.)

(b) Where was K used in the proof of the Scott-Curry theorem? Give a counterexample to the theorem for Church's λI-terms (Remark 1.36), using the fact that if $X =_\beta Y$ in the λI-system, then $FV(X) = FV(Y)$.

(c) Prove the Second Fixed-point theorem, namely that for each λ- or CL-term X there exists a P such that

$$X\ulcorner P \urcorner =_{\beta,w} P.$$

(Hint: modify J in the proof of 5.6, with X instead of F.) This theorem is Barendregt 1981, Theorem 6.5.9.

FURTHER READING 5.10.

Barendregt 1981, §§6.3 - 6.7 goes further into λ- definability, number systems and undecidability.

Visser 1980 treats some interesting deeper recursion-theoretic results.

CHAPTER SIX

THE FORMAL THEORIES $\lambda\beta$ AND CLw

6A THE DEFINITIONS OF THE THEORIES

The relations $\rhd_\beta$, $=_\beta$, $\rhd_w$ and $=_w$ were defined in Chapters 1 and 2 via contractions of redexes. The present chapter gives alternative definitions via formal theories with axioms and rules of inference.

These theories will give us a new view of the four relations, and will help to make the distinction between syntax and semantics clearer in the chapters on models. They will also give us more technical power, for example Church's undecidability proof depends on them (Remark 5.7). And they will give a more direct meaning to such phrases as 'add the equation $M = N$ as a new axiom to the definition of $=_\beta$ or $=_w$' (cf. Corollary 3.10.1).

NOTATION 6.1 (Formal theories). In this book, 'formal theory' will be used in the sense of Mendelson 1964, Chapter 1, §4, except that deductions will be viewed as trees, not linear sequences, and rules with a countable infinity of premises will be allowed. Here is a summary of the definition, with some slight changes of notation.

A formal theory $\mathcal{T}$ has three sets: formulas, axioms and rules (of inference). Rules will be written here with premises above a horizontal line and conclusion below (see Definition 6.2 below for example). A deduction in $\mathcal{T}$, of a formula F from a set $\mathcal{B}$ of formulas, is a tree of formulas, with those at the tops of branches being axioms or members of $\mathcal{B}$, the others being deduced from those immediately above them by a rule, and the bottom one being F. Non-axioms at the tops of branches are called assumptions. Iff such a deduction exists, we say

$$\mathcal{T}, \mathcal{B} \vdash F, \quad \text{or} \quad \mathcal{B} \vdash_{\mathcal{T}} F.$$

Iff $\mathcal{B}$ is empty, we call F a provable formula or theorem of $\mathcal{T}$, call the deduction a proof, and say

$$\mathcal{T} \vdash F, \quad \text{or} \quad \vdash_{\mathcal{T}} F.$$

If $\mathcal{T}$ has an infinite-premise rule, then a deduction is a tree in which an infinity of branches may grow from one point, but each linear path upward has only finite length.

Finally, an axiom-scheme will simply be a set of axioms which all conform to some given pattern.

DEFINITION 6.2 ($\lambda\beta$, the formal theory of β-equality). The formulas of $\lambda\beta$ are just equations $M = N$, for all λ-terms M, N. The axioms are the particular cases of the three axiom-schemes below, for all λ-terms M, N and all variables x, y.

Axiom-schemes:

(α) $\lambda x.M = \lambda y.[y/x]M$ if $y \notin FV(M)$;

(β) $(\lambda x.M)N = [N/x]M$;

(ρ) $M = M$.

Rules of inference:

$$(\mu)\ \frac{M = M'}{NM = NM'} \qquad (\tau)\ \frac{M = N \quad N = P}{M = P}$$

$$(\nu)\ \frac{M = M'}{MN = M'N} \qquad (\sigma)\ \frac{M = N}{N = M}.$$

$$(\xi)\ \frac{M = M'}{\lambda x.M = \lambda x.M'}$$

Iff the equation $M = N$ is provable in $\lambda\beta$, we say

$$\lambda\beta \vdash M = N.$$

(The names of the axiom-schemes and rules are from Curry and Feys 1958, and have now become fairly standard; similarly for other rule-names later.)

DEFINITION 6.3 ($\lambda\beta$, <u>the formal theory of</u> β-<u>reduction</u>). This theory is called $\lambda\beta$ like that above (the context will always make it clear which theory the name '$\lambda\beta$' means). Its <u>formulas</u> are expressions $M \rhd N$, for all λ-terms M, N. Its <u>axiom-schemes</u> and <u>rules</u> are the same as in 6.2, but with '=' changed to '$\rhd$' and rule (σ) omitted. Iff the expression $M \rhd N$ is provable in $\lambda\beta$, we say

$$\lambda\beta \vdash M \rhd N.$$

LEMMA 6.4.

(a) $M \rhd_\beta N \iff \lambda\beta \vdash M \rhd N$;

(b) $M =_\beta N \iff \lambda\beta \vdash M = N$.

Proof. Straightforward. □

DEFINITION 6.5 (CLw, <u>the formal theory of weak equality</u>). The <u>formulas</u> of CLw are equations $X = Y$, for all CL-terms X, Y. The <u>axioms</u> are the particular cases of the three axiom-schemes below, for all CL-terms X, Y, Z.

<u>Axiom-schemes</u>:

(K) $KXY = X$;

(S) $SXYZ = XZ(YZ)$;

(ρ) $X = X$.

<u>Rules of inference</u>:

$$(\mu)\ \frac{X = X'}{ZX = ZX'} \qquad (\tau)\ \frac{X = Y \quad Y = Z}{X = Z}$$

$$(\nu)\ \frac{X = X'}{XZ = X'Z} \qquad (\sigma)\ \frac{X = Y}{Y = X}.$$

Iff $X = Y$ is provable in CLw, we say

$$\text{CLw} \vdash X = Y.$$

DEFINITION 6.6 (CLw, the formal theory of weak reduction). The formulas of CLw are expressions $X \rhd Y$, for all CL-terms X, Y. The axiom-schemes and rules are the same as in 6.5, but with '=' changed to '$\rhd$' and rule (σ) omitted. Iff $X \rhd Y$ is provable in CLw, we say

$$\text{CLw} \vdash X \rhd Y.$$

LEMMA 6.7.

(a) $X \rhd_w Y \iff \text{CLw} \vdash X \rhd Y$;

(b) $X =_w Y \iff \text{CLw} \vdash X = Y$.

REMARK 6.8. By the Church-Rosser theorems and Lemmas 6.4 and 6.7, the theories $\lambda\beta$ and CLw are consistent in the sense that not all their formulas are provable.

6B FIRST-ORDER THEORIES AND DERIVABLE RULES

This section contains material from logic and proof-theory that will be used as background later.

DISCUSSION 6.9 (First-order theories). Most logic textbooks define a concept of first-order theory (e.g. Mendelson 1964, Chapter 2, §3). Such a theory $\mathcal{T}$ is a special kind of formal theory. Its formulas are built up from terms using predicate-symbols and the usual connectives and quantifiers. Its axioms are divided into two classes: logical axioms common to all first-order theories, and proper axioms peculiar to the theory $\mathcal{T}$. Its rules are the usual rules of classical first-order logic (e.g. Mendelson 1964, Chapter 2, §3), and are the same for all first-order theories.

In this book, 'first-order theory' will always mean 'first-order theory with equality', i.e. with the symbol '=' and the usual axioms for it (e.g. Mendelson 1964, Chapter 2, §8).

A model of a first-order theory will always mean a model in which '=' is interpreted as the identity relation (often called a normal model in logic textbooks).

REMARK 6.10. Neither of the equality-theories $\lambda\beta$ and CLw is a first-order theory.

For $\lambda\beta$, this is because it has no connectives and quantifiers, and because its terms contain λ; term-formation operators that bind variables are not allowed in first-order theories.

CLw also has no connectives, so it is not a first-order theory. But this lack is trivial, and CLw can easily be converted into a first-order theory CLw^+ as follows.

DEFINITION 6.11 (The first-order theory CLw^+). The terms of CLw^+ are CL-terms. The formulas are built from equations $X = Y$ using connectives and quantifiers in the normal way. The rules of inference and logical axioms are the usual ones for classical first-order logic with equality (e.g. Mendelson 1964, Chapter 2, §§2 and 8). The proper axioms are the following three:

(a) $(\forall x,y)(Kxy = x)$,

(b) $(\forall x,y,z)(Sxyz = xz(yz))$,

(c) $\neg(S = K)$.

LEMMA 6.12. CLw^+ *is a conservative extension of* CLw, *i.e. both theories have the same set of provable equations.*

Proof. Barendregt 1973, Theorem 2.12. (Note that axiom (c) can be included in CLw^+ because by the Church-Rosser theorem, $S = K$ is not provable in CLw.) □

DEFINITION 6.13 (Admissible and derivable rules). Suppose we have a formal theory $\mathcal{T}$, and we consider extending $\mathcal{T}$ by adding a new rule $\mathcal{R}$. It is natural to ask first whether $\mathcal{R}$ is already derivable in $\mathcal{T}$. But what exactly does '$\mathcal{R}$ is derivable' mean? Two answers will be given here, both well-known in proof-theory (e.g. Troelstra 1973, §1.11.1).

But first, what does 'a new rule $\mathcal{R}$' mean? Let $\mathcal{F}$ be the set of all formulas of $\mathcal{T}$, and let $n \geq 1$. Then every partial function ϕ from $\mathcal{F}^n$ into $\mathcal{F}$ determines a rule $\mathcal{R}(\phi)$ thus: each n-tuple $\langle A_1,\ldots,A_n\rangle$ in the domain of ϕ may be called a sequence of premises, and if

$\phi(A_1,\ldots,A_n) = B$, then B is the corresponding <u>conclusion</u>, and the expression

$$\frac{A_1,\ldots,A_n}{B}$$

is called an <u>instance</u> of $\mathcal{R}(\phi)$.

Similarly, if $\mathcal{F}^{\omega}$ is the set of all infinite sequences of formulas, each partial function $\phi : \mathcal{F}^{\omega} \to \mathcal{F}$ determines an infinite-premise rule $\mathcal{R}(\phi)$.

A rule $\mathcal{R}$ determined by a partial function as above is said to be <u>derivable in</u> $\mathcal{T}$ iff, for each instance of $\mathcal{R}$, there is a deduction in $\mathcal{T}$ of its conclusion from its premises;

$$\mathcal{T}, A_1, A_2,\ldots \vdash B.$$

On the other hand, $\mathcal{R}$ is said to be <u>admissible in</u> $\mathcal{T}$ iff, for each instance of $\mathcal{R}$, if all the premises are provable in $\mathcal{T}$ then so is the conclusion; i.e. iff

$$(\mathcal{T} \vdash A_1),\ (\mathcal{T} \vdash A_2),\ldots \Rightarrow (\mathcal{T} \vdash B).$$

Finally, a formula C, for example a proposed new axiom, is called both <u>admissible</u> and <u>derivable</u> in $\mathcal{T}$ iff $\mathcal{T} \vdash C$.

LEMMA 6.14.

(a) $\mathcal{R}$ *is admissible in* $\mathcal{T}$ *iff adding* $\mathcal{R}$ *to* $\mathcal{T}$ *as a new rule will not change* $\mathcal{T}$*'s set of theorems.*

(b) *Derivability implies admissibility, but not vice-versa.*

(c) *If* $\mathcal{R}$ *is derivable in* $\mathcal{T}$, *then* $\mathcal{R}$ *is derivable in any extension of* $\mathcal{T}$ *obtained by adding new axioms or rules.*

Proof. Trivial, except to prove admissibility weaker than derivability. Here is an example to prove this. (There will be another in Remark 11.2.) Add to $\lambda\beta$ four new constant atoms a, b, c, d, and let $\mathcal{R}$ be a one-premise rule with only one instance, namely

$$\frac{a = b}{c = d}.$$

This rule is trivially admissible in $\lambda\beta$, because $a = b$ cannot be proved in $\lambda\beta$ (by the Church-Rosser theorem). But it is not derivable. (Exercise: prove this via a modified Church-Rosser theorem after reading Appendix 1.) □

DEFINITION 6.15. Let $\mathcal{T}$, $\mathcal{T}'$ be formal theories with the same set of formulas. We shall call $\mathcal{T}$, $\mathcal{T}'$ theorem-equivalent iff every rule and axiom of $\mathcal{T}$ is admissible in $\mathcal{T}'$ and vice-versa, and rule-equivalent iff every rule and axiom of $\mathcal{T}$ is derivable in $\mathcal{T}'$ and vice-versa.

Clearly, theorem-equivalence is weaker than rule-equivalence. The following easy lemma shows why it is called 'theorem-equivalence'.

LEMMA 6.16. *Let $\mathcal{T}$, $\mathcal{T}'$ be formal theories with the same set of formulas: then $\mathcal{T}$, $\mathcal{T}'$ are theorem-equivalent iff they have the same set of theorems.*

Admissibility and theorem-equivalence will be more important than derivability and rule-equivalence in this book. But lack of derivability of a rule will cause trouble in at least one place, so the difference between the two pairs of concepts is worth remembering.

DEFINITION 6.17. Let $\mathcal{T}$ be a formal theory and let some of its formulas be equations $X = Y$, where X and Y are terms according to some definition. The equality relation determined by $\mathcal{T}$ is called $=_{\mathcal{T}}$ and is defined by

$$X =_{\mathcal{T}} Y \iff \mathcal{T} \vdash X = Y.$$

LEMMA 6.18. *Let $\mathcal{T}$, $\mathcal{T}'$ be formal theories with the same set of formulas, and let this set include some equations: if $\mathcal{T}$, $\mathcal{T}'$ are theorem-equivalent, then they determine the same equality-relation.*

CHAPTER SEVEN

EXTENSIONALITY IN λ-CALCULUS

7A EXTENSIONAL EQUALITY

The concept of function-equality used in most of mathematics is commonly called '<u>extensional</u>': that is, it includes the assumption that for functions f and g with the same domain,

$$((\forall x)\ f(x) = g(x)) \quad \Rightarrow \quad f = g.$$

In contrast, the formal theory $\lambda\beta$ is '<u>intensional</u>': there exist terms F, G such that for all terms X

$$\lambda\beta \ \vdash \ FX = GX,$$

but not $\lambda\beta \vdash F = G$. For example take any variable y and define

$$F \equiv y, \quad G \equiv \lambda x.yx.$$

This chapter is about adding an extensionality rule to $\lambda\beta$. For combinatory logic there is also an extensional theory, but it is not just a straight translation of the λ-theory, so it will be left to the next chapter.

After the discussion of extensionality the relations between λ and CL will become much clearer, and these will be examined in Chapter 9.

NOTATION 7.1. In this chapter, 'term' will mean 'λ-term'. Recall that a <u>closed term</u> is one without free variables. Recall also the formal theory of β-equality, (Definition 6.2).

The following three rules and one axiom-scheme have been proposed in the literature to formalize the extensionality concept in λ-calculus.

$$(ext)\quad \frac{MP = NP \quad \text{for all terms}\ \ P}{M = N}$$

$$(\omega)\quad \frac{MQ = NQ \quad \text{for all closed}\ \ Q}{M = N}$$

$$(\zeta)\quad \frac{Mx = Nx}{M = N} \qquad (\text{if}\ \ x \notin FV(MN))$$

$$(\eta)\quad \lambda x.Mx = M \qquad (\text{if}\ \ x \notin FV(MN)).$$

The first two rules have an infinity of premises; for example (*ext*) has one premise $MP = NP$ for each λ-term P.

Which of the above rules expresses extensionality best is open to discussion; what we shall do here is look at the relations between them.

DEFINITION 7.2. Let $\lambda\beta$ be the theory of equality in 6.2. Four new formal theories of equality are defined as follows:

$\lambda\beta$+*ext*: add rule (*ext*) to $\lambda\beta$;
$\lambda\beta\omega$: add rule (ω) to $\lambda\beta$;
$\lambda\beta\zeta$: add rule (ζ) to $\lambda\beta$;
$\lambda\beta\eta$: add axiom-scheme (η) to $\lambda\beta$.

(Adding axiom-scheme (η) means adding all equations $\lambda x.Mx = M$ as new axioms, for all terms M and all $x \notin FV(M)$.)

REMARK 7.3. Rule (ω) was first postulated by Paul Rosenbloom in his 1950, Chapter 3, Rule E5. It clearly implies (*ext*); more precisely, (*ext*) is admissible in $\lambda\beta\omega$ in the sense of Definition 6.13. But the converse is false. Gordon Plotkin has shown that (ω) is not admissible in $\lambda\beta$+*ext*, by constructing closed M, N such that $MQ = NQ$ is provable in $\lambda\beta$+*ext* (even in $\lambda\beta$) for all closed Q, yet $M = N$ is not provable in $\lambda\beta$+*ext*. (Plotkin 1974.)

For more on (ω), see Barendregt 1981 §§17.3, 17.4, and Hindley and Longo §5 ff. The focus here will be on (*ext*), (ζ) and (η) from now on.

THEOREM 7.4. *The theories* $\lambda\beta$+*ext*, $\lambda\beta\zeta$, $\lambda\beta\eta$ *are theorem-equivalent in the sense of* 6.15, *and hence determine the same equality-relation on the set of all* λ-*terms. They are also rule-equivalent.*

Proof. It is enough to prove rule-equivalence. First, rule (*ext*) is derivable in $\lambda\beta\zeta$. Because if we have an infinity of premises $MP = NP$, one of these will be $Mx = Nx$ for some $x \notin FV(MN)$, and from this one premise we can deduce $M = N$ by (ζ).

Second, rule (ζ) is derivable in $\lambda\beta\eta$. Because from a premise $Mx = Nx$ with $x \notin FV(MN)$, we can deduce

$$\begin{aligned} M &= \lambda x.Mx && \text{by } (\eta) \\ &= \lambda x.Nx && \text{by the premise, and rule } (\xi) \text{ in 6.2} \\ &= N && \text{by } (\eta). \end{aligned}$$

Finally, each (η)-axiom $\lambda x.Mx = M$ is provable in $\lambda\beta$+*ext*. Because by axiom-scheme (β),

$$(\lambda x.Mx)P = MP$$

for all P, so by (*ext*), $\lambda x.Mx = M$ follows. □

By Theorem 7.4, (*ext*), (ζ) and (η) have turned out equivalent in the strongest sense. Since axiom-schemes are simpler than rules, (η) is usually taken as the 'canonical' definition of extensionality in λ-calculus.

DEFINITION 7.5 ($\lambda\beta\eta$-equality). The equality-relation determined by the theory $\lambda\beta\eta$ is called $=_{\beta\eta}$; that is, we define

$$M =_{\beta\eta} N \iff \lambda\beta\eta \vdash M = N.$$

The main tool for proving results about $\lambda\beta\eta$-equality is a corresponding theory of reduction; its main definitions and theorems will be listed in the next section.

7B λβη-REDUCTION

DEFINITION 7.6 (η-reduction). An η-redex is any λ-term $\lambda x.Mx$ with $x \notin FV(M)$. Its contractum is M. The phrases 'P η-contracts to Q' and 'P η-reduces to Q' are defined like 'β-contracts', β-reduces' in 1.22; notation

$$P \rhd_{1\eta} Q, \quad P \rhd_{\eta} Q.$$

DEFINITION 7.7 (βη-reduction). A βη-redex is a β-redex or an η-redex. The phrases 'P βη-contracts to Q' and 'P βη-reduces to Q' are defined as in 1.22; notation

$$P \rhd_{1\beta\eta} Q, \quad P \rhd_{\beta\eta} Q.$$

Warning: βη-reduction is often called η-reduction in the literature.

DEFINITION 7.8 (βη-normal forms). A λ-term Q containing no βη-redexes is said to be in βη-normal form (βη-nf), and Q is a βη-normal form of P iff $P \rhd_{\beta\eta} Q$. (This is the same as Definition 3.6.)

DEFINITION 7.9 (The formal theory λβη of βη-reduction). This is defined by adding the axiom-scheme

$$(\eta) \qquad \lambda x.Mx \rhd M \quad \text{if} \quad x \notin FV(M)$$

to the theory of β-reduction, Definition 6.3.

LEMMA 7.10. $\quad P \rhd_{\beta\eta} Q \iff \lambda\beta\eta \vdash P \rhd Q.$

LEMMA 7.11.

(a) $P \rhd_{\beta\eta} Q \implies (v \notin FV(P) \implies v \notin FV(Q));$

(b) $P \rhd_{\beta\eta} Q \implies [P/x]M \rhd_{\beta\eta} [Q/x]M;$

(c) $P \rhd_{\beta\eta} Q \implies [N/x]P \rhd_{\beta\eta} [N/x]Q.$

Proof. Like the substitution lemma for β-reduction, 1.28. For (c), note that if $\lambda y.My$ is an η-redex, then so is $[N/x](\lambda y.My)$. □

THEOREM 7.12 (<u>Church-Rosser theorem for</u> $\beta\eta$-<u>reduction</u>). *If* $P \rhd_{\beta\eta} M$ *and* $P \rhd_{\beta\eta} N$, *then there exists* T *such that*

$$M \rhd_{\beta\eta} T \quad \textit{and} \quad N \rhd_{\beta\eta} T.$$

Proof. See Appendix 1, Theorem A1.9. □

The consequences of the Church-Rosser theorem are the same as for β, in particular, the uniqueness of normal forms modulo congruence.

THEOREM 7.13. *A λ-term has a $\beta\eta$-normal form iff it has a β-normal form.*

Proof. Barendregt 1981, Corollary 15.1.5 or Curry et al. 1972, §11E, Lemma 13.1. □

THEOREM 7.14. *In a $\beta\eta$-reduction all the η-contractions can be postponed to the end; that is, if* $M \rhd_{\beta\eta} N$ *then there exists a* P *such that*

$$M \rhd_{\beta} P \rhd_{\eta} N.$$

Proof. Barendregt 1981, Corollary 15.1.6. (The theorem was first stated in Curry and Feys 1958, §4D2, but the proof given there contains a gap.) □

The following theorem connects $\beta\eta$-reduction with the relation of $\lambda\beta\eta$-equality defined in 7.5. Its proof is straightforward.

THEOREM 7.15. $P =_{\beta\eta} Q$ *iff* Q *is obtained from* P *by a finite (perhaps empty) series of $\beta\eta$-contractions and reversed $\beta\eta$-contractions and changes of bound variables.*

COROLLARY 7.15.1. *If* $P =_{\beta\eta} Q$, *then there exists* T *such that*

$$P \rhd_{\beta\eta} T \quad \textit{and} \quad Q \rhd_{\beta\eta} T.$$

Proof. By Theorem 7.15 and the Church-Rosser theorem, 7.12. □

COROLLARY 7.15.2. *The theory* $\lambda\beta\eta$ *is consistent in the sense that not all equations are provable.*

REMARK 7.16. The above results show that the $\beta\eta$-system is very well-behaved, almost as easy to use as β. In fact, it is a bit surprising that we get extensionality with so little effort. However, in the deeper theory of reductions this good behaviour breaks down, and difficulties begin to appear for $\beta\eta$ that β does not have, and which have taken considerable ingenuity to overcome (Klop 1980, Chapter IV). For example, though there is an analogue of the Quasi-leftmost-reduction theorem, its proof is much harder than for β. (Klop 1980, Chapter IV, Corollary 5.13).

CHAPTER EIGHT

EXTENSIONALITY IN COMBINATORY LOGIC

8A EXTENSIONAL EQUALITY

In this chapter we shall look at axioms and rules to add to the theory of combinatory weak equality to make it extensional.

NOTATION 8.1. 'Term' will mean 'CL-term' in this chapter. Recall that $I \equiv SKK$.

The following rules will be studied; the first two are extensionality-rules from the last chapter, and the third is a defining-rule for $\lambda\beta$ but not for CLw. The first has an infinity of premises.

$$(ext) \qquad \frac{XZ = YZ \quad \text{for all terms } Z}{X = Y}$$

$$(\zeta) \qquad \frac{Xx = Yx}{X = Y} \qquad (\text{if } x \notin FV(XY))$$

$$(\xi) \qquad \frac{X = Y}{\lambda^{*}x.X = \lambda^{*}x.Y} \ .$$

(Rule (ω) will not be treated; it is well covered in Barendregt 1981, §§17.3, 17.4.) We shall also discuss the following analogue of the λ-calculus axiom-scheme (η):

$$(\eta) \qquad \lambda^{*}x.Ux = U \qquad (\text{if } x \notin FV(U)).$$

DEFINITION 8.2. Let CLw be the theory of weak equality, 6.5. Three new formal theories of equality are defined as follows:

CL+*ext*: add rule (*ext*) to CLw;
CLζ : add (ζ) to CLw;
CLξ : add (ξ) to CLw.

EXERCISE 8.3. Prove that none of (*ext*), (ζ), (ξ) is admissible in CLw. Hence all the above theories are proper extensions of CLw.

THEOREM 8.4. CL+*ext and* CLζ *are theorem-equivalent in the sense of* 6.15, *and hence determine the same equality-relation on the set of all* CL-*terms*.

Proof. Exercise: prove that (*ext*) is derivable in CLζ and (ζ) is admissible in CL+*ext*. ∎

The equality relation determined by both CL+*ext* and CLζ is the obvious candidate for the name 'extensional equality'. In the literature, it is usually called $\beta\eta$-equality or just η-equality to stress the analogy with $\lambda\beta\eta$, and the definition below will follow this convention; though we shall see later that (η) cannot be regarded as expressing extensionality in combinatory logic.

DEFINITION 8.5. <u>Combinatory extensional equality</u>, or $\beta\eta$-<u>equality</u> or $=_{c\beta\eta}$, is the equality-relation determined by the theory $\mathrm{CL}_{\mathrm{ext}}$:

$$X =_{c\beta\eta} Y \iff \mathrm{CL}+ext \vdash X = Y.$$

REMARK 8.6 (<u>The role of</u> (η)). In λ-calculus, (η) ranked as an extensionality-principle equivalent to (*ext*) and (ζ). But in combinatory logic it is not an extensionality-principle at all. All it says is that every term is equal to an abstraction, and by itself this has nothing to do with extensionality. Indeed, (η) is actually true in the weak theory CLw as an identity

$$\lambda^*x.Ux \equiv U \quad \text{if} \quad x \notin FV(U)$$

by Definition 2.14(c); but despite this, CLw is not extensional. In fact, none of the rules defining CLw mentions λ^*, so the truth or falsehood of (η) is irrelevant to properties of CLw.

To get extensionality from (η) we must combine it with (ξ). (In λ-calculus too, (ξ) was needed to derive (ζ) from (η) in Theorem 7.4, but (ξ) is hardly noticed when it is used in λ-calculus.) In combinatory logic (ξ) can be regarded as a weak extensionality-principle, because given (ξ) it is easy to see that

$$X = \lambda^*x.U, \quad Y = \lambda^*x.V, \quad Xx = Yx \quad \vdash \quad X = Y.$$

The role of (η) is to make the equations $X = \lambda^*x.U$ and $Y = \lambda^*x.V$ provable for all X, Y (with $U \equiv Xx$ and $V \equiv Yx$ for some $x \notin FV(XY)$), and thus to strengthen weak extensionality to full extensionality.

THEOREM 8.7. *CLξ is rule-equivalent to CLζ; hence CLξ, CLζ and CL+ext all determine the same equality-relation,* $=_{c\beta\eta}$.

Proof. (Cf. Theorem 7.4.) First we derive (ζ) in CLξ. Given an equation $Xx = Yx$ with $x \notin FV(XY)$, apply (ξ) to give

$$\lambda^*x.Xx = \lambda^*x.Yx,$$

and this is $X = Y$ by definition of λ^* (2.14(c)).

Next we derive (ξ) in CLζ. Given $X = Y$, deduce

$$\begin{aligned}(\lambda^*x.X)x &= X && \text{by Theorem 2.15}\\ &= Y && \text{given}\\ &= (\lambda^*x.Y)x && \text{by Theorem 2.15;}\end{aligned}$$

then $\lambda^*x.X = \lambda^*x.Y$ follows by (ζ). □

REMARK 8.8. It is possible to define $\lambda^*x.M$ by 2.14(a), (b), (f) only, without the clause (c) that makes (η) true. If we did this, the corresponding CLξ would not admit (*ext*), and to get (*ext*) we would have to add (η) as a new axiom-scheme just as in λ-calculus.

DISCUSSION 8.9. We now have three theories that determine $=_{c\beta\eta}$, namely CL+*ext*, CLζ and CLξ. Unfortunately, none of these is particularly easy to handle. However, Curry discovered a fourth theory in the 1920's that improves this situation enormously (Curry 1930). He showed that rule (*ext*) could be replaced by just a finite set of axioms. (In λ-calculus (*ext*) was proved equivalent to just one axiom-scheme (η), but (η) has an infinity of axioms as its instances; here the total number of axioms, not just of axiom-schemes, is finite.) Each of Curry's axioms is just an equation $A = B$ where A and B are closed terms.

A suitable set of axioms will be given below. To motivate them, suppose we have already chosen them and they form a finite set of equations

$$A_1 = B_1, \ldots, A_n = B_n$$

in which the A_i, B_i contain no variables, and let

$$CL_{ax}$$

be the formal theory obtained by adding these to CLw as new axioms.

The axioms are to be chosen so that CL_{ax} determines $=_{c\beta\eta}$, and for this it is enough that CL_{ax} be theorem-equivalent to $CL\xi$. (Note that the equivalence required is only theorem-equivalence not rule-equivalence, and this is all that Curry's construction of the axioms will give.)

LEMMA 8.10. *If the following were provable in* CL_{ax}

Ax 1. $\lambda^*xy.S(Kx)(Ky) = \lambda^*xy.K(xy)$,

Ax 2. $\lambda^*x.S(Kx)I = \lambda^*x.x$,

then for all X, Y, x,

$$CL_{ax} \vdash \lambda^*x.XY = S(\lambda^*x.X)(\lambda^*x.Y).$$

Proof. By the definition of λ^*, 2.14, the desired equation is an identity unless either (a) $x \notin FV(XY)$, or (c) $x \notin FV(X)$ and $Y \equiv x$.

In case (a):

$$\begin{aligned} \lambda^*x.XY &\equiv K(XY) && \text{by 2.14(a)} \\ &=_w (\lambda^*xy.K(xy))XY && \text{by 2.20} \\ &= (\lambda^*xy.S(Kx)(Ky))XY && \text{by Ax 1} \\ &=_w S(KX)(KY) && \text{by 2.20} \\ &\equiv S(\lambda^*x.X)(\lambda^*x.Y) && \text{by 2.14(a).} \end{aligned}$$

In case (c):

$$\begin{aligned} \lambda^*x.XY &\equiv X && \text{by 2.14(c)} \\ &= S(KX)I && \text{by Ax 2} \\ &\equiv S(\lambda^*x.X)(\lambda^*x.x) && \text{by 2.14(a),(b).} \end{aligned}$$ □

DISCUSSION 8.11. The axioms will be constructed so that rule (ξ) is admissible in CL_{ax} i.e. so that

$$CL_{ax} \vdash U = V \quad\Rightarrow\quad CL_{ax} \vdash \lambda^*v.U = \lambda^*v.V.$$

In particular, taking the special case that the equation $U = V$ is a (K) or (S) axiom, we must choose the axioms so that for all X, Y, Z, x,

(i) $CL_{ax} \vdash \lambda^*v.KXY = \lambda^*v.X,$

(ii) $CL_{ax} \vdash \lambda^*v.SXYZ = \lambda^*v.XZ(YZ).$

If we assume Ax 1 and Ax 2, then by Lemma 8.10 the left and right sides of the equations in (i) and (ii) can be analyzed as follows.

$$\text{(iii)}\quad \lambda^*v.KXY = S(\lambda^*v.KX)Y^* \qquad \text{where } Y^* \equiv \lambda^*v.Y$$
$$= S(S(KK)X^*)Y^* \qquad \text{where } X^* \equiv \lambda^*v.X$$

$$\text{(iv)}\quad \lambda^*v.SXYZ = S(\lambda^*v.SXY)(\lambda^*v.Z)$$
$$= S(S(\lambda^*v.SX)Y^*)Z^* \qquad \text{where } Z^* \equiv \lambda^*v.Z$$
$$= S(S(S(KS)X^*)Y^*)Z^*.$$

$$\text{(v)}\quad \lambda^*v.XZ(YZ) = S(\lambda^*v.XZ)(\lambda^*v.YZ)$$
$$= S(SX^*Z^*)(SY^*Z^*).$$

We see, then, that (i) and (ii) will follow from Ax 1, Ax 2, and the following:

Ax 3. $\lambda^*xy.S(S(KK)x)y = \lambda^*xy.x,$

Ax 4. $\lambda^*xyz.S(S(S(KS)x)y)z = \lambda^*xyz.S(Sxz)(Syz).$

For example, (ii) follows from (iv), (v) and Ax 4, thus:

$$\begin{aligned} \lambda^*v.SXYZ &= S(S(S(KS)X^*)Y^*)Z^* && \text{by (iv)} \\ &= (\lambda^*xyz.S(S(S(KS)x)y)z)X^*Y^*Z^* && \text{by 2.20} \\ &= (\lambda^*xyz.S(Sxz)(Syz))X^*Y^*Z^* && \text{by Ax 4} \\ &= S(SX^*Z^*)(SY^*Z^*) && \text{by 2.20} \\ &= \lambda^*v.XZ(YZ) && \text{by (v).} \end{aligned}$$

The proof that (i) follows is similar.

DEFINITION 8.12. The <u>combinatory</u> βη-<u>axioms</u> are:

βη-<u>ax</u> 1. $\lambda^*xy.S(Kx)(Ky) = \lambda^*xy.K(xy),$

βη-<u>ax</u> 2. $\lambda^*x.S(Kx)I = \lambda^*x.x,$

βη-<u>ax</u> 3. $\lambda^*xy.S(S(KK)x)y = \lambda^*xy.x,$

βη-<u>ax</u> 4. $\lambda^*xyz.S(S(S(KS)x)y)z = \lambda^*xyz.S(Sxz)(Syz).$

The formal theory obtained by adding these axioms to CLw will be called $CL\beta\eta_{ax}$.

Of course, this definition depends on the definition of λ^*. But note that the axioms do not actually contain 'λ^*'; when written out in full, they are simply equations between combinations of S and K.

There is a slightly different set of axioms in Barendregt 1981, Corollary 7.3.15 (the set $A_{\beta\eta}$). There is a set for the system whose atoms are I, K, S in Curry and Feys 1958, p. 203 (the set ω_η). Different axiomatizations are compared in Curry et al. 1972, §11D.

THEOREM 8.13. $CL\beta\eta_{ax}$ *is theorem-equivalent to* $CL\xi$, $CL\zeta$ *and* CL+*ext*, *and hence determines the same equality as they do, namely* $=_{c\beta\eta}$.

Proof. It is enough to prove $CL\beta\eta_{ax}$ theorem-equivalent to $CL\xi$. First, every theorem of $CL\beta\eta_{ax}$ is a theorem of $CL\xi$, because each $\beta\eta$-axiom can easily be proved in $CL\xi$, and the other axioms and rules of $CL\beta\eta_{ax}$ are also rules of $CL\xi$.

For the converse, we must prove (ξ) admissible in $CL\beta\eta_{ax}$; that is, we must show that if an equation X = Y is provable in $CL\beta\eta_{ax}$ by a proof with, say, n steps, then $\lambda^*x.X = \lambda^*x.Y$ is also provable in $CL\beta\eta_{ax}$. This we shall do by induction on n. Recall the axioms and rules of $CL\beta\eta_{ax}$ (Definitions 8.12 and 6.5).

If n = 1 and the equation X = Y is an instance of axiom-scheme (K) in 6.5, then $\lambda^*x.X = \lambda^*x.Y$ follows from (i) in Discussion 8.11, which is provable in $CL\beta\eta_{ax}$ by 8.10 and 8.11. Similarly for axiom-scheme (S), using 8.11(ii).

If n = 1 and X = Y is a $\beta\eta$-axiom, then X and Y contain no variables, so the equation $\lambda^*x.X = \lambda^*x.Y$ is just KX = KY, which follows from X = Y by rule (μ) in 6.5.

Suppose now that n ≥ 2 and X = Y is a consequence by rule (μ) in 6.5. Then X ≡ ZU, Y ≡ ZV, and

$$CL\beta\eta_{ax} \vdash U = V$$

by a proof with n-1 steps. By the induction-hypothesis,

$$CL\beta\eta_{ax} \vdash \lambda^*x.U = \lambda^*x.V.$$

Hence in $CL\beta\eta_{ax}$ we can prove

$$\begin{aligned}\lambda^*x.X &= S(\lambda^*x.Z)(\lambda^*x.U) && \text{by Lemma 8.10}\\ &= S(\lambda^*x.Z)(\lambda^*x.V) && \text{by induction-hypothesis and } (\mu)\\ &= \lambda^*x.Y && \text{by Lemma 8.10.}\end{aligned}$$

Rule (ν) is handled like (μ), and the other rules in 6.5 are trivial. □

Warning. Rules (ξ) and (ζ) are not derivable in $CL\beta\eta_{ax}$ (to be proved in §11.2), although by the above theorem they are admissible (§6.13).

EXERCISE 8.14. Write out the $\beta\eta$-axioms in detail. (The abbreviations $B_X \equiv S(KX)$ and $B_X^2 \equiv S(KB_X)$ might help.)

8B $\beta\eta$-STRONG REDUCTION

DISCUSSION 8.15. In the previous chapters, each equality-relation has had a corresponding reduction with a Church-Rosser theorem and other helpful properties. So now the natural next step is to look for a reduction for $=_{c\beta\eta}$.

Ideally, it should have the following properties (in which '$X \succ Y$' means 'X reduces to Y').

(a) All the rules and axioms of CLw hold true for $\succ$; in particular, $\succ$ is transitive, reflexive and has this substitution property:

$$X \succ Y \quad\Rightarrow\quad [X/v]Z \succ [Y/v]Z.$$

(b) The other two 'substitution' properties hold:

$$X \succ Y \quad\Rightarrow\quad (v \notin FV(X) \quad\Rightarrow\quad v \notin FV(Y)),$$
$$X \succ Y \quad\Rightarrow\quad [Z/v]X \succ [Z/v]Y.$$

(c) The equivalence relation generated by $\succ$ is the same as $=_{c\beta\eta}$; that is, $X =_{c\beta\eta} Y$ iff X goes to Y by a finite series of reductions and reversed reductions.

(d) The Church-Rosser theorem holds for $\succ$; i.e.

$$(U \succ X,\ U \succ Y) \quad\Rightarrow\quad (\exists Z)(X \succ Z,\ Y \succ Z).$$

(e) If we call X <u>irreducible</u> iff $(\forall Y)(X \succ Y \Rightarrow Y \equiv X)$, then the set of all irreducibles can be characterized in some reasonably neat way.

DEFINITION 8.16 (βη-<u>strong reduction</u>). The formal theory of βη-<u>strong reduction</u> is obtained from the theory CLw of weak reduction defined in 6.6 by changing '$\rhd$' to '$\succ$' and adding this rule:

$$(\xi)\quad \frac{X \succ Y}{\lambda^* x.X \succ \lambda^* x.Y}$$

Iff $X \succ Y$ is provable in this theory, we say that X βη-<u>strongly reduces to</u> Y, or $X \succ_{\beta\eta} Y$. (The 'βη' is usually omitted in the literature.)

THEOREM 8.17. *βη-strong reduction has properties* 8.15(a)-(d).

Proof. (a)-(c) are easy. For (d), see Exercise 9.16(a). □

DISCUSSION 8.18. What about property (e)? It has been stated only vaguely, and one very attractive way of making it precise is to demand that the irreducibles be exactly the 'strong normal forms' of Definition 3.7, which played a role in Böhm's theorem and also corresponded closely to normal forms in λ-calculus.

Unfortunately, one of the simplest strong normal forms, I, is not irreducible. (<u>Exercise</u>: prove that $SK \succ_{\beta\eta} KI$, and deduce

$$I \succ_{\beta\eta} KIK \succ_{\beta\eta} K(KIK)K \succ_{\beta\eta} \ldots\ .)$$

However, if I was not defined as SKK but assumed to be a new atom with an axiom-scheme

$$IX \succ X,$$

this anomaly would disappear and the irreducibles would be exactly the strong normal forms (Lercher 1967a). And all previous results on weak reduction and equality would stay valid, provided the extra axiom-schemes

$$IX = X, \qquad IX \rhd X$$

were added to Definitions 6.5 and 6.6. (The proof is routine.) For βη-equality, an extra axiom $S(KI) = I$ would have to be added to 8.12.

So if I is assumed to be an atom, $\succ_{\beta\eta}$ seems to behave very nicely. But when we look deeper, we find other snags. The proof of the Church-Rosser theorem for $\succ_{\beta\eta}$ proceeds by deducing it from the theorem for $\lambda\beta\eta$-reduction, and no more direct proof is known. And the proof that the irreducibles are exactly the strong normal forms is by no means easy. In fact, the metatheory of $\succ_{\beta\eta}$ is quite messy, and this has limited its use.

But nevertheless, all its major properties are known. There is a clear short account in Hindley et al. 1972, Chapter 7, and a more detailed one in Curry et al. 1972, §11E. The latter reference contains a standardization theorem. Also rule (ξ) can be replaced by an infinite but recursive set of axioms (Hindley 1967, Lercher 1967b), and this fact is used in Hindley and Lercher 1970 to simplify the characterization proof for the irreducibles.

DEFINITION 8.19. A CL-term X <u>has a strong normal form</u> X^* iff X^* is in strong nf (Definition 3.7) and

$$X \succ_{\beta\eta} X^*.$$

LEMMA 8.20.

(a) *A term* X *cannot have more than one strong nf.*

(b) *If* X^* *is in strong nf and* $U =_{c\beta\eta} X^*$, *then* $U \succ_{\beta\eta} X^*$.

(c) X^* *is the strong nf of* X *iff* X^* *is in strong nf and*

$$X =_{c\beta\eta} X^*.$$

Proof. Clearly, (c) follows from (b). To prove (a) and (b) would be easy if I was an atom, because then every strong nf would be irreducible, as noted above, and a Church-Rosser argument would suffice (cf. Theorem 1.35 and its corollaries). But I is not an atom here, so we must use a more roundabout method; see Exercise 9.16(d). □

Further results on strong nf's will be given in Exercises 9.16.

CHAPTER NINE

THE CORRESPONDENCE BETWEEN λ AND CL

9A INTRODUCTION

Everything done so far has emphasized the close correspondence between λ-calculus and combinatory logic, in both motivation and results. But only now do we have the tools to make this correspondence precise. This is the aim of the present chapter.

The correspondence between the 'extensional' equalities will be described first, in §9B.

The non-extensional equalities are less straightforward. We have $=_\beta$ in λ-calculus and $=_w$ in combinatory logic, and despite their many parallel properties, these differ crucially in that λβ admits (ξ) and CLw does not. To get a close correspondence in any reasonable sense, we must define a new relation called combinatory β-equality, and relate it to λβ. This will be done in §9C.

(It is also possible to define a new 'weak λ-equality' corresponding to CLw, but this has at present only specialist interest; the definition is due to W. Howard, and appears in Hindley 1977, §2, and Hindley and Longo 1980, §8.15.)

NOTATION 9.1. This chapter is about both λ- and CL-terms, so 'term' will never be used without 'λ-' or 'CL-'.

Congruent λ-terms will be regarded as identical, and 'M is congruent to N' will be written as 'M ≡ N'. (So in effect, the word 'λ-term' will mean 'congruence-class of λ-terms', i.e. the class of all λ-terms congruent to a given term.) Define

$$\Lambda \;=\; \text{the class of all (congruence-classes of) λ-terms,}$$
$$\mathcal{C} \;=\; \text{the class of all CL-terms.}$$

It will be assumed that the variables in $\mathcal{C}$ are the same as those in Λ (cf. Definitions 1.1, 2.1).

For CL-terms: in this chapter only, the $\lambda^*x.M$ of Definition 2.14 will be called

$$\lambda^\eta x.M;$$

this is to distinguish it from a new definition of $\lambda^*x.M$ in §9C.

The letters 'I', 'K', 'S' will denote CL-terms only, not any λ-terms. Note that $I \equiv SKK$.

Recall the four main equality-relations; two in λ-calculus:

$=_\beta$ (determined by the theory $\lambda\beta$ in 6.2),

$=_{\beta\eta}$ (determined by any of $\lambda\beta\eta$, $\lambda\beta\zeta$, $\lambda\beta+ext$ in 7.2),

and two in combinatory logic:

$=_w$ (determined by the theory CLw in 6.5),

$=_{c\beta\eta}$ (determined by any of $CL\zeta$, $CL\xi$, $CL+ext$, $CL\beta\eta_{ax}$ (8.2, 8.5, 8.12)).

DEFINITION 9.2 (The λ-transformation). This very natural mapping is the basis of the correspondence between λ and CL. To each CL-term X, associate a λ-term X_λ by induction on X, thus:

(a) $x_\lambda \equiv x$,

(b) $K_\lambda \equiv \lambda xy.x$, $S_\lambda \equiv \lambda xyz.xz(yz)$,

(c) $(XY)_\lambda \equiv X_\lambda Y_\lambda$.

NOTE 9.3. Clearly X_λ is unique (modulo congruence). Also $X \not\equiv Y$ implies $X_\lambda \not\equiv Y_\lambda$, so the λ-transformation is one-to-one. It maps $\mathcal{C}$ onto a proper subset of Λ which we shall call $\mathcal{C}_\lambda$:

$$\mathcal{C}_\lambda = \{X_\lambda : X \in \mathcal{C}\}.$$

Note also that $FV(X_\lambda) = FV(X)$, and that

(9.4) $$([Z/x]X)_\lambda \equiv [Z_\lambda/x](X_\lambda).$$

LEMMA 9.5. *For CL-terms* X, Y:

(a) $X \triangleright_w Y \Rightarrow X_\lambda \triangleright_\beta Y_\lambda$,

(b) $X =_w Y \Rightarrow X_\lambda =_\beta Y_\lambda$,

(c) $X =_{c\beta\eta} Y \Rightarrow X_\lambda =_{\beta\eta} Y_\lambda$.

Proof. For (a) and (b), use induction on the definitions of CLw (6.6, 6.5). Cases (ρ), (μ), (ν), (τ), (σ) are trivial.

Case (S): $(SXYZ)_\lambda \equiv (\lambda xyz.xz(yz))X_\lambda Y_\lambda Z_\lambda$

$\equiv_\alpha (\lambda uvw.uw(vw))X_\lambda Y_\lambda Z_\lambda$ where $u, v, w \notin FV(XYZ)$

$\rhd_\beta X_\lambda Z_\lambda(Y_\lambda Z_\lambda)$

$\equiv (XZ(YZ))_\lambda$.

Case (K): $(KXY)_\lambda \equiv (\lambda xy.x)X_\lambda Y_\lambda \rhd_\beta X_\lambda$.

For (c), note that CLζ determines $=_{c\beta\eta}$ by 8.4, and use induction on the axioms and rules of CLζ. Rule (ζ) is admissible in λβη by 7.4. □

9B THE EXTENSIONAL EQUALITIES

DEFINITION 9.6 (<u>The induced combinatory</u> βη-<u>equality</u>). λβη-conversion induces the following relation on CL-terms via the λ-transformation:

$$X =_{\beta\eta\,\underline{\text{induced}}} Y \iff X_\lambda =_{\beta\eta} Y_\lambda.$$

The aim of this section is to prove that this induced relation is the same as the $=_{c\beta\eta}$ defined in 8.5, and to describe the correspondence between it and λβη in detail.

The main tool is the following mapping from Λ into 𝒞.

DEFINITION 9.7 (<u>The H_η-transformation</u>). To each λ-term M we associate a CL-term called M_{H_η} (or usually just M_H), thus:

(a) $x_H \equiv x$,

(b) $(MN)_H \equiv M_H N_H$,

(c) $(\lambda x.M)_H \equiv \lambda^\eta x.(M_H)$.

(As mentioned earlier, '$\lambda^\eta x$' denotes the $\lambda^* x$ of Definition 2.14.)

LEMMA 9.8. *For CL-terms* X: $(X_\lambda)_{H_\eta} \equiv X$.

Proof. Induction on X. The cases $X \equiv x$, $X \equiv YZ$ are trivial.

If $X \equiv K$: $K_{\lambda H} \equiv (\lambda xy.x)_H \equiv \lambda^\eta xy.x$

$\equiv K$ by Example 2.18(a).

If $X \equiv S$: $S_{\lambda H} \equiv \lambda^\eta xyz.xz(yz)$

$\equiv S$ by Example 2.18(b). □

REMARK 9.9. Lemma 9.8 implies that every CL-term is in the range of H_η. Also, although H_η is fairly obviously not one-to-one, it becomes one-to-one when its domain is restricted to $\mathcal{C}_\lambda$, and in fact becomes the inverse of the λ-transformation; because

$$M \in \mathcal{C}_\lambda \Rightarrow M \equiv X_\lambda \quad \text{for some } X$$
$$\Rightarrow M_{H\lambda} \equiv X_{\lambda H \lambda} \equiv X_\lambda \quad \text{by Lemma 9.8}$$
$$\equiv M.$$

In the language of Chapter 11, H_η is a 'left inverse' of λ (Notation 11.1, Figure 11.1).

LEMMA 9.10. *For λ-terms* M, N: *if* H *is* H_η, *then*

(a) $FV(M_H) = FV(M)$;

(b) $M \equiv_\alpha N \Rightarrow M_H \equiv N_H$;

(c) $([N/x]M)_H \equiv [N_H/x](M_H)$.

Proof. (a) is obvious. Before proving (b), it is best to prove (c) in the special case that no variable bound in M is free in N. This is done by induction on M. The cases $x \notin FV(M)$, $M \equiv x$, $M \equiv PQ$ are routine. Now suppose $M \equiv \lambda y.P$ and $y \notin FV(xN)$: then

$$([N/x](\lambda y.P))_H \equiv (\lambda y.[N/x]P)_H \quad \text{by 1.11(e)}$$

$$\equiv \lambda^\eta y.([N/x]P)_H \quad \text{by 9.7(c)}$$

$$\equiv \lambda^\eta y.[N_H/x](P_H) \quad \text{by induction-hypothesis}$$

$$\equiv [N_H/x](\lambda^\eta y.P_H) \quad \text{by 2.21(c)}$$

$$\equiv [N_H/x](\lambda y.P)_H \quad \text{by 9.7(c).}$$

Next, we prove (b). It is enough to prove that

$$(\lambda x.P)_H \equiv (\lambda y.[y/x]P)_H \quad (\text{if } y \notin FV(xP)).$$

This is deduced from the above special case of (c), thus:

$$\begin{aligned}(\lambda x.P)_H &\equiv \lambda^{\eta}x.(P_H) \equiv \lambda^{\eta}y.[y/x](P_H) && \text{by 2.21(b)}\\ &\equiv \lambda^{\eta}y.([y/x]P)_H && \text{by (c) above}\\ &\equiv (\lambda y.[y/x]P)_H.\end{aligned}$$

Finally, we deduce (c) from the above special case, by first changing all clashing bound variables and using (b). □

LEMMA 9.11. *For* λ-*terms* M, N: *if* H *is* H_η, *then*

$$(\lambda\beta\eta \vdash M = N) \Rightarrow (\text{CL}+ext \vdash M_H = N_H).$$

Proof. Induction on λβη (Definitions 7.2 and 6.2). Case (α) is done by 9.10(b). Cases (ρ), (σ), (τ), (μ), (ν) are trivial.

Case (β). Let $M \equiv (\lambda x.P)Q$ and $N \equiv [Q/x]P$. Then

$$\begin{aligned}M_H &\equiv (\lambda^{\eta}x.P_H)Q_H \rhd_w [Q_H/x]P_H\\ &\equiv N_H \quad \text{by 9.10(c)}.\end{aligned}$$

Case (η). Let $M \equiv \lambda x.Nx$ and $x \notin FV(N)$. Then

$$M_H \equiv \lambda^{\eta}x.N_H x \equiv N_H \quad \text{by 2.14(c)}.$$

Case (ξ). Rule (ξ) is admissible in CL+*ext*, by 8.7. □

THEOREM 9.12. *The equality relation induced on* CL-*terms by* λβη *(Definition* 9.6*) is the same as* $=_{c\beta\eta}$ *(Definition* 8.5*); i.e. for all* CL-*terms* X, Y,

$$(\lambda\beta\eta \vdash X_\lambda = Y_\lambda) \Leftrightarrow (\text{CL}+ext \vdash X = Y).$$

Proof. '⇐' is 9.5(c), and '⇒' comes from 9.8 and 9.11. □

LEMMA 9.13. *For* CL-*terms* Z: $\lambda\beta\eta \vdash (\lambda^{\eta}x.Z)_\lambda = \lambda x.(Z_\lambda)$.

Proof. $(\lambda^{\eta}x.Z)_{\lambda} =_{\eta} \lambda x.(\lambda^{\eta}x.Z)_{\lambda}x$ by (η)

$\equiv \lambda x.((\lambda^{\eta}x.Z)x)_{\lambda}$ by 9.2(c), (a)

$=_{\beta} \lambda x.Z_{\lambda}$ by 9.5(b) since $(\lambda^{\eta}x.Z)x =_{w} Z$. □

The next theorem describes the correspondence between the λ- and combinatory βη-equalities in detail.

THEOREM 9.14 (Equivalence theorem for βη). *For all* CL-*terms* X, Y *and* λ-*terms* M, N: *if* H *is* H_{η}, *then*

(a) $X_{\lambda H} \equiv X$,

(b) $M_{H\lambda} =_{\beta\eta} M$,

(c) $X =_{c\beta\eta} Y \iff X_{\lambda} =_{\beta\eta} Y_{\lambda}$,

(d) $M =_{\beta\eta} N \iff M_{H} =_{c\beta\eta} N_{H}$.

Proof. (a) is 9.8 and (c) is 9.12.

For (b): use induction on M. For example, if $M \equiv \lambda x.P$, then

$$\begin{aligned} M_{H\lambda} &\equiv (\lambda^{\eta}x.P_{H})_{\lambda} \\ &=_{\beta\eta} \lambda x.(P_{H\lambda}) \quad \text{by 9.13} \\ &=_{\beta\eta} \lambda x.P \quad \text{by induction hypothesis and } (\xi). \end{aligned}$$

For (d): '⇒' is 9.11, and '⇐' comes from (c) and (b). □

REMARK 9.15 (Reduction). The correspondence between λ and CL is nowhere near as neat for reduction as for equality. In λ-calculus, the 'extensional' reduction is $\triangleright_{\beta\eta}$ (Definition 7.8), and in combinatory logic it is $\succ_{\beta\eta}$ (Definition 8.16). It is easy to prove that

(a) $$M \triangleright_{\beta\eta} N \implies M_{H_{\eta}} \succ_{\beta\eta} N_{H_{\eta}}$$

(cf. the proof of 9.11). But if we ask for the converse of (a), we shall be disappointed, because $X \succ_{\beta\eta} Y$ does not imply $X_{\lambda} \triangleright_{\beta\eta} Y$, but only $X_{\lambda} =_{\beta\eta} Y_{\lambda}$ (Exercise 9.16(b) below). Some of the problems involved are discussed in Hindley 1977, §3.

The lack of a two-way correspondence between the two reductions is no great drawback, however. Equality is the main thing, and all we should ask of a reduction is that it helps in the study of equality (and, perhaps, that it behaves in some sense like the informal process of calculating the value of a function). Reduction's main use is to prove the consistency of the theory of equality, and other results of the form 'X = Y is not provable'.

EXERCISES 9.16 (<u>Strong reduction and strong nf's</u>).

(a) Using 9.15(a) and the Church-Rosser theorem for $\lambda\beta\eta$ reduction (7.12), prove the Church-Rosser theorem for $\succ_{\beta\eta}$ (8.17(d)). (Answer: Curry et al. 1958, §6F Theorem 3, or Hindley et al. 1972, Theorem 7.6.)

(b) Find CL-terms X, Y such that $X \succ_{\beta\eta} Y$ but not $X_\lambda \rhd_{\beta\eta} Y_\lambda$. (Answer: Curry et al. 1958, p. 221.)

(c) Prove that, for any CL-term X,

$$X \text{ is in strong nf} \iff (\exists M \text{ in } \beta\eta\text{-nf})(X \equiv M_{H_\eta})$$
$$\iff (\exists M \text{ in } \beta\text{-nf})(X \equiv M_{H_\eta}).$$

(d) Using (c) and 9.14 and 9.15(a), prove Lemma 8.20(a), (b).

(e) Using (c) and 8.20(c) and 9.14, prove that for all CL-terms X and λ-terms M,

(i) X has a strong nf $\iff$ X_λ has a λβη-nf;

(ii) M has a λβη-nf $\iff$ M_{H_η} has a strong nf.

(f) Using (e) and 3.20, prove that the Y-combinators in CL have no strong nf. (Though, as remarked in 3.20, they are in weak nf.)

(g) Do not confuse <u>having</u> a normal form with <u>being in</u> normal form! For example, every CL-term in strong nf is also in weak nf (by 3.9), but there are terms X which have a strong nf but do not have a weak nf. Prove that SK(YK) is one such X.

9C COMBINATORY β-EQUALITY

DEFINITION 9.17. λβ-conversion induces the following relation on CL-terms, which we shall call $=_{c\beta}$:

$$X =_{c\beta} Y \iff X_\lambda =_\beta Y_\lambda.$$

The aim of this section is to find a formal theory entirely within combinatory logic that determines this relation, and to prove a detailed equivalence-theorem like 9.14.

EXERCISE 9.18.* The relation $=_{c\beta}$ is intermediate in strength between $=_w$ and $=_{c\beta\eta}$. It is easy to prove that

(a) $$X =_w Y \Rightarrow X =_{c\beta} Y \Rightarrow X =_{c\beta\eta} Y.$$

Show that these implications cannot be reversed, by proving

(b) $$SK =_{c\beta} KI, \qquad SK \neq_w KI;$$

(c) $$S(KI) =_{c\beta\eta} I, \quad S(KI) \neq_{c\beta} I.$$

NOTE 9.19. The proof of the equivalence theorem will need an H-transformation from Λ into $\mathcal{C}$. But H_η is not suitable, because it does not satisfy $M_{H\lambda} =_\beta M$ but only $M_{H\lambda} =_{\beta\eta} M$. (For example, let $M \equiv \lambda x.yx$.) The trouble is that the definition of λ^* on which H_η is based contains the clause

(c) $$\lambda^*x.Ux \equiv U \quad \text{if} \quad x \notin FV(U),$$

which corresponds to (η). For β-equality, we shall need a new definition of λ^* with (c) omitted, or at least severely restricted. It is simplest to omit it, as follows.

DEFINITION 9.20 (Weak abstraction, λ^w). *For* CL-*terms* X, *define* $\lambda^w x.X$ *thus*:

(a) $\lambda^w x.X \equiv KX$ if $x \notin FV(X)$,

(b) $\lambda^w x.x \equiv I$,

(f) $\lambda^w x.UV \equiv S(\lambda^w x.U)(\lambda^w x.V)$ if $x \in FV(UV)$.

This definition is the same as Curry and Feys 1958, §6A Algorithm (abf), and as the λ^* in Barendregt 1981, Definition 7.1.5. It has the same basic properties as λ^η, namely, 2.15, 2.20 and 2.21, with essentially the same proofs.

(Why then was it not introduced in Chapter 2 instead of λ^η, whose definition is longer? One reason is practical: calculate and compare

$$\lambda^w xyz.xz(yz), \qquad \lambda^\eta xyz.xz(yz),$$

for example. Another reason is that one or two of the less basic but nonetheless useful properties of λ^η (e.g. $X_{\lambda H} \equiv X$) fail for λ^w, and there is an alternative definition of λ^w for which these properties do not fail. This alternative is a little more complicated to define, however (Definition 9.34), so we shall stay with the present one for the moment.)

DEFINITION 9.21 (The H_w-transformation). For each λ-term M, define M_{H_w} just like M_{H_η} in Definition 9.7, but using λ^w instead of λ^η.

LEMMA 9.22. *The three parts of Lemma* 9.10 *hold for* H_w.

Proof. Like Lemma 9.10. □

DEFINITION 9.23 (Functional terms and the theory $CL\zeta_\beta$). A CL-term S, K, SX, KX or SXY (for any X, Y) is called functional or fnl. The theory $CL\zeta_\beta$ is obtained by adding to CLw the rule

$$(\zeta_\beta)\ \frac{Ux = Vx}{U = V} \quad (\text{if } x \notin FV(UV) \text{ and } U, V \text{ fnl}).$$

LEMMA 9.24.

(a) *For all* CL-*terms* X: $\lambda^w x.X$ *is functional.*

(b) *If* U *is fnl, then* $U_\lambda \triangleright_\beta \lambda x.M$ *for some* λ-*term* M.

(A converse to Lemma 9.24(b) also holds, see Remark 9.37.)

LEMMA 9.25. *For* CL-*terms* X, Y *and* λ-*terms* M, N:

(a) $(CL\zeta_\beta \vdash X = Y) \Rightarrow (\lambda\beta \vdash X_\lambda = Y_\lambda)$;

(b) $(\lambda\beta \vdash M = N) \Rightarrow (CL\zeta_\beta \vdash M_{H_w} = N_{H_w})$.

Proof.

(a): Induction on the rules defining $CL\zeta_\beta$ in 9.23 and 6.5. For (K) and (S) see the proof of 9.5. Rules (ρ), (μ), (ν), (τ), (σ) are trivial. For rule (ζ_β): let U, V be fnl CL-terms and suppose $U_\lambda x =_\beta V_\lambda x$. Then by 9.24(b),

$$U_\lambda \rhd_\beta \lambda x.M, \quad V_\lambda \rhd_\beta \lambda x.N$$

for some λ-terms M, N. Hence

$$\begin{aligned} U_\lambda \rhd_\beta \lambda x.M &=_\beta \lambda x.(\lambda x.M)x && \text{since } (\lambda x.M)x \rhd_\beta M \\ &=_\beta \lambda x.U_\lambda x && \text{since } U_\lambda =_\beta \lambda x.M \\ &=_\beta \lambda x.V_\lambda x && \text{by hypothesis and } (\xi) \\ &=_\beta V_\lambda && \text{similarly.} \end{aligned}$$

(b): Induction on λβ, Definition 6.2. For cases (α) and (β), use Lemma 9.22. Cases (ρ), (μ), (ν), (τ), (σ) are trivial. For (ξ): suppose that

$$CL\zeta_\beta \vdash M_H = N_H,$$

where H is H_w. We want to deduce $(\lambda x.M)_H = (\lambda x.N)_H$. Now

$$(\lambda x.M)_H x \equiv (\lambda^\beta x.M_H)x =_w M_H,$$

and similarly for N, so

$$CL\zeta_\beta \vdash (\lambda x.M)_H x = (\lambda x.N)_H x.$$

But $(\lambda x.M)_H$ and $(\lambda x.N)_H$ are fnl by 9.24(a), so (ζ_β) can be applied, giving

$$CL\zeta_\beta \vdash (\lambda x.M)_H = (\lambda x.N)_H. \qquad \square$$

LEMMA 9.26. *For* CL-*terms*: $CL\zeta_\beta \vdash (X_\lambda)_{H_w} = X$.

Proof. Induction on X. For example, let X ≡ S. (The cases X ≡ x, X ≡ YZ are trivial, and X ≡ K is like X ≡ S). Let H be H_w. Then

$$\begin{aligned} S_{\lambda H}xyz &\equiv (\lambda^\beta xyz.xz(yz))xyz =_w xz(yz) \\ &=_w Sxyz. \end{aligned}$$

Now Sxy, Sx, S and $S_{\lambda H}$ are fnl, and $S_{\lambda H}xy$, $S_{\lambda H}x$ can easily be seen to be weakly equal to fnl terms, so (ζ_β) can be applied three times to give

$$CL\zeta_\beta \vdash S_{\lambda H} = S. \qquad \square$$

THEOREM 9.27. *The equality relation* $=_{c\beta}$ *induced on* CL-*terms by* $\lambda\beta$ *(Definition* 9.17*) is the same as the equality determined by* $CL\zeta_\beta$; *i.e. for all* CL-*terms* X, Y,

$$(\lambda\beta \vdash X_\lambda = Y_\lambda) \iff (CL\zeta_\beta \vdash X = Y).$$

Proof. '⇐' is 9.25(a); and '⇒' comes from 9.25(b) and 9.26. □

THEOREM 9.28 (Equivalence theorem for β). *For all* CL-*terms* X, Y *and* λ-*terms* M, N: *if* H *is* H_w, *then*

(a) $X_{\lambda H} =_{c\beta} X$,

(b) $M_{H\lambda} =_\beta M$,

(c) $X =_{c\beta} Y \iff (\lambda\beta \vdash X_\lambda = Y_\lambda) \iff (CL\zeta_\beta \vdash X = Y)$,

(d) $M =_\beta N \iff M_H =_{c\beta} N_H$.

Proof. Like Theorem 9.14. For (b), first prove

(9.29) $$\lambda\beta \vdash (\lambda^w x.Z)_\lambda = \lambda x.(Z_\lambda).$$ □

REMARK 9.30. The theory $CL\zeta_\beta$ gives the purely combinatory characterization of $=_{c\beta}$ that we wanted.

But we can go further, just as we did for βη, and determine $=_{c\beta}$ by a finite set of axioms (added to CLw). There is no space to give details here, so a suitable set will be simply stated below without proof. It is the set A_β from Barendregt 1981, §7.3, p. 154, with two redundant members omitted.

The first published axioms for β are in Rosser 1935, though his system corresponds to λI-calculus and hence has no K (cf. 1.36). There is also a set of axioms in Curry and Feys 1958 p. 203 for the system with basic combinators I, K, S. Various methods of axiomatization are analyzed in Curry et al. 1972, §11D.

DEFINITION 9.31. The <u>combinatory</u> β-<u>axioms</u> are:

β-<u>ax</u> 1. $\lambda^w xy.Kxy = K$,

β-<u>ax</u> 2. $\lambda^w xyz.Sxyz = S$,

β-<u>ax</u> 3. $\lambda^w xy.S(Kx)(Ky) = \lambda^w xy.K(xy)$,

β-<u>ax</u> 4. $\lambda^w xy.S(S(KK)x)y = \lambda^w xyz.xz$,

β-<u>ax</u> 5. $\lambda^w xyz.S(S(S(KS)x)y)z = \lambda^w xyz.S(Sxz)(Syz)$.

The theory obtained by adding these axioms to CLw will be called

$$CL\beta_{ax}.$$

THEOREM 9.32. *The equality determined by* $CL\beta_{ax}$ *is* $=_{c\beta}$, *i.e.*

$$CL\beta_{ax} \vdash X = Y \quad iff \quad \lambda\beta \vdash X_\lambda = Y_\lambda.$$

Proof. Barendregt 1981 Theorem 7.3.10 proves the result for his set A_β, Definition 7.3.6. Two members of A_β are redundant, namely A.1′ and A.2′, as he states in the proof of 7.3.15. (They follow from A.1, A.2 and the corresponding weak axioms by Lemma 7.3.5(iii).) The above set is A_β without A.1′ and A.2′. □

COROLLARY 9.32.1. $CL\beta_{ax}$ *is theorem-equivalent to* $CL\zeta_\beta$.

We will see in Remark 11.2 that $CL\beta_{ax}$ is not rule-equivalent to $CL\zeta_\beta$.

REMARK 9.33 (<u>Rule</u> (ξ)). In the last chapter, CLζ was proved equivalent to CLξ. Can this be done here with $CL\zeta_\beta$?

The answer appears to be 'no'. When abstraction is λ^w, the rule (ξ) becomes

$$(\xi_w)\ \frac{X = Y}{\lambda^w x.X = \lambda^w x.Y}\ .$$

There is no problem deriving this in $CL\zeta_\beta$ by a proof like that of 9.25(b); however, we the authors are unable to prove the converse and we conjecture that it is false.

On the other hand, (ξ) is not an absolute concept like (ζ) or (ζ_β); it depends on the definition of abstraction. This raises the possibility that there might be some new definition of abstraction which makes (ξ) equivalent to (ζ_β). And in fact there is. It is due to Curry and is more complicated to define than either λ^η or λ^w, but has some of the virtues of both.

DEFINITION 9.34 (β-abstraction, λ^β). For CL-terms, define

(a) $\lambda^\beta x.X \equiv KX$ if $x \notin FV(X)$,

(b) $\lambda^\beta x.x \equiv I$,

(c_β) $\lambda^\beta x.Ux \equiv U$ if U is fnl and $x \notin FV(U)$,

(f_β) $\lambda^\beta x.UV \equiv S(\lambda^\eta x.U)(\lambda^\eta x.V)$ if (a)-(c_β) do not apply.

NOTE 9.35. The η's in (f_β) are not misprints; together with (c_β) they give λ^β a property that λ^η has but λ^w does not, namely

$$(K_\lambda)_{H_\beta} \equiv K, \quad (S_\lambda)_{H_\beta} \equiv S,$$

where H_β is the H defined by λ^β. On the other hand, (c_β) gives λ^β a property that λ^w has but λ^η does not, namely

$$\lambda^\beta x.X \ \underline{\text{is fnl for all}}\ X.$$

Routine proofs show that λ^β satisfies 2.15, 2.20, 2.21(a), (b), and

$$[U/x](\lambda^\beta y.V) =_{c\beta} \lambda^\beta y.[U/x]V \quad \text{if } y \notin FV(xU)$$

(a weak form of 2.21(c)). The analogue of (ξ) for λ^β is the rule

$$(\xi_\beta)\ \frac{X = Y}{\lambda^\beta x.X = \lambda^\beta x.Y},$$

and if we add this to CLw we get a theory rule-equivalent to $CL\zeta_\beta$.

The proof of the equivalence theorem (Theorem 9.28) can easily be carried out with λ^β in place of λ^w, and 9.28(a) strengthens to

$$X_{\lambda H_\beta} \equiv X.$$

Thus H_β is a left-inverse of the λ-transformation, just like H_η (Remark 9.9). (Of course this does not imply $H_\beta = H_\eta$, but only that they would become equal if restricted to $\mathcal{C}_\lambda$.)

Thus it seems λ^β is the abstraction best suited to formalizing combinatory β-equality.

REMARK 9.36 (β-strong reduction). Several definitions of reduction have been proposed for $=_{c\beta}$, but unfortunately none is very easy to work with. If we simply changed abstraction to λ^{w} in the definition of βη-strong reduction (Definition 8.16), we would get a reduction that does not have a Church-Rosser theorem. (Lercher, correspondence, discussed in Hindley 1977, §3). The problem is that some of the key results in the theory of βη-strong reduction depend on the fact that abstraction is λ^{η}, and clause (c) in the definition of λ^{η} is incompatible with $=_{c\beta}$ (cf. Note 9.19).

Towards the end of his life, Curry investigated the reduction obtained by replacing λ^* in Definition 8.16 by λ^{β}, and this does have some of the properties desired in a β-strong reduction. An improved candidate which has more of the desired properties is described in Mezghiche 1984.

REMARK 9.37 (Fnl terms and λ-terms $\lambda x.M$). For CL-terms X, call X an O_1-CL-term iff $X =_w U$ for some fnl U. For λ-terms P, call P an O_1-λ-term iff $P =_\beta \lambda x.M$ for some M. Then we have

(a) X *is an* O_1*-CL-term iff* X_λ *is an* O_1*-λ-term.*

('Only if' comes from Lemma 9.24(b) and the Church-Rosser theorems. The converse is a good exercise for the reader who knows some reduction theory, in particular the standardization theorem. The authors do not remember seeing a proof in print, however.)

EXERCISE 9.38. A result that will be needed later is the lemma of invariance of form: if a, b are non-redex atoms, and

$$aX_1\ldots X_k =_{c\beta} bY_1\ldots Y_m,$$

then $a \equiv b$, $k = m$, and $X_i =_{c\beta} Y_i$ for $i = 1,\ldots,k$. Prove this lemma for combinatory β-equality, using the β-strong reduction in Mezghiche 1984, Definition 4.2. (The clause '$U \not\equiv [x]_\beta.Y$' in that definition can be omitted without affecting any of Mezghiche's results; indeed, its presence seems to make their proofs more difficult.) Alternatively, deduce the lemma from the corresponding lemma for λβ-equality, Corollary 1.35.5.

EXERCISE 9.39. Show that $S(K(Kx)) =_{c\beta} K(Kx)$, so $K(Kx)$ is a fixed-point of S. Show that this fails for weak equality.

CHAPTER TEN

MODELS OF CLw

10A APPLICATIVE STRUCTURES

In ordinary first-order logic, a natural question to ask about a theory is 'what are its models like?'. For $\lambda\beta$ and CLw the first person to ask this was Dana Scott in the 1960's, during his studies of computability for functions of higher type; and it was he who constructed the first non-trivial model, D_∞. Since then about a dozen other models have been discovered, and a beginning has been made on a general theory of models of $\lambda\beta$ and CLw.

But this is only a beginning, though combinators and λ-calculus are now over 60 years old. Why has growth been so slow? There are two main reasons.

First, the origins of $\lambda\beta$ and CL. Both Church and Curry viewed these theories, not from within the semantics that most mathematical logicians are trained in nowadays, but from the alternative viewpoint described in Discussion 3.24. Their aim was to formalize a concept of operator which is independent of the concept of set, and does not correspond to the function-concept in the usual set-theories (e.g. Zermelo-Fraenkel set-theory, ZF). In contrast, the logicians' 'standard' semantics presupposes the set-concept, and to ask for a model of CLw is really asking for an interpretation of CLw in ZF. From the Church-Curry point of view this question was interesting, but it was definitely not primary.

Second, the complexity of the models. The problem of constructing set-theoretic objects which behave like combinators was far from easy, its solution was a considerable achievement, and the resulting structures are not simple (though simpler models have since been found).

The present chapter discusses a few basic general properties of models of CLw. Models of $\lambda\beta$ are a more complicated concept and will be left to the next chapter. Chapter 12 will describe the construction of specific models (of both CLw and $\lambda\beta$), and give references for others.

NOTATION 10.1. In this chapter, 'term' will mean 'CL-term'. The theories whose models will be studied are:

CLw (Definition 6.5), which determines $=_w$;

$CL\beta_{ax}$ (Definition 9.31), which determines $=_{c\beta}$;

$CL\beta\eta_{ax}$ (Definition 8.12), which determines $=_{c\beta\eta}$.

Identity will, as usual, be '≡' for terms, and '=' for all other objects, in particular for members of a model's domain.

Vars will be the class of all variables. As usual, variables will be denoted by 'x', 'y', 'z', 'u', 'v', 'w'.

In contrast, 'a', 'b', 'c', 'd', 'e' will denote arbitrary members of a given set D (see later).

If • is a mapping from D^2 into D, expressions such as

$$(((a \cdot b) \cdot c) \cdot d)$$

will be shortened to $a \cdot b \cdot c \cdot d$ (the convention of 'association to the left').

If D is a set, any mapping ρ: Vars → D will be called a valuation. For $d \in D$ and $x \in$ Vars,

$$[d/x]\rho$$

will denote the valuation ρ' which is the same as ρ except that $\rho'(x) = d$. (If $\rho(x) = d$, then $[d/x]\rho = \rho$.)

The reader is assumed to know the usual definition of a model of a first-order theory with equality (e.g. see Mendelson 1964, Chapter 2, §§2, 3, 8.) A model is often called 'normal' when '=' is interpreted as identity; here 'model' will always mean 'normal model'.

Models will be defined in the usual informal set-theory in which mathematics is commonly written (and which has been used throughout this book for the metatheory of CLw, λβ, etc.). If desired, this could be formalized in ZF + the axiom of choice. In particular, recall that, as usual, 'function', 'mapping' and 'map' mean 'a set of ordered pairs such that no two pairs have the same first member'.

DEFINITION 10.2. An <u>applicative structure</u> is a pair $\langle D, \cdot \rangle$ where D is a set (called the <u>domain</u> of the structure) with at least 2 members, and $\cdot$ is any mapping from D^2 to D.

A model of CLw or $\lambda\beta$ will be an applicative structure with extra items, and such that $\cdot$ has some of the properties of function-application. The 2-member condition is just to prevent triviality.

DEFINITION 10.3. Let $\langle D, \cdot \rangle$ be an applicative structure and let $n \geq 1$. A function $\theta: D^n \to D$ is <u>representable in</u> D iff D has a member a such that

$$(\forall d_1, \ldots, d_n \in D) \qquad a \cdot d_1 \cdot d_2 \cdot \ldots \cdot d_n = \theta(d_1, \ldots, d_n).$$

(By the association-to-the-left convention this equation really says

$$(\ldots((a \cdot d_1) \cdot d_2) \cdot \ldots \cdot d_n) = \theta(d_1, \ldots, d_n).)$$

Each such a is called a <u>representative</u> of θ. The set of all representable functions from D^n to D is called

$$(D^n \to D)_{rep}.$$

In general very few functions on D are representable, because there are far more functions than members to represent them. On the other hand, each $a \in D$ represents a function of each finite number of arguments.

DEFINITION 10.4. For each $a \in D$, <u>Fun</u>(a) is the unique one-place function that a represents:

$$(\forall d \in D) \qquad \underline{\mathrm{Fun}}(a)(d) = a \cdot d.$$

For $\theta \in (D \to D)_{rep}$, <u>Reps</u>$(\theta)$ is the set of all θ's representatives in D.

DEFINITION 10.5. For $a, b \in D$; a <u>is extensionally equivalent to</u> b $(a \sim b)$ iff

$$(\forall d \in D)\ a \cdot d = b \cdot d.$$

For $a \in D$, the <u>extensional-equivalence-class</u> containing a is

$$\tilde{a} = \{b \in D: b \sim a\}.$$

The set of all these classes is called $D/\sim$:

$$D/\sim\ = \{\tilde{a}: a \in D\}.$$

LEMMA 10.6. *Let* $\langle D, \cdot \rangle$ *be an applicative structure. Then*

(a) $a \sim b \iff \underline{\mathrm{Fun}}(a) = \underline{\mathrm{Fun}}(b)$,

(b) $a \sim b \iff \tilde{a} = \tilde{b}$,

(c) *the members of* $D/\sim$ *are non-empty, disjoint, and their union is* D.

(d) $(D \to D)_{rep}$ *corresponds one-to-one with* $D/\sim$ *by the map* $\underline{\mathrm{Reps}}$.

DEFINITION 10.7. An applicative structure $\langle D, \cdot \rangle$ is extensional iff, for all $a, b \in D$,

$$(\forall d \in D)(a \cdot d = b \cdot d) \implies a = b.$$

LEMMA 10.8. *Extensionality is equivalent to any one of:*

(a) $(\forall a, b \in D)\ a \sim b \implies a = b$,

(b) $(\forall a \in D)\ \tilde{a}$ *is a singleton.*

(c) $(\forall \theta \in (D \to D)_{rep})\ \underline{\mathrm{Reps}}(\theta)$ *is a singleton.*

(d) D *corresponds one-to-one with* $(D \to D)_{rep}$ *by the map* $\underline{\mathrm{Fun}}$.

10B COMBINATORY ALGEBRAS

The definition of model of CLw is essentially the same as that of model in first-order logic, cf. Mendelson 1964, Chapter 2, §§3 and 8. Here it is, in two slightly different forms.

DEFINITION 10.9. A combinatory algebra is a pair $\mathfrak{D} = \langle D, \cdot \rangle$ where D is a set with at least 2 members, $\cdot$ maps D^2 to D, and D has members k and s such that

(a) $(\forall a, b \in D)\quad k \cdot a \cdot b = a$,

(b) $(\forall a, b, c \in D)\quad s \cdot a \cdot b \cdot c = a \cdot c \cdot (b \cdot c)$.

A model of CLw is a quadruple $\langle D, \cdot, k, s \rangle$ such that $\langle D, \cdot \rangle$ is a combinatory algebra and k, s satisfy (a) and (b).

(The name 'combinatory algebra' is from Barendregt 1981, §5.1. Models of CLw have sometimes been called 'Schönfinkel algebras', Lambek 1980.)

Exercise. Prove that $k \neq s$ in a combinatory algebra.

DEFINITION 10.10 (Interpretations, $[\![X]\!]_\rho^{\mathcal{D}}$). Let $\mathcal{D} = \langle D, \cdot, k, s \rangle$ where $\langle D, \cdot \rangle$ is an applicative structure and $k, s \in D$. For each valuation ρ, define

(a) $[\![x]\!]_\rho^{\mathcal{D}} = \rho(x)$,

(b) $[\![K]\!]_\rho^{\mathcal{D}} = k$, $\quad [\![S]\!]_\rho^{\mathcal{D}} = s$,

(c) $[\![XY]\!]_\rho^{\mathcal{D}} = [\![X]\!]_\rho^{\mathcal{D}} \cdot [\![Y]\!]_\rho^{\mathcal{D}}$.

Clearly $[\![X]\!]_\rho^{\mathcal{D}} \in D$; it is called the interpretation of X determined by ρ. When no confusion is likely, it will be called just

$$[\![X]\!]_\rho \quad \text{or} \quad [\![X]\!].$$

Example. If $\rho(x) = a$ and $\rho(y) = b$, then

$$[\![Sx(yK)]\!]_\rho = s \cdot a \cdot (b \cdot k), \in D.$$

LEMMA 10.11. *If* $\rho(x) = \sigma(x)$ *for all* $x \in FV(X)$, *then*

$$[\![X]\!]_\rho = [\![X]\!]_\sigma .$$

COROLLARY 10.11.1. *For closed* X, $[\![X]\!]_\rho$ *is independent of* ρ.

LEMMA 10.12. *Interpretation commutes with substitution; i.e.*

$$[\![[N/x]M]\!]_\rho = [\![M]\!]_{[b/x]\rho} \quad \textit{where } b = [\![N]\!]_\rho .$$

DEFINITION 10.13 (Satisfaction). Let $\mathcal{D} = \langle D, \cdot, k, s \rangle$ where $\langle D, \cdot \rangle$ is an applicative structure and $k, s \in D$. Let ρ be any valuation. Define satisfies ($\models$) thus: for any equation $X = Y$,

$$\mathcal{D}, \rho \models X = Y \iff [\![X]\!]_\rho = [\![Y]\!]_\rho$$

$$\mathcal{D} \models X = Y \iff (\forall \rho)(\mathcal{D}, \rho \models X = Y).$$

Note. '=' is used in two senses here; as a formal symbol in CLw (e.g. '$X = Y$'), and as a meta-language symbol for identity (e.g. '$[\![X]\!]_\rho = [\![Y]\!]_\rho$'),

DEFINITION 10.14. A combinatory β-model, or model of $CL\beta_{ax}$, is a model $\langle D,\cdot,k,s\rangle$ of CLw that satisfies the β-axioms, Definition 9.31.

(What are essentially models of $CL\beta_{ax}$ with a little extra structure are called λ-algebras in Barendregt 1981 and pseudo-models of λβ in Hindley and Longo 1980; cf. Hindley and Longo 1980, Proposition 8.9.)

DEFINITION 10.15. A combinatory βη-model, or model of $CL\beta\eta_{ax}$, is a model $\langle D,\cdot,k,s\rangle$ of CLw that satisfies the βη-axioms, Definition 8.12. (Lambek 1980 calls these models Curry algebras.)

LEMMA 10.16. *Each model of* CLw, $CL\beta_{ax}$ *or* $CL\beta\eta_{ax}$ *satisfies all the provable equations of the corresponding theory.*

Proof. The axioms are satisfied, by definition. And each rule of inference is a property of identity. □

REMARK 10.17. CLw is not quite a first-order theory in the usual sense (cf. 6.9), but the difference is trivial and Definition 6.11 showed how to change it into a first-order theory CLw^+ without changing its set of provable equations. The models of CLw are exactly the normal models of CLw^+ in the usual first-order-logic sense.

A similar remark holds for $CL\beta_{ax}$ and $CL\beta\eta_{ax}$.

REMARK 10.18. We now have enough material to build a simple model. First, CLw^+ is consistent, because by the Church-Rosser theorem there are equations that have no proofs in CLw, and hence none in CLw^+. And every consistent first-order theory has a model (e.g. Mendelson 1964, Proposition 2.12). The usual proofs of this fact construct a model which may be defined directly as follows.

DEFINITION 10.19 (Term models). Let $\mathcal{J}$ be CLw or $CL\beta_{ax}$ or $CL\beta\eta_{ax}$. For each CL-term X, define

$$[X] = \{Y: \mathcal{J} \vdash X = Y\}.$$

The term model of $\mathcal{J}$, called $\mathcal{M}(\mathcal{J})$, is $\langle D,\cdot,k,s\rangle$, where

$$D = \{[X]: X \text{ is a CL-term}\},$$

$$[X]\cdot[Y] = [XY],$$

$$k = [K], \quad s = [S].$$

REMARK 10.20. It is routine to prove that $\cdot$ is well-defined, i.e. that

$$[X] = [X'],\ [Y] = [Y'] \quad \Rightarrow \quad [XY] = [X'Y'];$$

and that $\mathcal{M}(\mathcal{T})$ is indeed a model of $\mathcal{T}$. It is also routine to prove that interpretation is the same as substitution in this model; i.e. that if $FV(X) = \{x_1,\dots,x_n\}$ and $\rho(x_i) = [Y_i]$ for each i, then

$$[\![X]\!]_\rho = [[Y_1/x_1,\dots,Y_n/x_n]X].$$

Thus $\mathcal{M}(\mathcal{T})$ is in a sense trivial; it is really just a reflection of the syntax of $\mathcal{T}$ and tells us very little new about combinators. The models in Chapter 12 will be much deeper.

REMARK 10.21. The following theorem is a standard result from the model-theory of algebra. It is almost trivial for combinatory logic but will fail for λ-calculus, and is an interesting way of expressing the difference between them (cf. 11.25).

THEOREM 10.22 (<u>The Submodel theorem</u>). *Let* $\mathcal{T}$ *be* CLw *or* $CL\beta_{ax}$ *or* $CL\beta\eta_{ax}$, *and let* $\langle D,\cdot,k,s\rangle$ *be a model of* $\mathcal{T}$. *Then for every* $D' \subseteq D$ *which contains* k,s *and is closed under* $\cdot$, $\langle D',\cdot,k,s\rangle$ *is a model of* $\mathcal{T}$.

Proof. By assumption, D' has at least two members. And if the axioms 6.11(a)-(b) hold in D, then they hold in D', a fortiori. Similarly for the β- and $\beta\eta$-axioms. (This proof depends on the very simple form of 6.11(a)-(b); if they contained existential quantifiers the proof would fail.) □

DEFINITION 10.23. The <u>interior</u>, $\mathcal{D}^o$, of a model $\mathcal{D} = \langle D,\cdot,k,s\rangle$ of CLw or $CL\beta_{ax}$ or $CL\beta\eta_{ax}$ is $\langle D^o,\cdot,k,s\rangle$, where

$$D^o = \{[\![X]\!] : X \text{ closed}\}.$$

(This is a model of the corresponding theory, by Theorem 10.22.)

REMARK 10.24 (Extensionality). How does the extensionality concept defined in 10.7 relate to the theories of extensional equality in Chapter 8? For example, it would be nice to prove that a model of CLw is extensional iff it is a model of $CL\beta\eta_{ax}$.

Part of this conjecture is easy. (Exercise: prove that every extensional model of CLw is a model of $CL\beta\eta_{ax}$.) But unfortunately the converse part is false.

For a counterexample take $(\mathcal{M}(CL\beta\eta_{ax}))^{o}$, the interior of the term model of $CL\beta\eta_{ax}$. This is a model of $CL\beta\eta_{ax}$ by above. But extensionality demands that for all closed X and Y,

$$\{(\forall \text{ closed CL-terms } Z)([XZ] = [YZ])\} \Rightarrow [X] = [Y].$$

That is

$$\{(\forall \text{ closed CL-terms } Z)(CL\beta\eta_{ax} \vdash XZ = YZ)\} \Rightarrow \{CL\beta\eta_{ax} \vdash X = Y\}.$$

By Theorem 9.12 this is equivalent to

$$\{(\forall \text{ closed } \lambda\text{-terms } Q)(X_\lambda Q =_{\beta\eta} Y_\lambda Q)\} \Rightarrow \{X_\lambda =_{\beta\eta} Y_\lambda\}.$$

And this is false, by Plotkin's counterexample in Remark 7.3.

Thus the 'extensionality' expressed by $CL\beta\eta_{ax}$ is weaker than the extensionality-concept in Definition 10.7.

For a fuller discussion of extensionality see Hindley and Longo 1980 and Barendregt 1981, Chapter 20; but first read Chapter 11 below.

REMARK 10.25. Definition 10.9 defined two concepts, 'combinatory algebra' and 'model of CLw'. The latter is tied to a particular formalization of combinatory logic; for example, if I was added as a new atomic combinator, a model of CLw would have to be re-defined as a quintuple $\langle D, \cdot, k, s, i\rangle$.

In contrast, 'combinatory algebra' is independent of the formalism. This might not be immediately obvious from Definition 10.9, so let us try to rewrite the definition to avoid mentioning s and k.

The characteristic property of combinatory algebras is called combinatory completeness; it is defined as follows.

DEFINITION 10.26. A <u>combination</u> of $x_1,\ldots,x_n$ is any CL-term X whose only atoms are $x_1,\ldots,x_n$. (X need not contain all of $x_1,\ldots,x_n$, but X must not contain K or S.) Any such X can be interpreted in any applicative structure $\langle D,\cdot\rangle$ in the natural way; we simply define $[\![X]\!]_\rho$ by 10.10(a) and (c). (Lemmas 10.11 and 10.12 still hold.)

DEFINITION 10.27. An applicative structure $\langle D,\cdot\rangle$ is <u>combinatorially complete</u> iff: for every sequence $u, x_1,\ldots,x_n$, and every combination X of $x_1,\ldots,x_n$ only, the formula

$$(\exists u)(\forall x_1,\ldots,x_n)(ux_1\ldots x_n = X)$$

is true in D. That is, iff there exists $a \in D$ such that

$$(\forall d_1,\ldots,d_n \in D)\quad a\cdot d_1\cdot\ldots\cdot d_n = [\![X]\!]_{[d_1/x_1]\ldots[d_n/x_n]\rho}$$

where ρ is arbitrary.

(This definition is equivalent to Barendregt 1981, Definition 5.1.3, though very slightly simpler.)

THEOREM 10.28. *An applicative structure* $\langle D,\cdot\rangle$ *is combinatorially complete iff it is a combinatory algebra.*

Proof. Immediate. □

So combinatory completeness is a way of defining combinatory algebras without mentioning k and s. But it is not so easy a property to handle as the axioms for k and s. For example, in practice, the quickest way to show that a particular structure is combinatorially complete is usually to find members k and s which satisfy these axioms. And standard results in the model-theory of algebra (e.g. Theorem 10.22) are harder to deduce directly from the combinatory completeness definition.

CHAPTER ELEVEN

MODELS OF $\lambda\beta$

11A THE DEFINITION OF λ-MODEL

The discussion of models in the last chapter was almost trivial, so simple was the theory CLw. In contrast the theory $\lambda\beta$ has bound variables and rule (ξ), and these make the concept of model much deeper. Indeed, its definition has only been arrived at after some disagreement in the literature.

This chapter will consider the underlying problems and present three equivalent definitions, each highlighting a different part of the overall picture.

The first definition will demand the least prior insight from the reader; in fact its clauses will correspond very closely to the axioms and rules of $\lambda\beta$. The second will define 'model' entirely in terms of internal structure and will not mention $\lambda\beta$ at all. The third will be by far the simplest, though to see why it is a good definition both the first and the second will be needed.

NOTATION 11.1. This chapter will use the notation of 10.1, except that 'term' will now mean 'λ-term'.

The composition of functions ϕ and ψ is defined as usual by

$$(\psi \circ \phi)(a) = \psi(\phi(a)),$$

and its domain is $\{a : \phi(a)$ is defined and in the domain of $\psi\}$.

The identity-function on a set C will be called I_C.

If C and D are sets, and functions $\phi: C \to D$ and $\psi: D \to C$ satisfy

(a) $$\psi \circ \phi = I_C,$$

then ψ is called a left inverse of ϕ and C is called a retract of D by ϕ and ψ. (Figure 11.1.) The pair $\langle \phi, \psi \rangle$ will be called a retraction.

For example, both H_η and H_β are left inverses of the λ-transformation; see Remark 9.9 and Note 9.35.

From (a) it easily follows that

(b) $(\phi \circ \psi) \circ (\phi \circ \psi) = \phi \circ \psi$,

(c) ψ is onto C (i.e. its range is the whole of C),

(d) ϕ is one-to-one.

Thus a function ϕ that has a left inverse is a one-to-one embedding of C onto a subset $\phi(C)$ of D.

Incidentally, (b) - (d) together imply (a).

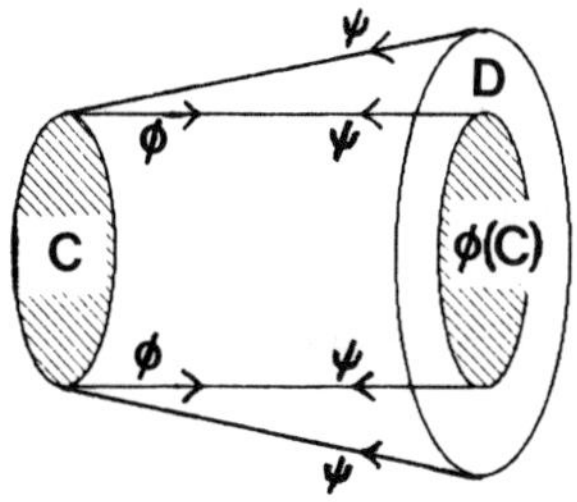

Figure 11.1

REMARK 11.2. Before starting, let us look at one temptation and dismiss it. Why not simply identify 'model of λβ' with 'model of $CL\beta_{ax}$'? After all, λβ and $CL\beta_{ax}$ have the same set of provable equations (modulo the λ- and H_β-transformations), so why should they not have the same models?

The snag is as follows. In λβ we can make deductions by rule (ξ), for example

$$\lambda x.yx = y \ \vdash\ \lambda yx.yx = \lambda y.y.$$

So any reasonable definition of 'model of λβ' should have the property that if $\mathcal{D}$ is such a model, then

$$(\mathcal{D} \models \lambda x.yx = y) \quad \Rightarrow \quad (\mathcal{D} \models \lambda yx.yx = \lambda y.y).$$

But this property fails for models of $CL\beta_{ax}$. (Hindley and Longo 1980 §7 gives two counterexamples; one is the interior of $\mathcal{M}(\lambda\beta\eta)$, and comes from Plotkin's example mentioned in Remark 7.3.)

This failure is only surprising if we think of $CL\beta_{ax}$ as satisfying (ξ) in some sense. But it does not. True, it is equivalent to a theory in which a form of (ξ) is derivable (the $CL\zeta_\beta$ of Corollary 9.32.1 and Note 9.35), but the equivalence is only theorem-equivalence, and the above counterexamples show it cannot be rule-equivalence.

So 'model of $\lambda\beta$' is a more complicated concept than 'model of $CL\beta_{ax}$'. Here is the first definition.

DEFINITION 11.3. A λ-<u>model</u> or <u>model of</u> $\lambda\beta$ is a triple $\mathcal{D} = \langle D, \cdot, [\![\ ]\!] \rangle$, where $\langle D, \cdot \rangle$ is an applicative structure and $[\![\ ]\!]$ is a mapping which assigns, to each λ-term M and each valuation ρ, a member $[\![M]\!]_\rho \in D$ such that

(a) $[\![x]\!]_\rho = \rho(x)$;

(b) $[\![PQ]\!]_\rho = [\![P]\!]_\rho \cdot [\![Q]\!]_\rho$;

(c) $[\![\lambda x.P]\!]_\rho \cdot d = [\![P]\!]_{[d/x]\rho}$ $(\forall d \in D)$;

(d) $[\![M]\!]_\rho = [\![M]\!]_\sigma$ if $\sigma(x) = \rho(x)$ for all $x \in FV(M)$;

(e) $[\![\lambda x.M]\!]_\rho = [\![\lambda y.[y/x]M]\!]_\rho$ if $y \notin FV(M)$;

(f) if $(\forall d \in D)\{[\![M]\!]_{[d/x]\rho} = [\![N]\!]_{[d/x]\rho}\}$,

then $[\![\lambda x.M]\!]_\rho = [\![\lambda x.N]\!]_\rho$.

($[\![M]\!]_\rho$ may also be called $[\![M]\!]_\rho^{\mathcal{D}}$, or simply $[\![M]\!]$ when it is known to be independent of ρ.)

COMMENTS 11.4. Each of (a) - (f) is a condition we would intuitively expect a model of $\lambda\beta$ to satisfy. First, a model of $\lambda\beta$, no matter how it is defined, should allow us to define $[\![M]\!]_\rho$ for all M and ρ. And by analogy with first-order logic, a change in the mapping $[\![\ ]\!]$ should be regarded as creating a new model. So we may as well include $[\![\ ]\!]$ in the definition of model.

Clauses (a) and (b) are in fact the definition of $[\![M]\!]_\rho$ in the cases $M \equiv x$ and $M \equiv PQ$. The case $M \equiv \lambda x.P$ is more difficult, which is why we need (c) - (f).

Clause (c) expresses in model-theoretic language the intuitive meaning behind the λ-notation; it says that $[\![\lambda x.P]\!]_\rho$ acts like a function whose value is calculated by interpreting x as d. More precisely, suppose $[\![P]\!]_{[d/x]\rho}$ has already been defined, for all $d \in D$. Then we can define a function $\psi_{P,x,\rho}$ from D to D thus:

(11.5) $$\psi_{P,x,\rho}(d) = [\![P]\!]_{[d/x]\rho} \qquad (\forall x \in D).$$

And (c) says that $[\![\lambda x.P]\!]_\rho$ is a representative of $\psi_{P,x,\rho}$. Of course this implies that for all P, x, ρ, the function $\psi_{P,x,\rho}$ is representable (Definition 10.3).

Clause (d) is a standard property in first-order model theory, cf. Lemma 10.11.

Clause (e) corresponds to axiom-scheme (α), and is needed to make congruent terms have the same interpretation.

Clause (f) is the interpretation of rule (ξ). In the ψ-notation above, it says that

$$\psi_{M,x,\rho} = \psi_{N,x,\rho} \quad \Rightarrow \quad [\![\lambda x.M]\!]_\rho = [\![\lambda x.N]\!]_\rho .$$

Together with (c) it implies a property called <u>weak extensionality</u>, namely

(11.6) $$[\![\lambda x.M]\!]_\rho \sim [\![\lambda x.N]\!]_\rho \quad \Rightarrow \quad [\![\lambda x.M]\!]_\rho = [\![\lambda x.N]\!]_\rho .$$

(Here $\sim$ is extensional equivalence.) This says that objects of form $[\![\lambda x.M]\!]_\rho$ satisfy the extensionality property, 10.7 (cf. Remark 8.6).

Finally, (a) - (f) together ensure that every λ-model satisfies all the provable equations of $\lambda\beta$, as the following three lemmas and theorem will show.

LEMMA 11.7. *Let* $\langle D,\cdot,[\![\ ]\!]\rangle$ *be a* λ*-model.*

(a) *If* $\rho(y) = \rho(x)$ *and* $y \notin FV(M)$, *then*

$$[\![[y/x]M]\!]_\rho = [\![M]\!]_\rho .$$

(b) *If* $FV(M) \subseteq \{x_1,\ldots,x_n\}$ *and* $y_1,\ldots,y_n,x_1,\ldots,x_n$ *are distinct and* $\sigma(y_i) = \rho(x_i)$ *for* $i=1,\ldots,n$, *then*

$$[\![[y_1/x_1]\ldots[y_n/x_n]M]\!]_\sigma = [\![M]\!]_\rho .$$

Proof.

(a) Let $d = \rho(y) = \rho(x)$. Then $[d/x]\rho = [d/y]\rho = \rho$, and

$$\begin{aligned} [\![M]\!]_\rho &= [\![M]\!]_{[d/x]\rho} = [\![\lambda x.M]\!]_\rho \cdot d && \text{by 11.3(c)} \\ &= [\![\lambda y.[y/x]M]\!]_\rho \cdot d && \text{by 11.3(e)} \\ &= [\![[y/x]M]\!]_\rho && \text{by 11.3(c).} \end{aligned}$$

(b) Let $d_i = \rho(x_i) = \sigma(y_i)$. Let $\tau = [d_1/y_1]\ldots[d_n/y_n]\rho$.

Then

$$\llbracket M \rrbracket_\rho = \llbracket M \rrbracket_\tau \quad \text{by 11.3(d)}$$
$$= \llbracket [y_1/x_1]\ldots[y_n/x_n]M \rrbracket_\tau \quad \text{by (a) repeated}$$
$$= \llbracket [y_1/x_1]\ldots[y_n/x_n]M \rrbracket_\sigma \quad \text{by 11.3(d)}. \qquad \square$$

The second lemma of the three is a small but significant strengthening of the weak extensionality property (11.6), and will be the key to the syntax-free analysis of models given later.

LEMMA 11.8 (Berry's extensionality property). *Let* $\langle D, \cdot, \llbracket\ \rrbracket \rangle$ *be a* λ-*model. Then for all* P, Q, x, y, ρ, σ,

(a) $(\forall d \in D)(\llbracket P \rrbracket_{[d/x]\rho} = \llbracket Q \rrbracket_{[d/y]\sigma}) \Rightarrow \llbracket \lambda x.P \rrbracket_\rho = \llbracket \lambda y.Q \rrbracket_\sigma$,

(b) $\llbracket \lambda x.P \rrbracket_\rho \sim \llbracket \lambda y.Q \rrbracket_\sigma \Rightarrow \llbracket \lambda x.P \rrbracket_\rho = \llbracket \lambda y.Q \rrbracket_\sigma$.

Proof. To prove (a), assume that $\llbracket P \rrbracket_{[d/x]\rho} = \llbracket Q \rrbracket_{[d/y]\sigma}$ for all $d \in D$. Suppose

$$FV(P)-\{x\} = \{x_1,\ldots,x_m\}, \quad FV(Q)-\{y\} = \{y_1,\ldots,y_n\}.$$

(The x's need not be distinct from the y's.) Let $a_i = \rho(x_i)$ and $b_j = \sigma(y_j)$ for $1 \le i \le m$ and $1 \le j \le n$. Choose distinct new variables $z, u_1, \ldots, u_m, v_1, \ldots, v_n$ not in any term mentioned above. Define

$$P' \equiv [u_1/x_1]\ldots[u_m/x_m][z/x]P,$$
$$Q' \equiv [v_1/y_1]\ldots[v_n/y_n][z/y]Q,$$
$$\tau = [a_1/u_1]\ldots[a_m/u_m][b_1/v_1]\ldots[b_n/v_n]\rho.$$

Then for all $d \in D$,

$$\llbracket P' \rrbracket_{[d/z]\tau} = \llbracket P \rrbracket_{[d/x]\rho} \quad \text{by 11.7(b)}$$
$$= \llbracket Q \rrbracket_{[d/y]\sigma} \quad \text{by assumption}$$
$$= \llbracket Q' \rrbracket_{[d/z]\tau} \quad \text{by 11.7(b)}.$$

Hence by 11.3(f),

(i) $$\llbracket \lambda z.P' \rrbracket_\tau = \llbracket \lambda z.Q' \rrbracket_\tau .$$

Then

$$\llbracket \lambda x.P \rrbracket_\rho = \llbracket \lambda z.[z/x]P \rrbracket_\rho \qquad \text{by 11.3(e)}$$
$$= \llbracket \lambda z.P' \rrbracket_\tau \qquad \text{by 11.7(b)}$$
$$= \llbracket \lambda z.Q' \rrbracket_\tau \qquad \text{by (i) above}$$
$$= \llbracket \lambda y.Q \rrbracket_\sigma \qquad \text{by 11.7(b) and 11.3(e).}$$

This proves (a). Part (b) follows by 11.3(c). □

EXERCISE 11.9. Prove that (d) - (f) in Definition 11.3 could have been replaced by Berry's extensionality property; i.e. prove that 11.3(a) - (c) and 11.8(b) imply 11.3(d) - (f).

LEMMA 11.10. *Let* $\langle D, \cdot, \llbracket\ \rrbracket \rangle$ *be a λ-model. Then*

(a) $\llbracket [N/x]M \rrbracket_\rho = \llbracket M \rrbracket_{[b/x]\rho}$ *where* $b = \llbracket N \rrbracket_\rho$;

(b) $\llbracket (\lambda x.M)N \rrbracket_\rho = \llbracket [N/x]M \rrbracket_\rho$.

Proof.

(a) Induction on M. The only non-trivial case is $M \equiv \lambda y.P$, $y \not\equiv x$. By 11.3(e) we can assume $y \notin FV(N)$. The induction hypothesis implies that for all $d \in D$,

$$\llbracket [N/x]P \rrbracket_{[d/y]\rho} = \llbracket P \rrbracket_{[b/x][d/y]\rho} .$$

Hence, by Lemma 11.8(a) applied with $\sigma = [b/x]\rho$,

$$\llbracket \lambda y.[N/x]P \rrbracket_\rho = \llbracket \lambda y.P \rrbracket_{[b/x]\rho} .$$

And $[N/x](\lambda y.P) \equiv \lambda y.[N/x]P$, so (a) holds in the case $M \equiv \lambda y.P$.

(b) $\llbracket (\lambda x.M)N \rrbracket_\rho = \llbracket \lambda x.M \rrbracket_\rho \cdot b = \llbracket M \rrbracket_{[b/x]\rho}$ by 11.3(c)

$= \llbracket [N/x]M \rrbracket_\rho$ by (a). □

DEFINITION 11.11. <u>Satisfies</u> (⊨) is defined in a λ-model $\mathfrak{D}$ in the usual way, thus:

$$\mathfrak{D},\rho \models M = N \iff \llbracket M \rrbracket_\rho = \llbracket N \rrbracket_\rho ;$$

$$\mathfrak{D} \models M = N \iff (\forall\rho)(\mathfrak{D},\rho \models M = N).$$

THEOREM 11.12. *Every λ-model satisfies all provable equations of* λβ.

Proof. Induction on $\lambda\beta$ (Definition 6.2). Case (α) is 11.3(e), case (β) is 11.10(b), case (ξ) is 11.3(f). The rest are trivial. □

COROLLARY 11.12.1. *If* $\langle D, \cdot, [\![\ ]\!] \rangle$ *is a* λ*-model then* $\langle D, \cdot \rangle$ *is a combinatory algebra, and hence is combinatorially complete.*

Proof. Define $k = [\![\lambda xy.x]\!]$, $s = [\![\lambda xyz.xz(yz)]\!]$. □

COROLLARY 11.12.2. *If* $\langle D, \cdot, [\![\ ]\!] \rangle$ *is a* λ*-model then* D *contains* k *and* s *such that* $\langle D, \cdot, k, s \rangle$ *is a model of* $CL\beta_{ax}$.

Proof. Define k, s as above; then use the correspondence between $\lambda\beta$ and $CL\beta_{ax}$, Theorem 9.27. □

REMARK 11.13. The converses to both these corollaries are false; there exist combinatory algebras, even models of $CL\beta_{ax}$, which cannot be made into λ-models by any definition of $[\![\ ]\!]$. (Barendregt and Koymans 1980, §3.)

DEFINITION 11.14 (<u>Models of</u> $\lambda\beta\eta$). A <u>model of</u> $\lambda\beta\eta$ is a λ-model that satisfies the equation $\lambda x.Mx = M$ for all M and all $x \notin FV(M)$.

It is easy to see that every model of $\lambda\beta\eta$ satisfies all provable equations of $\lambda\beta\eta$; cf. 11.12.

THEOREM 11.15. *A* λ*-model* $\mathfrak{D}$ *is extensional iff it is a model of* $\lambda\beta\eta$.

Proof. Exercise. (Cf. the proof of Theorem 7.4, especially the derivability of (ζ) in $\lambda\beta\eta$.) □

This theorem contrasts with combinatory algebras: as Remark 10.24 showed, a combinatory algebra can be a model of $CL\beta\eta_{ax}$ without being extensional.

DEFINITION 11.16 (<u>Term models</u>). Let $\mathcal{T}$ be $\lambda\beta$ or $\lambda\beta\eta$. For each λ-term M, define

$$[M] = \{N : \mathcal{T} \vdash M = N\}.$$

The <u>term model</u> of $\mathcal{T}$, called $M(\mathcal{T})$, is $\langle D, \cdot, [\![\ ]\!] \rangle$, where

$$D = \{[M] : M \text{ is a } \lambda\text{-term}\},$$
$$[M] \cdot [N] = [MN],$$
$$[\![M]\!]_\rho = [\,[N_1/x_1,\ldots,N_n/x_n]M],$$

where $\{x_1,\ldots,x_n\} = FV(M)$ and $\rho(x_i) = [N_i]$, and $[N_1/x_1,\ldots,N_n/x_n]$ is simultaneous substitution, cf. Exercise 2.6(b).

REMARK 11.17. It is routine to prove that $\cdot$ and $[\![\ ]\!]$ are well-defined and that $\mathcal{M}(\mathcal{T})$ is a model of $\mathcal{T}$. As noted in Remark 10.20 term models are just reflections of syntax so they are in a sense trivial.

But in fact this very triviality makes them one of the tests for a good definition of 'model'; if the term model of $\lambda\beta$ had not satisfied the conditions of Definition 11.3, those conditions would have had to be changed. Fortunately they pass the test.

11B SYNTAX-FREE DEFINITIONS

DISCUSSION 11.18 (The mapping Λ). Let us now try to characterize λ-models by their internal structure without mentioning the theory $\lambda\beta$.

One way to do this starts with the extensional-equivalence relation $\sim$ (Definition 10.5). Let $\langle D, \cdot, [\![\ ]\!] \rangle$ be any λ-model. For each extensional-equivalence-class $\tilde{a} \subseteq D$, there exist M, x, ρ such that $[\![\lambda x.M]\!]_\rho \in \tilde{a}$. For example, take

$$[\![\lambda x.ux]\!]_{[a/u]\sigma}$$

for any σ; this is extensionally equivalent to a, because for all $d \in D$,

$$\begin{aligned} [\![\lambda x.ux]\!]_{[a/u]\sigma} \cdot d &= [\![ux]\!]_{[d/x][a/u]\sigma} && \text{by 11.3(c)} \\ &= a \cdot d && \text{by 11.3(a),(b).} \end{aligned}$$

Of course there are an infinity of other examples. But it is important to note that whatever example we take, the value of $[\![\lambda x.M]\!]_\rho$ is always the same (by Berry's extensionality property, Lemma 11.8(b)). So one member of $\tilde{a}$ is chosen to be the value of $[\![\lambda x.M]\!]_\rho$ for all M, x, ρ such that $[\![\lambda x.M]\!]_\rho \in \tilde{a}$, and all other members are ignored. Call this member $\Lambda(a)$:

(i) $$\Lambda(a) = [\![\lambda x.ux]\!]_{[a/u]\sigma} \quad \text{for any } \sigma.$$

(In the literature $\Lambda(a)$ is often called '$\lambda x.ax$', mixing the formal λ-language with its meta-language.) The basic properties of the map Λ are:

(ii) $\Lambda(a) \sim a$ (by above),

(iii) $\Lambda(a) \sim \Lambda(b) \Rightarrow \Lambda(a) = \Lambda(b)$ (by 11.8(b)),

(iv) $a \sim b \Leftrightarrow \Lambda(a) = \Lambda(b)$ (by (ii), (iii)),

(v) $\Lambda(\Lambda(a)) = \Lambda(a)$ (by (ii), (iv)).

Moreover, the map Λ is representable in D; i.e. there exists $e \in D$ such that

(vi) $$(\forall a \in D)\ \ e \cdot a = \Lambda(a).$$

One suitable such e is the member of D corresponding to the Church numeral $\bar{1}$:

$$e = [\![\bar{1}]\!] = [\![\lambda xy.xy]\!]_\rho$$

(for any ρ); this e works because

$$\begin{aligned} [\![\bar{1}]\!] \cdot a &= [\![\lambda y.xy]\!]_{[a/x]\rho} && \text{by 11.3(c)} \\ &= \Lambda(a) && \text{by (i) above.} \end{aligned}$$

The definition of λ-model can be written in terms of Λ independently of the λ-syntax, as follows.

DEFINITION 11.19. A (<u>syntax-free</u>) <u>λ-model</u> is a triple $\langle D, \cdot, \Lambda \rangle$ where $\langle D, \cdot \rangle$ is an applicative structure, Λ maps D to D, and

(a) $\langle D, \cdot \rangle$ is combinatorially complete,

(b) $(\forall a \in D)\ \ \Lambda(a) \sim a$,

(c) $(\forall a,b \in D)\ \ a \sim b \Rightarrow \Lambda(a) = \Lambda(b)$,

(d) $(\exists e \in D)(\forall a \in D)\ \ e \cdot a = \Lambda(a)$.

('Syntax-free' will be omitted when there is no chance of confusion.)

THEOREM 11.20. *$\langle D, \cdot, \Lambda \rangle$ is a syntax-free λ-model iff $\langle D, \cdot, [\![\ ]\!] \rangle$ is a λ-model in the sense of* 11.3. *Here $[\![\ ]\!]$ is defined in terms of Λ by*

(a) $[\![x]\!]_\rho = \rho(x)$,

(b) $[\![PQ]\!]_\rho = [\![P]\!]_\rho \cdot [\![Q]\!]_\rho$,

(c) $[\![\lambda x.P]\!]_\rho = \Lambda(a)$, *where* $(\forall d \in D)(a \cdot d = [\![P]\!]_{[d/x]\rho})$.

Conversely Λ is defined in terms of $[\![\ ]\!]$ by

(d) $$\Lambda(a) = [\![\lambda x.ux]\!]_{[a/u]\sigma} \ \textit{for any}\ \sigma.$$

Proof. For 'if', see Discussion 11.18.

For 'only if': let $\langle D, \cdot, \Lambda \rangle$ be a syntax-free λ-model. We must first prove that (a) - (c) define $[\![M]\!]_\rho$ for all M and ρ, and to do this the only problem is to prove that a exists in (c). In fact we shall prove the following simultaneously:

(i) Clauses (a) - (c) define $[\![M]\!]_\rho$ for all ρ;

(ii) $[\![M]\!]_\rho$ is independent of $\rho(z)$ if $z \notin FV(M)$;

(iii) for each sequence $y_1, \ldots, y_n \supseteq FV(M)$ there exists $b \in D$ such that for all $d_1, \ldots, d_n \in D$,

$$b \cdot d_1 \cdot \ldots \cdot d_n = [\![M]\!]_{[d_1/y_1]\ldots[d_n/y_n]\rho}.$$

Proof of (i) - (iii): Induction on M.

<u>Case</u> 1: M is a combination of variables. Then (i) and (ii) are trivial and (iii) holds by combinatory completeness.

<u>Case</u> 2: $M \equiv PQ$. Then (i) and (ii) are trivial. For (iii), let b_P, b_Q satisfy (iii) for P, Q and the given $y_1, \ldots, y_n$. Define

$$g = [\![\lambda uvy_1 \ldots y_n.(uy_1 \ldots y_n)(vy_1 \ldots y_n)]\!].$$

(g exists by combinatory completeness.) Then $b = g \cdot b_P \cdot b_Q$ satisfies (iii).

<u>Case</u> 3: $M \equiv \lambda x.P$. By induction-hypothesis (iii) applied to P and $y_1, \ldots, y_n, x$, there exists $b_P \in D$ such that for all $d_1, \ldots, d_n, d \in D$,

$$b_P \cdot d_1 \cdot \ldots \cdot d_n \cdot d = [\![P]\!]_{[d/x][d_1/y_1]\ldots[d_n/y_n]\sigma}$$

where σ is arbitrary. (The right side is independent of σ by induction hypothesis (ii).)

To prove (i), it is enough to show that the a in (c) exists. Take any ρ, let $d_i = \rho(y_i)$, and define

$$a = b_P \cdot d_1 \cdot \ldots \cdot d_n.$$

Then $a \cdot d = [\![P]\!]_{[d/x]\rho}$ by the equation for b_P. This proves (i). Part (ii) is obvious from the proof of (c).

To prove (iii), take any e that represents Λ; then define

$$f = [\![\lambda uvy_1 \ldots y_n . u(vy_1 \ldots y_n)]\!].$$

(f exists by combinatory completeness.) Define $b = f \cdot e \cdot b_P$. For any $d_1, \ldots, d_n \in D$, define

$$a = b_P \cdot d_1 \cdot \ldots \cdot d_n;$$

then for all $d \in D$, $a \cdot d = [\![P]\!]_{[d/x]\rho}$, where ρ is defined by $\rho(y_i) = d_i$. Hence by (c),

$$[\![\lambda x.P]\!]_\rho = \Lambda(a).$$

So

$$\begin{aligned} b \cdot d_1 \cdot \ldots \cdot d_n &= e \cdot a && \text{by definition of } b \text{ and } f \\ &= \Lambda(a) && \text{since } e \text{ represents } \Lambda \\ &= [\![\lambda x.P]\!]_\rho && \text{by above.} \end{aligned}$$

This ends the proof of (i) - (iii). To complete the theorem, just check that $[\![\;]\!]$ satisfies 11.3(a) - (f). This is easy. □

COROLLARY 11.20.1. *The constructions of* $[\![\;]\!]$ *and* Λ *in Theorem* 11.20 *are mutual inverses, so the two definitions of* λ*-model are equivalent in the strongest possible sense. More precisely: let* $\langle D, \cdot, [\![\;]\!]' \rangle$ *be a* λ*-model in the sense of* 11.3; *if we first define* Λ *from* $[\![\;]\!]'$ *by* 11.20(d) *and then define* $[\![\;]\!]$ *by* 11.20(a) - (c), *we shall get* $[\![\;]\!] = [\![\;]\!]'$. *Conversely, if we start with a syntax-free* λ*-model* $\langle D, \cdot, \Lambda' \rangle$ *and define first* $[\![\;]\!]$ *and then* Λ, *we shall get* $\Lambda = \Lambda'$.

Proof. Easy. □

REMARK 11.21. Definition 11.19 is clearly independent of the λ-syntax. Even better, in contrast to the earlier definition it is essentially just a finite set of first-order axioms. In fact, its clause (a) is equivalent to

(a′) $(\exists k,s \in D)(\forall a,b,c \in D)(k \cdot a \cdot b = a \wedge s \cdot a \cdot b \cdot c = a \cdot c \cdot (b \cdot c))$,

and (b) and (c) are equivalent to

(b′) $(\forall a,b \in D)((\Lambda(a)) \cdot b = a \cdot b)$,

(c′) $(\forall a,b \in D)((\forall d \in D)(a \cdot d = b \cdot d) \Rightarrow \Lambda(a) = \Lambda(b))$.

But the following definition is simpler still. It is due to Dana Scott and Albert Meyer, and instead of focussing on Λ it focusses on one of its representatives in D. By this means it avoids the need for the function-symbol 'Λ'.

DEFINITION 11.22 (Scott-Meyer λ-models). Let $\langle D, \cdot \rangle$ be an applicative structure and let $e \in D$. The triple $\langle D, \cdot, e \rangle$ is called a loose Scott-Meyer λ-model iff

(a) $\langle D, \cdot \rangle$ is combinatorially complete,

(b) $(\forall a,b \in D)\quad e \cdot a \cdot b = a \cdot b,$

(c) $(\forall a,b \in D)\quad (\forall d \in D)(a \cdot d = b \cdot d) \quad \Rightarrow \quad e \cdot a = e \cdot b,$

and a strict Scott-Meyer λ-model (or just a Scott-Meyer λ-model) iff also

(d) $$e \cdot e = e.$$

(Cf. Scott 1980b pp. 421-5, Meyer 1982 Definition 1.3a,b 'combinatory models'.)

DISCUSSION 11.23. If we take a Scott-Meyer model $\langle D, \cdot, e \rangle$, strict or loose, and let Λ be the function that e represents (i.e.

(i) $$\Lambda = \underline{\text{Fun}}(e)$$

in the notation of 10.4), then $\langle D, \cdot, \Lambda \rangle$ is clearly a λ-model in the sense of Definition 11.19.

Conversely, if we take a λ-model $\langle D, \cdot, \Lambda \rangle$ and let e be any representative of Λ, then $\langle D, \cdot, e \rangle$ is a loose Scott-Meyer model. So the loose Scott-Meyer definition of model is essentially equivalent to the earlier definitions, 11.19 and 11.3.

However, one Λ may have many representatives, so one model in the sense of 11.19 may give rise to many loose Scott-Meyer models. But only one of these is strict. To find it, choose

(ii) $$e_o = [\![\bar{1}]\!] = [\![\lambda xy.xy]\!]$$

where $[\![\ ]\!]$ is defined from Λ by 11.20(a) - (c). This e_o represents Λ, by the end of Discussion 11.18, and we also have

$$e_o \cdot e_o = [\![\bar{1}\bar{1}]\!] = [\![\bar{1}]\!] = e_o,$$

so $\langle D, \cdot, e_o \rangle$ is a strict model.

No other representative of Λ gives a strict model. Because if e' represents Λ and $\langle D, \cdot, e' \rangle$ is a strict model, we get $e' = e_o$; in detail,

$$
\begin{aligned}
e' &= e' \cdot e' && \text{by 11.22(d) for } e' \\
&= \Lambda(e') && \text{since } e' \text{ represents } \Lambda \\
&= e_o \cdot e' && \text{since } e_o \text{ represents } \Lambda \\
&= e_o \cdot e_o && \text{by 11.22(c) for } e_o \\
&= e_o && \text{by 11.22(d) for } e_o.
\end{aligned}
$$

This discussion can be summed up as follows.

THEOREM 11.24. *The constructions* $\Lambda = \underline{\text{Fun}}(e)$ *and* $e_o = [\![\bar{1}]\!]$ *are mutual inverses. More precisely, if* $\langle D, \cdot, \Lambda \rangle$ *is a λ-model and* $e_o = [\![\bar{1}]\!]$, *then* $\langle D, \cdot, e_o \rangle$ *is a strict Scott-Meyer λ-model and* $\underline{\text{Fun}}(e_o) = \Lambda$. *Conversely, if* $\langle D, \cdot, e \rangle$ *is a strict Scott-Meyer λ-model and* $\Lambda = \underline{\text{Fun}}(e)$, *then* $\langle D, \cdot, \Lambda \rangle$ *is a λ-model in the sense of* 11.19, *and* $[\![\bar{1}]\!] = e$.

REMARK 11.25. The Scott-Meyer definition is clearly simpler than both the previous definitions, in fact it is almost as simple as the definition of 'combinatory algebra'. All its clauses except (c) can be expressed in the form

(i) $$(\exists x_1, \ldots, x_m)(\forall y_1, \ldots, y_n)(X = Y) \qquad (m, n \geq 0).$$

But the exception is very important. If every clause had form (i), then an analogue of the Submodel theorem would hold (Theorem 10.22), and in particular the interior of every λ-model would be a λ-model. But the latter is not true. (Hindley and Longo 1980, §6 - 7 give two counter-examples; one is the interior of $\mathcal{M}(\lambda\beta\eta)$.) So we have gone about as far as we can go in simplifying the definition of λ-model.

This fact can be seen from another point of view: every combinatory algebra contains members e satisfying 11.22(b) and (d), but there need not be one satisfying (c) as well. (Cf. Remark 11.13.)

11C GENERAL PROPERTIES

DISCUSSION 11.26. By concentrating on Λ and its representatives we came very quickly to a simple, almost algebraic, definition of λ-model. But λ-calculus is function-theory, not algebra, so a more function-oriented view seems also desirable. Here is one such view.

Consider an arbitrary λ-model $\langle D, \cdot, \Lambda \rangle$. In the notation of 10.3 and 10.4, $(D \to D)_{rep}$ is the set of all its representable one-place functions, Reps(θ) is the set of all representatives of a function $\theta \in (D \to D)_{rep}$, and Fun(a) is the one-place function represented by $a \in D$.

Of course Reps(θ) may have many members. But Λ gives us a way of choosing a 'canonical' one, which will be called Rep(θ):

(i) $$\text{Rep}(\theta) = \Lambda(a) \quad \text{for any} \quad a \in \text{Reps}(\theta).$$

(This definition is independent of a by 10.6 and 11.18.) We clearly have

(ii) $$\text{Fun}(\text{Rep}(\theta)) = \theta.$$

Thus Fun is a left inverse of Rep, so by 11.1, Rep is a one-to-one embedding of $(D \to D)_{rep}$ into D, and $(D \to D)_{rep}$ is a retract of D. (Figure 11.2.)

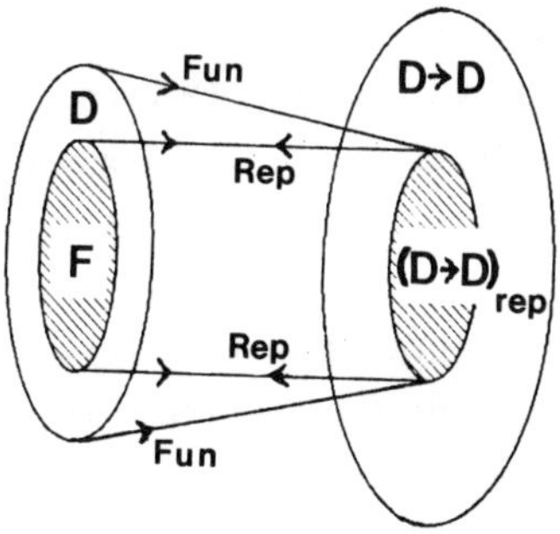

Figure 11.2.

It is possible to reverse the above discussion and define application in terms of representability. Let D be any set, C be any set of one-place functions from D to D, and $\underline{Rep}: C \to D$, $\underline{Fun}: D \to C$ be any pair of functions that form a retraction (Notation 11.1). That is, let

(iii) $$\underline{Fun} \circ \underline{Rep} = I_C.$$

Then we can define application for all $a, b \in D$ thus:

(iv) $$a \cdot b = (\underline{Fun}(a))(b).$$

It is easy to show that C becomes exactly the set of all functions representable in $\langle D, \cdot \rangle$ when $\cdot$ is defined in this way. Next, define

(v) $$\Lambda = \underline{Rep} \circ \underline{Fun}.$$

This Λ is easily seen to satisfy (b) and (c) in the definition of λ-model, 11.19. Now, in the present language the other conditions in that definition say:

(vi) D has at least two members;

(vii) if $\cdot$ is defined by (iv), then $\langle D, \cdot \rangle$ is combinatorially complete;

(viii) $\underline{Rep} \circ \underline{Fun} \in C$.

Hence any retraction that satisfies (vi) - (viii) gives rise to a λ-model.

DISCUSSION 11.27. The above conditions can be simplified greatly by using the language of category theory (MacLane 1971).

Let $\mathcal{C}$ be a cartesian closed category (MacLane 1971, Chapter IV, §6, or Barendregt 1981 (II), Definition 5.5.1). Suppose also that the objects of $\mathcal{C}$ are sets, the arrows of $\mathcal{C}$ are functions, and that the cartesian product, exponentiation, etc. in the definition of 'cartesian closed' are the usual set-theoretic constructions, except that not every function from an object A to an object B need be an arrow in $\mathcal{C}$, and the object B^A corresponding to the set of all arrows from A to B may be a proper sub-set of the set of all functions from A to B. (Such a $\mathcal{C}$ is called 'strictly concrete' in Barendregt 1981 (II), Definition 5.5.8.)

Suppose $\mathcal{C}$ has an object D with at least two members, and suppose there are two arrows

$$\underline{Fun}: D \to D^D, \qquad \underline{Rep}: D^D \to D,$$

such that $\langle \underline{Rep}, \underline{Fun} \rangle$ is a retraction (i.e. $\underline{Fun} \circ \underline{Rep} = I_{(D^D)}$). Then all the conditions in the preceding discussion are satisfied. In fact, (iii) and (vi) are given assumptions, (viii) follows from the fact that categories are always 'closed' with respect to composition, and (vii) comes from the definition of 'cartesian closed' (Barendregt 1981 (II), Proposition 5.5.7(ii) and the note after 5.5.8).

So every retraction in a strictly concrete cartesian closed category gives rise to a λ-model, provided its domain has at least two members. Conversely, every λ-model can be described as a retraction in some strictly concrete cartesian closed category.

Further, the λ-model is extensional iff the retraction is an isomorphism (between D and D^D).

For more on a category-theoretic view of λ-models, see Plotkin 1978b, Lambek 1980, Koymans 1982, and Barendregt 1981 (II), §5.5.

REMARK 11.28 (<u>The set</u> F). If $\langle D, \cdot, \Lambda \rangle$ is a λ-model in the sense of Definition 11.19, the range of Λ is called F. By Discussion 11.18, F has exactly one member in each extensional-equivalence-class in D, and corresponds one-to-one with $(D \to D)_{rep}$ by the map <u>Rep</u>. (Figure 11.2.) An alternative characterization of F is

(a) $$F = \{d \in D: (\exists M, x, \rho)(d = [\![\lambda x.M]\!]_\rho)\}.$$

(Cf. 11.18.) Finally, by Lemma 10.8(b) a λ-model is extensional iff $F = D$.

REMARK 11.29 (<u>Combinatory algebras and</u> λ-<u>models</u>). Suppose we have a combinatory algebra $\langle D, \cdot \rangle$: how many maps Λ exist such that $\langle D, \cdot, \Lambda \rangle$ is a λ-model? The answer depends on the given algebra.

(a) There are examples with none (Remark 11.13).

(b) There are examples with just one. (A $\langle D, \cdot \rangle$ with just one is called <u>lambda-categorical</u>.) For example, by the next theorem every extensional $\langle D, \cdot \rangle$ is λ-categorical; and by Bruce and Longo 1984, §2, there is also a non-extensional $\langle D, \cdot \rangle$ that is lambda-categorical. (It is $P\omega$, cf. §12.66 later.) In an extensional algebra, not only is Λ unique but its representative e is unique as well. On the other hand, in P_ω the unique Λ has an infinity of representatives e, and each of these gives a loose Scott-Meyer model (though of course only one e gives a strict Scott-Meyer model).

(c) There are examples with an infinity, and examples with finite numbers >1 (Longo 1983, Theorem 4.1 and the remarks after 4.5).

Further results on changing combinatory algebras into λ-models are in Barendregt and Koymans 1980, Meyer 1982, Longo 1983, §4, Bruce and Longo 1984, and are summarized in Barendregt 1981(II), §5.2.

If the combinatory algebra is extensional, the task reduces to triviality, as the following theorem shows.

THEOREM 11.30. *Every extensional combinatory algebra* $\langle D,\cdot\rangle$ *can be made into a λ-model* $\langle D,\cdot,\Lambda\rangle$ *in exactly one way, namely by defining* $\Lambda(a) = a$. *And* $\langle D,\cdot,\Lambda\rangle$ *is a model of* $\lambda\beta\eta$.

Proof. By extensionality, each $\tilde{a}$ has only one member, so $\Lambda(a) = a$ is the only way to define Λ such that $\Lambda(a) \in \tilde{a}$. And this Λ does satisfy Definition 11.19(a) - (d). □

REMARK 11.31. In an extensional combinatory algebra, how do we extend the definition of $[\![\ ]\!]$ from combinatory terms to λ-terms? The above theorem and Theorem 11.20 give an implicit method, but that is not much use in practice. Here is an explicit method. (In fact, by extensionality, all definitions of $[\![\ ]\!]$ that satisfy the conditions of Definition 11.3 will give the same value to $[\![M]\!]_\rho$.)

First, find $k, s \in D$ satisfying the axioms of CLw, and define $[\![\ ]\!]$ for combinatory terms in the usual way (10.10). Then define $[\![\ ]\!]$ for λ-terms via the H_η-transformation:

$$[\![M]\!]_\rho = [\![M_H]\!]_\rho .$$

This definition can easily be checked to satisfy 11.3(a) - (f). For example, here is the proof of (c):

$$\begin{aligned}[\![\lambda x.P]\!]_\rho \cdot d &= [\![\lambda^* x.(P_H)]\!]_\rho \cdot d \\ &= [\![(\lambda^* x.P_H)x]\!]_{[d/x]\rho} && \text{by 10.10(c),(a)} \\ &= [\![P_H]\!]_{[d/x]\rho} && \text{by 2.15} \\ &= [\![P]\!]_{[d/x]\rho} .\end{aligned}$$

SUMMARY 11.32. Five main classes of models have been defined in this chapter and the previous one; they are

$\overline{CLw}$: combinatory algebras/ models of CLw,

$\overline{CL\beta}$: models of $CL\beta_{ax}$ (cf. Barendregt's 'λ-algebras'),

$\overline{CL\beta\eta}$: models of $CL\beta\eta_{ax}$,

$\overline{\lambda\beta}$: λ-models,

$\overline{\lambda\beta\eta}$: extensional combinatory algebras.

By Theorem 11.30 we have $\overline{\lambda\beta\eta} \subseteq \overline{\lambda\beta}$, in the sense that every member of $\overline{\lambda\beta\eta}$ can be made into a member of $\overline{\lambda\beta}$ by adding some extra structure (namely Λ). In a similar sense, by Theorem 11.12 and its corollaries we can say

(i) $\overline{\lambda\beta\eta} \subseteq \overline{CL\beta\eta} \subseteq \overline{CL\beta} \subseteq \overline{CLw}$,

(ii) $\overline{\lambda\beta\eta} \subseteq \overline{\lambda\beta} \subseteq \overline{CL\beta}$.

All these inclusions except the second one in (i) are known to be proper (see Remark 10.24 for the first in (i), Barendregt and Koymans 1980 for the rest). To prove the second inclusion in (i) proper would mean giving a model of $CL\beta_{ax}$ that could not be made into a model of $CL\beta\eta_{ax}$ by changing k and s.

In the literature the two classes $\overline{\lambda\beta}$ and $\overline{CL\beta}$ have been to some extent rivals for the definition of λ-model. $\overline{CL\beta}$ has some nice properties that $\overline{\lambda\beta}$ does not, typically the submodel theorem. But models of $CL\beta_{ax}$ need not satisfy (ξ), so $\overline{\lambda\beta}$ has been generally adopted.

Nevertheless, $\overline{CL\beta}$ has its own interest. For example Gérard Berry in his 1981 argues that the definition of λ-model given earlier is not flexible enough to capture all the aspects of an interpretation in a denotational semantics, and that a definition like that of 'model of $CL\beta_{ax}$' with some extra structure would be better.

FURTHER READING 11.33. The first published definition of λ-model is in Henkin 1950 (p. 83 'standard model', p. 84 'general model'). But Henkin's theory contains an extensionality axiom and this simplifies the definition of model considerably, as we have seen. (And Henkin's theory is a type-theory, which complicates the definition again, but in another direction.)

For the basic ideas behind the λ-model concept the best-written account is Meyer 1982, and the most comprehensive is Hindley and Longo 1980. Further interconnections are described in Barendregt and Koymans 1980, Koymans 1984, Bruce and Longo 1984, and Barendregt 1981(II), Chapter 5. (In Barendregt 1981(I) Chapter 5 the account is not quite as 'clean' as Barendregt's accounts usually are, being based on his 1977; but the account in his 1981(II) is based on his 1984 and is much neater.)

Further references are postponed to the end of the next chapter as most of them are about particular models, or else assume detailed knowledge of particular models.

CHAPTER TWELVE

D_∞ AND OTHER MODELS

12A INTRODUCTION: COMPLETE PARTIAL ORDERS

Having looked at the abstract definition of 'model' in the last two chapters, let us now study a particular concrete model in detail. The one presented here is a variant of Dana Scott's D_∞. This was the first non-trivial model (dating from 1969), and the one whose influence on the semantics of λ-calculus and programming languages has been greatest. It is a model of both CLw and λβ, and is also extensional.

The description below will owe much to Scott's accounts, published and unpublished, and to the very elegant account of D_∞ in Barendregt 1981, but it will give more details and assume the reader has a less mathematical background.

At the end of the chapter some other models will be defined in outline, with references. These are simpler than D_∞, and the reader who only wishes to see a model defined without looking any deeper should go straight to the end of the chapter. (But be warned: much of the language used to describe models in the current literature was derived from work on D_∞, and if you do not understand the D_∞-construction you will not get very far with the literature on other models either.)

Scott originally built D_∞ as a model for his theory of computable functions of higher type. (The theory and the model are both explained together in Scott 1970a, 1972 and 1973.) Scott's theory has had a considerable impact on the semantics of programming languages, as can be seen by reading J. Stoy's textbook 1977 for example. Cf. Scott and Strachey 1971.

On the other hand, the model has had just as great an effect on λ-calculus semantics, and even on the study of pure syntax. For example, although D_∞ is a model of type-free λ-calculus, it induces on λ-terms a type-structure which, with modifications, has turned out very useful in analysing reductions (Barendregt 1981, Chapter 14). With hindsight, this type-structure could have been discovered by contemplating pure syntax, but as an historical fact it was not; it was found via D_∞.

D_∞ has even had an impact on abstract lattice theory, having sparked off the study of a new class, the 'continuous lattices'; Scott 1972, 1976, Gierz et al. 1980.

NOTATION 12.1. In this chapter, $\mathbb{N}$ will be the set of all natural numbers as usual. The following notation will be new:

D, D', D'', X, Y, J :	arbitrary sets;
$a,\ldots,h$:	members of these sets;
ϕ, ψ, χ :	functions;
$\sqsubseteq, \sqsubseteq', \sqsubseteq''$:	partial orderings (Definition 12.2) on D, D', D'' respectively;
$\sqsupseteq, \sqsupseteq', \sqsupseteq''$:	the reversed orderings ($a \sqsupseteq b$ iff $b \sqsubseteq a$, etc.);
$\bot, \bot', \bot''$:	the least members of D, D', D'' respectively ($\bot$ is called <u>'bottom'</u>);
$(D \to D')$:	the set of all functions $\phi: D \to D'$;
$[D \to D']$:	the set of all continuous $\phi: D \to D'$ (Definition 12.10);
$\phi(X)$:	$\{\phi(d) : d \in X\}$, where X is any given set;
$\bigsqcup X$:	the least upper bound (supremum) of X, Definition 12.3;
$\bigsqcup_{n \geq p} \ldots$:	$\bigsqcup \{\ldots : n \geq p\}$;
$\bigsqcup X \simeq \bigsqcup Y$:	$\bigsqcup X$ exists iff $\bigsqcup Y$ exists, and $\bigsqcup X = \bigsqcup Y$ when they exist.

<u>An informal</u> λ-<u>notation</u> '$\boldsymbol{\lambda}$' will be used to define functions. For example, suppose sets D, D' are given, and $a_1,\ldots,a_n \in D$ and $\phi: D^{n+1} \to D'$. Then there is a function $\psi: D \to D'$ such that

$$(\forall d \in D)\ \psi(d) = \phi(a_1,\ldots,a_n,d).$$

This ψ will be called

$$\boldsymbol{\lambda} d \varepsilon D.\phi(a_1,\ldots,a_n,d).$$

Other examples of the λ-notation are

$$\lambda d\varepsilon D.\phi(\chi(d)) \quad \text{for} \quad \phi \circ \chi,$$

$$\lambda d\varepsilon D.b \quad \text{for} \quad \psi \quad \text{such that} \quad (\forall d \in D)(\psi(d) = b).$$

The notation has the following properties:

$$(\lambda a\varepsilon D.\phi(a))(b) = \phi(b),$$

$$\lambda a\varepsilon D.\phi(a) = \phi.$$

But note that this notation is not a new formal language. It will only be used to denote functions that are obviously definable without it (though their definitions without it may be very tedious). The '=' in the above two equations is identity as usual, and means that both sides denote the same set-theoretic function, i.e. the same set of ordered pairs.

DEFINITION 12.2. A partially ordered set (or poset) is a pair $\langle D, \sqsubseteq \rangle$ where D is a set and $\sqsubseteq$ is a binary relation on D, such that

(a) $a \sqsubseteq b$ and $b \sqsubseteq c \Rightarrow a \sqsubseteq c$ (transitivity);

(b) $a \sqsubseteq b$ and $b \sqsubseteq a \Rightarrow a = b$ (anti-symmetry);

(c) $a \sqsubseteq a$ (reflexivity).

The least member of D (if D has one) is called $\bot$, or bottom; then

$$(\forall d \in D) \quad \bot \sqsubseteq d.$$

DEFINITION 12.3. Let $\langle D, \sqsubseteq \rangle$ be a partially ordered set. An upper bound (u.b.) of a subset X of D is any $b \in D$ such that

(a) $$(\forall a \in X) \quad a \sqsubseteq b.$$

And $\bigsqcup X$ (the least upper bound, or l.u.b. or supremum of X) is any u.b. of X, call it b, such that

(b) $$(\forall c \in D)(c \text{ is an u.b. of } X \Rightarrow b \sqsubseteq c).$$

Of course $\bigsqcup X$ might not exist; and if it does, it need not be in X. But if it exists, it is clearly unique. Hence, to prove an equation $b = \bigsqcup X$ we need only prove that b satisfies (a) and (b) above.

EXERCISE 12.4. For any partially ordered set $\langle D, \sqsubseteq \rangle$, prove that:

(a) D has a bottom $\perp$ iff the empty set $\emptyset$ has a l.u.b.; and

$$\perp = \bigsqcup \emptyset.$$

(b) If $X, Y \subseteq D$ and every member of X is $\sqsubseteq$ a member of Y and vice versa, then

$$\bigsqcup X \simeq \bigsqcup Y.$$

(c) Let J be any set and $\{X_j : j \in J\}$ be a family of subsets of D, each X_j having an l.u.b. $\bigsqcup X_j$. If Y is the union of this family, then

$$\bigsqcup Y \simeq \bigsqcup \{\bigsqcup X_j : j \in J\}.$$

DEFINITION 12.5. Let $\langle D, \sqsubseteq \rangle$ be a partially ordered set. A subset $X \subseteq D$ is <u>directed</u> iff $X \neq \emptyset$ and

$$(\forall a, b \in X)(\exists c \in X)\ a \sqsubseteq c \text{ and } b \sqsubseteq c.$$

Directed sets are used in mathematics as index-sets in the theory of convergence on nets; cf. Kelley 1955, Chapter 2. The most important examples are finite or infinite increasing chains:

$$a_1 \sqsubseteq a_2 \sqsubseteq a_3 \sqsubseteq \ldots \ .$$

A simple example which is not a chain is the set of all partitions of an interval $[a,b]$, used in defining the Riemann integral.

DEFINITION 12.6 (<u>Complete partial orders</u>, <u>c.p.o's</u>). A c.p.o. is a partially ordered set $\langle D, \sqsubseteq \rangle$ such that

(a) D has a least member (called $\perp$);

(b) every directed $X \subseteq D$ has a l.u.b. (called $\bigsqcup X$);

<u>Notation</u>. 'The c.p.o. D' will be used to mean 'the c.p.o. $\langle D, \sqsubseteq \rangle$'. Similarly for D', D'', etc. It will always be assumed that $\sqsubseteq$ is the ordering on D and $\sqsubseteq'$ on D' and $\sqsubseteq''$ on D''. For example, the first line of Definition 12.10 below means 'Let $\langle D, \sqsubseteq \rangle$ and $\langle D', \sqsubseteq' \rangle$ be c.p.o's'.

REMARK 12.7. The above definitions might seem to be diverging from what we would expect the essential components of a λ-model to be, so let us look at the motivation for introducing partial orderings.

D_∞ was originally built as a model for Scott's theory of computability. In such a theory it seems natural to regard computable functions as partial functions over some set, for example the partial recursive functions over $\mathbb{N}$. But in Scott's theory the functions are total. Instead of $\mathbb{N}$, he works with

$$\mathbb{N}^+ = \mathbb{N} \cup \{\bot\} \qquad (\bot \notin \mathbb{N}),$$

where $\bot$ is an arbitrary object introduced to represent 'undefinedness'. Each partial function ϕ from $\mathbb{N}$ to $\mathbb{N}$ determines a total function $\phi^+ \in (\mathbb{N}^+ \to \mathbb{N}^+)$, defined thus:

$$\phi^+(n) = \begin{cases} \phi(n) & \text{if } \phi(n) \text{ is defined} \\ \bot & \text{otherwise} \end{cases} \qquad (\forall n \in \mathbb{N})$$

$$\phi^+(\bot) = \bot.$$

Introducing $\bot$ has several advantages. One is to allow us to define two kinds of constant-function. For each $p \in \mathbb{N}$, we can define

(i) ψ': $(\forall n \in \mathbb{N})(\psi'(n) = p)$, $\psi'(\bot) = \bot$;
(ii) ψ'': $(\forall n \in \mathbb{N})(\psi''(n) = p)$, $\psi''(\bot) = p$.

Now if ϕ is the constant-function $\phi(n) = p$ for all $n \in \mathbb{N}$, then $\psi' = \phi^+$. In contrast, ψ'' does not have form ϕ^+ for any ϕ, and theories of partial functions often omit it. Nevertheless it is programmable in practice, and Scott's theory therefore includes it.

The disadvantage of introducing $\bot$ is that if we are not careful, we might end up treating it as an output-value with the same status as a natural number. To prevent this, Scott defines the following partial order on $\mathbb{N}^+$; it corresponds to the intuition that an output $\phi(n) = \bot$ carries less information than an output $\phi(n) = m \in \mathbb{N}$.

DEFINITION 12.8. Choose any object $\bot \notin \mathbb{N}$, and define $\mathbb{N}^+ = \mathbb{N} \cup \{\bot\}$. For all $a, b \in \mathbb{N}^+$, define

$$a \sqsubseteq b \iff (a = \bot \text{ and } b \in \mathbb{N}) \text{ or } a = b.$$

(Cf. Figure 12.1.) The pair $\langle \mathbb{N}^+, \sqsubseteq \rangle$ will be called just $\mathbb{N}^+$.

Figure 12.1

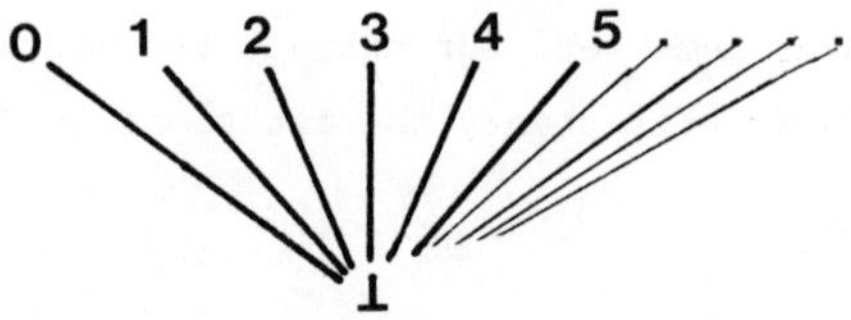

LEMMA 12.9. $\mathbb{N}^+$ *is a c.p.o.*

Proof. Clearly $\sqsubseteq$ is a partial order. The only directed sets in $\mathbb{N}^+$ are: (i) one-member sets, and (ii) pairs $\{\bot, n\}$ with $n \in \mathbb{N}$. Both these have obvious l.u.b's. □

The c.p.o. $\mathbb{N}^+$ will be used later. But now we return to definitions for arbitrary c.p.o's.

12B CONTINUOUS FUNCTIONS

DEFINITION 12.10. Let D, D' be c.p.o's, and let $\phi : D \to D'$. We say ϕ is <u>monotonic</u> iff

(a) $$a \sqsubseteq b \;\Rightarrow\; \phi(a) \sqsubseteq' \phi(b);$$

ϕ is <u>continuous</u> iff

(b) $$(\forall \text{ directed } X \subseteq D) \quad \phi(\bigsqcup X) = \bigsqcup(\phi(X)).$$

EXERCISE 12.11. Show that every continuous function is monotonic.

EXERCISE 12.12. Show that there are only two kinds of continuous function from $\mathbb{N}^+$ to $\mathbb{N}^+$; those of form ϕ^+ for ϕ a partial function from $\mathbb{N}$ to $\mathbb{N}$, and those of form ψ'' in 12.7(ii).
(Hint: for $\psi \in (\mathbb{N}^+ \to \mathbb{N}^+)$, prove that

$$\begin{aligned} \psi \text{ continuous} &\iff \psi \text{ monotonic} \\ &\iff \psi(\bot) = \bot \text{ or } (\exists p \in \mathbb{N})(\forall a \in \mathbb{N}^+)(\psi(a) = p).) \end{aligned}$$

EXERCISE 12.13. Show that if ϕ is monotonic, then

$$X \text{ directed } \subseteq D \quad \Rightarrow \quad \phi(X) \text{ directed } \subseteq D'.$$

Hence $\bigsqcup(\phi(X))$ exists for all directed $X \subseteq D$, because D' is a c.p.o.

REMARK 12.14. To prove a function ϕ continuous, one must prove that if X is directed then $\phi(\bigsqcup X)$ satisfies the two conditions for being the l.u.b. of $\phi(X)$: 12.3(a) and (b). This task is often easier if one already knows that $\bigsqcup(\phi(X))$ exists, and Exercise 12.13 shows that $\bigsqcup(\phi(X))$ always does exist if ϕ is monotonic.

Several proofs below will involve proving a function continuous. In all cases, it will be fairly obvious that the function is monotonic. So by Exercise 12.13 we will know that the l.u.b's involved exist, and the continuity proofs will reduce to straightforward calculations with l.u.b's. If we had to stop and check the conditions of the definition of l.u.b. at each stage, the proofs would become much longer.

REMARK 12.15. The word 'continuous' suggests topology, and indeed every c.p.o. has a topology called the Scott topology, whose continuous functions are exactly those in Definition 12.10 (Scott 1972, or Barendregt 1981, Definition 1.2.3.) In fact Scott's theory of computability is formulated in topological terms, and the account of D_∞ in Scott 1972 is written in these terms.

DEFINITION 12.16 (The function-set $[D \to D']$). For c.p.o's D, D', define $[D \to D']$ to be the set of all continuous functions from D to D'. For $\phi, \psi \in [D \to D']$, define

$$\phi \sqsubseteq \psi \quad \Leftrightarrow \quad (\forall d \in D)\ \phi(d) \sqsubseteq' \psi(d).$$

Intuitively, if we think of $a \sqsubseteq' b$ as meaning that a gives less or the same information as b, then $\phi \sqsubseteq \psi$ says that each output-value $\phi(d)$ carries less or the same information as $\psi(d)$.

Clearly $\langle [D \to D'], \sqsubseteq \rangle$ is a partially ordered set. It has a least member, namely the function $\bot$ defined by

$$(\forall d \in D) \qquad \bot(d) = \bot'.$$

In the special case $D = D'$, $[D \to D']$ contains the identity function I_D. Also, it is easy to see that

(12.17) $\quad \phi_1 \sqsubseteq \phi_2,\ d_1 \sqsubseteq d_2 \quad \Rightarrow \quad \phi_1(d_1) \sqsubseteq' \phi_2(d_2).$

LEMMA 12.18. *Let* D, D′ *be c.p.o's; then* [D→D′] *is a c.p.o., and for each directed* $Y \subseteq [D \to D']$, *we have*

$$(\forall d \in D) \qquad (\bigsqcup Y)(d) = \bigsqcup\{\phi(d) : \phi \in Y\}.$$

Proof. Let $Y \subseteq [D \to D']$ be directed. For each $d \in D$, define

$$Y(d) = \{\phi(d) : \phi \in Y\}.$$

Then Y(d) is directed $\subseteq D'$. (Proof: if $a, b \in Y(d)$, then $a = \phi(d)$, $b = \psi(d)$ for $\phi, \psi \in Y$. Since Y is directed it contains $\chi \sqsupseteq \phi, \psi$; then $\chi(d) \sqsupseteq a, b$.) So Y(d) has a l.u.b., $\bigsqcup(Y(d)) \in D'$, and the right-hand side of the equation in the lemma is meaningful.

Define a function $\psi: D \to D'$ thus:

$$\psi(d) = \bigsqcup(Y(d)).$$

We must prove that $\psi = \bigsqcup Y$. But first we must prove $\psi \in [D \to D']$, and this means proving that ψ is continuous. Let $X \subseteq D$ be directed; then

$$\begin{aligned}
\psi(\bigsqcup X) &= \bigsqcup(Y(\bigsqcup X)) && \text{by definition of } \psi\\
&= \bigsqcup\{\phi(\bigsqcup X) : \phi \in Y\} && \text{by definition of } Y(d)\\
&= \bigsqcup\{\bigsqcup(\phi(X)) : \phi \in Y\} && \text{by continuity of } \phi\\
&= \bigsqcup\{\phi(a) : a \in X \text{ and } \phi \in Y\} && \text{by Exercise 12.4(c)}\\
&= \bigsqcup\{\bigsqcup(Y(a)) : a \in X\} && \text{by Exercise 12.4(c)}\\
&= \bigsqcup\{\psi(a) : a \in X\} && \text{by definition of } \psi.
\end{aligned}$$

(It is easy to check that all the sets above are directed in D or D′, so all the l.u.b's mentioned above do exist.) Thus $\psi \in [D \to D']$.

Also ψ is an u.b. of Y; because for all $\phi \in Y$ and $d \in D$,

$$\phi(d) \sqsubseteq \bigsqcup(Y(d)) = \psi(d).$$

Finally, if χ is any other u.b. of Y, then $\psi \sqsubseteq \chi$; because for all $d \in D$, $\chi(d)$ will be an u.b. of Y(d) and hence $\sqsupseteq$ its least u.b., which is $\psi(d)$. □

LEMMA 12.19. *The composition of continuous functions is continuous. That is if* D, D′, D″ *are c.p.o's and* $\psi \in [D \to D']$ *and* $\phi \in [D' \to D'']$, *then* $\phi \circ \psi \in [D \to D'']$.

Proof. Routine. □

DEFINITION 12.20 (<u>Isomorphism</u>). Let D, D' be c.p.o's and let $\phi \in [D \to D']$ and $\psi \in [D' \to D]$ satisfy

$$\phi \circ \psi = I_{D'}, \qquad \psi \circ \phi = I_D;$$

then we say D <u>is imomorphic to</u> D', or $D \cong D'$.

(Note that any such ϕ and ψ must be one-to-one and onto. They are also continuous, and hence monotonic, i.e. order-preserving.)

DEFINITION 12.21 (<u>Projections</u>). Let D, D' be c.p.o's. A <u>projection from</u> D' <u>to</u> D is a pair of functions $\phi \in [D \to D']$ and $\psi \in [D' \to D]$, such that

$$\psi \circ \phi = I_D, \qquad \phi \circ \psi \sqsubseteq I_{D'}.$$

If such ϕ, ψ exist, we say D' <u>is projected onto</u> D <u>by</u> $\langle \phi, \psi \rangle$. (Cf. Figure 12.2.)

Every projection $\langle \phi, \psi \rangle$ is a retraction in the sense of 11.1, but with the extra properties that ϕ and ψ are continuous and

$$\phi \circ \psi \sqsubseteq I_{D'}.$$

Hence ϕ, ψ make D isomorphic to a subset $\phi(D)$ of D'. Moreover, it is easy to show that the bottom members of D and D' must correspond:

$$\phi(\bot) = \bot', \qquad \psi(\bot') = \bot.$$

Figure 12.2.

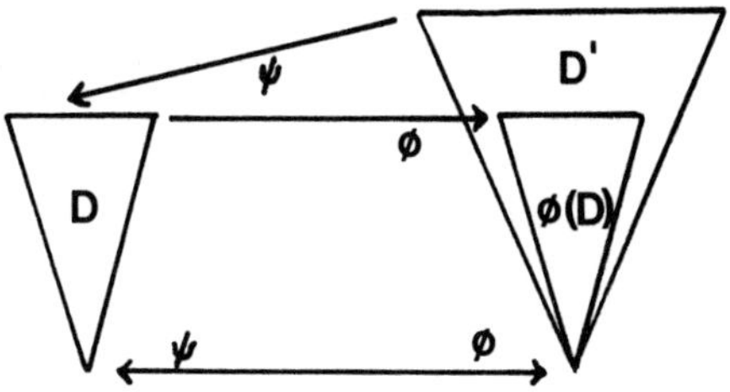

12C THE CONSTRUCTION OF D_∞

D_∞ will be the limit of a sequence $D_0, D_1, D_2, \ldots$ of c.p.o's, each of which is the function-set of the one before it. Their precise definition is as follows.

DEFINITION 12.22 (<u>The sequence</u> $\langle D_n \rangle$). Define $D_0 = \mathbb{N}^+$, and

$$D_{n+1} = [D_n \to D_n].$$

The $\sqsubseteq$-relation on D_n will be called just '$\sqsubseteq$'; it is defined in 12.16. The least member of D_n will be called $\perp_n$.

DISCUSSION 12.23. To construct a λ-model $\langle D_\infty, \cdot, [\![\;]\!] \rangle$ in ZF set-theory, we cannot take D_∞ to be a set of functions and $\cdot$ to be function-application because no ZF-function can be applied to itself.

Scott avoided this problem by a device which, in principle, is very simple. He took the members of D_∞ to be not just functions, but infinite sequences

$$\phi = \langle \phi_0, \phi_1, \phi_2, \ldots \rangle$$

with $\phi_n \varepsilon D_n$. Application was defined by

$$\phi \cdot \psi = \langle \phi_1(\psi_0), \phi_2(\psi_1), \phi_3(\psi_2), \ldots \rangle.$$

With this definition, self-application becomes immediately possible:

$$\phi \cdot \phi = \langle \phi_1(\phi_0), \phi_2(\phi_1), \phi_3(\phi_2), \ldots \rangle.$$

It is a long way from this simple idea to a λ-model, and the definition of application will become somewhat more complicated before it gets there, but the above is its motivation.

DEFINITION 12.24 (<u>The initial maps</u>). To begin the limit-construction, we embed D_0 into D_1 by a map ϕ_0 and define a reverse map ψ_0:

(a) $\phi_0(d) = \lambda a \varepsilon D_0 \,.\, d \qquad (\forall d \;\varepsilon\; D_0);$

(b) $\psi_0(g) = g(\perp_0) \qquad (\forall g \;\varepsilon\; D_1).$

For each $d \in D_0$, $\phi_0(d)$ is the constant-function with value d; this is obviously continuous, so $\phi_0(d) \in D_1$, and hence $\phi_0 \in (D_0 \to D_1)$. Also $\psi_0 \in (D_1 \to D_0)$.

LEMMA 12.25. *The pair* $\langle \phi_0, \psi_0 \rangle$ *is a projection from* D_1 *to* D_0; *i.e.*

(a) ϕ_0 *and* ψ_0 *are continuous;*

(b) $\psi_0 \circ \phi_0 = I_{D_0}$; *i.e.* $\psi_0(\phi_0(d)) = d$ *for all* $d \in D_0$;

(c) $\phi_0 \circ \psi_0 \sqsubseteq I_{D_1}$; *i.e.* $\phi_0(\psi_0(g)) \sqsubseteq g$ *for all* $g \in D_1$.

Proof.

(a) Exercise.

(b) $\psi_0(\phi_0(d)) = \phi_0(d)(\bot_0) = d$.

(c) Let $g \in D_1$. Then g is monotonic, so $g(\bot_0) \sqsubseteq g(d)$ for all $d \in D_0$. Then

$$\begin{aligned} \phi_o(\psi_o(g)) &= \lambda d\varepsilon D_0 \,.\, g(\bot_0) \\ &\sqsubseteq \lambda d\varepsilon D_0 \,.\, g(d) \qquad \text{by definition of } \sqsubseteq \text{ in } D_1 \\ &= g. \end{aligned}$$

□

REMARK 12.26. The above initial projection is going to be made to induce a projection $\langle \phi_n, \psi_n \rangle$ from D_{n+1} to D_n for each $n \geq 1$, in a very natural way. In category-theory language, D_∞ will be the inverse limit of this sequence of projections, in the category of c.p.o's and continuous functions (Figure 12.3).

For each n, we shall have

$$D_n \leftarrow [D_n \to D_n],$$

where '$\leftarrow$' denotes projection, and in the limit we shall have

$$D_\infty \cong [D_\infty \to D_\infty].$$

Figure 12.3.

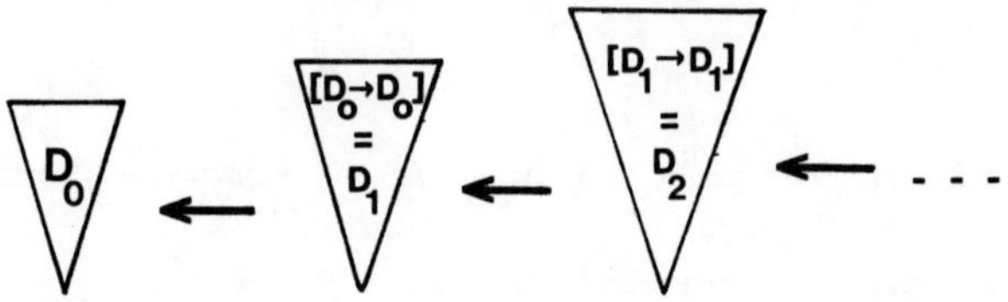

DEFINITION 12.27. Let $n \geq 1$. For all $f \in D_n$ and all $g \in D_{n+1}$, define

$$\phi_n(f) = \phi_{n-1} \circ f \circ \psi_{n-1},$$

$$\psi_n(g) = \psi_{n-1} \circ g \circ \phi_{n-1}.$$

LEMMA 12.28. $\langle \phi_n, \psi_n \rangle$ *is a projection from* D_{n+1} *to* D_n; *i.e.*

(a) $\phi_n \in [D_n \to D_{n+1}]$, $\psi_n \in [D_{n+1} \to D_n]$;

(b) $\psi_n \circ \phi_n = I_{D_n}$;

(c) $\phi_n \circ \psi_n \sqsubseteq I_{D_{n+1}}$.

Proof.

(a) Here, $\phi_n \in [D_n \to D_{n+1}]$ will be proved. (ψ_n is similar.) To do this, two things must be proved:

(a_1) $\phi_n(d)$ is continuous for all $d \in D_n$,
(this implies $\phi_n(d) \in D_{n+1}$, so $\phi_n \in (D_n \to D_{n+1})$);

(a_2) ϕ_n is continuous.

These will be proved together by induction on n. The basis $(n = 0)$ is 12.25. Now let $n \geq 1$. Then (a_1) holds by the induction-hypothesis (a_2) and Lemma 12.19.

To prove (a_2), let $X \subseteq D_n$ be directed. Now ϕ_n can easily be seen to be monotonic, so $\phi_n(X)$ is directed and hence $\bigsqcup(\phi_n(X))$ exists. Both $\bigsqcup(\phi_n(X))$ and $\phi_n(\bigsqcup X)$ are functions, so to prove them equal we only need prove

$$\phi_n(\bigsqcup X)(d) = (\bigsqcup(\phi_n(X))(d) \qquad (\forall d \in D_n).$$

But

$$\begin{aligned}
\phi_n(\bigsqcup X)(d) &= \phi_{n-1}((\bigsqcup X)(\psi_{n-1}(d))) && \text{by definition of } \phi_n\\
&= \phi_{n-1}(\bigsqcup\{f(\psi_{n-1}(d)) : f \in X\}) && \text{by 12.18}\\
&= \bigsqcup\{\phi_{n-1}(f(\psi_{n-1}(d))) : f \in X\} && \text{by continuity of } \phi_{n-1}\\
&= \bigsqcup\{\phi_n(f)(d) : f \in X\} && \text{by definition of } \phi_n\\
&= (\bigsqcup(\phi_n(X)))(d) && \text{by 12.18.}
\end{aligned}$$

(b) The proof is by induction on n. For $n = 0$, use 12.25(b). For $n \geq 1$, let $f \in D_n$. Then for all $d \in D_{n-1}$,

$$\begin{aligned}
\psi_n(\phi_n(f))(d) &= \psi_{n-1}(\phi_n(f)(\phi_{n-1}(d))) && \text{by definition of } \psi_n\\
&= \psi_{n-1}(\phi_{n-1}(f(\psi_{n-1}(\phi_{n-1}(d))))) && \text{by definition of } \phi_n\\
&= f(d) && \text{by induction hypothesis.}
\end{aligned}$$

So $\psi_n(\phi_n(f)) = f$.

(c) Like (b). □

LEMMA 12.29. *ϕ_n and ψ_n preserve application, in the following sense: for all* $a \in D_{n+1}$ *and* $b \in D_n$,

(a) $\psi_{n-1}(a(b)) \sqsupseteq \psi_n(a)(\psi_{n-1}(b))$ if $n \geq 1$;

(b) $\phi_n(a(b)) = \phi_{n+1}(a)(\phi_n(b))$ if $n \geq 0$.

Proof.

(a) By the definition of ψ_n,

$$\begin{aligned}
\psi_n(a)(\psi_{n-1}(b)) &= \psi_{n-1}(a(\phi_{n-1}(\psi_{n-1}(b))))\\
&\sqsubseteq \psi_{n-1}(a(b)) && \text{by 12.28(c)}
\end{aligned}$$

(b) Similar, using 12.28(b). □

EXERCISE 12.30. For $n \geq 2$, D_n contains the following analogue of K:

$$k_n = \lambda a \varepsilon D_{n-1} . \lambda b \varepsilon D_{n-2} . \psi_{n-2}(a).$$

Prove that

(a) $k_n \in D_n$ for $n \geq 2$;

(b) $\psi_1(k_2) = I_{D_o}$, $\psi_o(\psi_1(k_2)) = \perp_o$;

(c) $\psi_n(k_{n+1}) = k_n$ for $n \geq 2$.

(Hint: (a) has two parts, like 12.28(a); (a_1) for all $a \in D_{n-1}$, $k_n(a)$ is continuous, and (a_2) k_n is continuous.)

EXERCISE 12.31. For $n \geq 3$, D_n contains the following analogue of S:

$$s_n = \lambda a \varepsilon D_{n-1} . \lambda b \varepsilon D_{n-2} . \lambda c \varepsilon D_{n-3} . a(\phi_{n-3}(c))(b(c)).$$

Prove that

(a) $s_n \in D_n$ for $n \geq 3$;

(b) $\psi_2(s_3) = \lambda a \varepsilon D_1 . \lambda b \varepsilon D_o . a(\perp_o)$, $\psi_1(\psi_2(s_3)) = I_{D_o}$;

(c) $\psi_n(s_{n+1}) = s_n$ for $n \geq 3$.

DEFINITION 12.32. From ϕ_n and ψ_n we can define composite maps $\phi_{m,n}$ from D_m to D_n for all $m, n \geq 0$, thus:

$$\phi_{m,n} = \begin{cases} \phi_{n-1} \circ \phi_{n-2} \circ \ldots \circ \phi_{m+1} \circ \phi_m & \text{if } m < n, \\ I_{D_m} & \text{if } m = n, \\ \psi_n \circ \psi_{n+1} \circ \ldots \circ \psi_{m-2} \circ \psi_{m-1} & \text{if } m > n. \end{cases}$$

LEMMA 12.33.

(a) $\phi_{m,n} \in [D_m \to D_n]$;

(b) $\phi_{n,m} \circ \phi_{m,n} = I_{D_m}$ *if* $m \leq n$,

$\sqsubseteq I_{D_m}$ *if* $m > n$;

(c) $\phi_{k,n} \circ \phi_{m,k} = \phi_{m,n}$ *if* k *is between* m *and* n.

Proof. Lemmas 12.19 and 12.28. □

DEFINITION 12.34. D_∞ is the set of all infinite sequences

$$d = \langle d_0, d_1, d_2, \ldots \rangle$$

where $d_n \in D_n$ and $\psi_n(d_{n+1}) = d_n$ for all $n \geq 0$. For $d, d' \in D_\infty$, define

$$d \sqsubseteq d' \iff (\forall n \geq 0)(d_n \sqsubseteq d'_n).$$

NOTATION 12.35. In the rest of this chapter 'a_n', 'b_n', '$(a \cdot b)_n$', etc. will denote the n-th member of sequences a or b or $a \cdot b$, etc. in D_∞. For $X \subseteq D_\infty$, X_n is defined by

$$X_n = \{a_n : a \in X\}.$$

12D BASIC PROPERTIES OF D_∞

LEMMA 12.36. D_∞ *is a c.p.o. Its least member is*

$$\bot = \langle \bot_0, \bot_1, \bot_2, \ldots \rangle$$

where $\bot_n$ *is the least member of* D_n. *For directed* $X \subseteq D_\infty$,

$$\bigsqcup X = \langle \bigsqcup (X_0),\ \bigsqcup (X_1),\ \bigsqcup (X_2), \ldots \rangle.$$

Proof. Let $X \subseteq D_\infty$ be directed. Then each X_n is directed, so $\bigsqcup (X_n)$ exists $\in D_n$. Also

$$\begin{aligned} \psi_n(\bigsqcup (X_{n+1})) &= \bigsqcup (\psi_n(X_{n+1})) && \text{by continuity of } \psi_n \\ &= \bigsqcup \{\psi_n(a_{n+1}) : a \in X\} \\ &= \bigsqcup (X_n) && \text{since } \psi_n(a_{n+1}) = a_n. \end{aligned}$$

So the above sequence for $\bigsqcup X$ is in D_∞. And an easy proof shows that it satisfies the two conditions in the definition of l.u.b., 12.3. □

DEFINITION 12.37 (Embedding D_n into D_∞). Define mappings $\phi_{\infty,n} : D_\infty \to D_n$ and $\phi_{n,\infty} : D_n \to D_\infty$ thus: for all $d \in D_\infty$ and all $a \in D_n$,

$$\phi_{\infty,n}(d) = d_n;$$

$$\phi_{n,\infty}(a) = \langle \phi_{n,0}(a),\ \phi_{n,1}(a),\ \phi_{n,2}(a), \ldots \rangle.$$

(Note that $\phi_{n,n}(a) = a$ if $a \in D_n$.)

LEMMA 12.38. $\langle \phi_{n,\infty}, \phi_{\infty,n} \rangle$ *is a projection from* D_∞ *to* D_n, *i.e.*

(a) $\phi_{n,\infty} \in [D_n \to D_\infty]$, $\phi_{\infty,n} \in [D_\infty \to D_n]$;

(b) $\phi_{\infty,n} \circ \phi_{n,\infty} = I_{D_n}$;

(c) $\phi_{n,\infty} \circ \phi_{\infty,n} \sqsubseteq I_{D_\infty}$.

Also, for all $d \in D_m$,

(d) $\phi_{n,\infty}(\phi_{m,n}(d)) = \phi_{m,\infty}(d)$ *when* $m \le n$.

Proof. Straightforward. □

LEMMA 12.39. *For all* $a \in D_\infty$ *and all* $n, r \ge 0$:

(a) $\phi_{n+r,n}(a_{n+r}) = a_n$;

(b) $\phi_{n,\infty}(a_n) \sqsubseteq \phi_{n+1,\infty}(a_{n+1})$;

(c) $a = \bigsqcup_{n \ge 0} \phi_{n,\infty}(a_n)$

$= \bigsqcup_{n \ge r} \phi_{n,\infty}(a_n)$.

Proof.

(a) By the definitions of D_∞ and $\phi_{n+r,n}$.

(b) $\phi_{n,\infty}(a_n) = \phi_{n,\infty}(\psi_n(a_{n+1}))$ since $a \in D_\infty$

$= \phi_{n+1,\infty}(\phi_n(\psi_n(a_{n+1})))$ by 12.38(d)

$\sqsubseteq \phi_{n+1,\infty}(a_{n+1})$ by 12.28(c).

(c) Let $X = \{\phi_{n,\infty}(a_n) : n \geq 0\}$. By (b), X is an increasing chain. Hence X is directed, $\bigsqcup X$ exists, and for any r we have

$$\bigsqcup X = \bigsqcup_{n \geq r} \phi_{n,\infty}(a_n).$$

To prove (c), we must prove that $(\bigsqcup X)_p = a_p$ for all $p \geq 0$. But

$$(\bigsqcup X)_p = (\bigsqcup_{n \geq p} \phi_{n,\infty}(a_n))_p \quad \text{by above}$$

$$= \bigsqcup_{n \geq p} (\phi_{n,\infty}(a_n))_p \quad \text{by 12.36}$$

$$= \bigsqcup_{n \geq p} \phi_{n,p}(a_n) \quad \text{by definition of } \phi_{n,\infty}$$

$$= \bigsqcup \{a_p\} \quad \text{by (a)}$$

$$= a_p. \qquad \square$$

REMARK 12.40. By Lemma 12.38, $\phi_{n,\infty}$ embeds D_n isomorphically into D_∞. And by 12.38(d), we have, modulo isomorphism,

$$D_0 \subseteq D_1 \subseteq D_2 \subseteq \ldots \subseteq D_\infty.$$

So we can identify each $d \in D_n$ with $\phi_{n,\infty}(d) \in D_\infty$, and speak of d as if it was a member of D_∞. For example, with this convention, Lemma 12.39(b) implies that for each $a \in D_\infty$,

$$a_0 \sqsubseteq a_1 \sqsubseteq a_2 \sqsubseteq \ldots,$$

and 12.39(c) says that

$$a = \bigsqcup \{a_0, a_1, a_2, \ldots\}.$$

So $a_0, a_1, a_2, \ldots$ can be thought of as a sequence of better and better 'approximations' to a.

Identifying members of D_n with members of D_∞ may perhaps confuse a beginning reader a little, so it will not be done in this book; but it is normal practice in the literature.

DEFINITION 12.41 (Application in D_∞). For $a, b \in D_\infty$, define

$$a \cdot b = \bigsqcup_{n \geq 0} \phi_{n,\infty}(a_{n+1}(b_n)).$$

(The set $\{\phi_{n,\infty}(a_{n+1}(b_n)) : n \geq 0\}$ is an increasing chain by Lemma 12.42 below, so its l.u.b. does exist.)

Viewed in the light of Remark 12.40 above, $a \cdot b$ is the l.u.b. of an increasing sequence of approximations, $a_{n+1}(b_n)$. This is the modification of the simple application-definition in Remark 12.23, that is needed to make D_∞ into a λ-model.

LEMMA 12.42. *For* $a, b \in D_\infty$,

$$\phi_{n,\infty}(a_{n+1}(b_n)) \sqsubseteq \phi_{n+1,\infty}(a_{n+2}(b_{n+1})).$$

Proof. First,

$$\begin{aligned} \phi_n(a_{n+1}(b_n)) &= \phi_n(\psi_{n+1}(a_{n+2})(\psi_n(b_{n+1}))) && \text{since } a, b \in D_\infty \\ &\sqsubseteq \phi_n(\psi_n(a_{n+2}(b_{n+1}))) && \text{by 12.29(a)} \\ &\sqsubseteq a_{n+2}(b_{n+1}) && \text{by 12.28(c).} \end{aligned}$$

Now apply $\phi_{n+1,\infty}$ to both sides and use 12.38(d). □

COROLLARY 12.42.1. *For all* $a, b \in D_\infty$ *and* $r \geq 0$,

(a) $a \cdot b = \bigsqcup_{n \geq r} \phi_{n,\infty}(a_{n+1}(b_n))$;

(b) $(a \cdot b)_r = \bigsqcup_{n \geq r} \phi_{n,r}(a_{n+1}(b_n))$;

(c) $(a \cdot b)_r \sqsupseteq a_{r+1}(b_r)$.

Proof.

(a) By 12.41 and 12.42.

(b) By (a), 12.36, and the definition of $\phi_{n,\infty}$.

(c) By (a) and 12.42. □

DEFINITION 12.43. Although we are not yet ready to interpret terms containing 'λ' or combinators in D_∞, we can interpret combinations of variables. For any $\rho : \underline{\text{Vars}} \to D_\infty$, define

(a) $[\![x]\!]_\rho = \rho(x)$,

(b) $[\![PQ]\!]_\rho = [\![P]\!]_\rho \cdot [\![Q]\!]_\rho$.

DEFINITION 12.44. Let M be any combination of variables. Besides $[\![M]\!]_\rho \in D_\infty$, each $\rho : \underline{\text{Vars}} \to D_\infty$ also generates an 'approximate' interpretation $[\![M]\!]^n_\rho$ in each D_n, as follows:

(a) $[\![x]\!]^n_\rho = (\rho(x))_n$;

(b) $[\![PQ]\!]^n_\rho = [\![P]\!]^{n+1}_\rho ([\![Q]\!]^n_\rho)$.

EXAMPLES 12.45.

(a) Let $M \equiv xz(yz)$ and $\rho(x) = a$, $\rho(y) = b$, $\rho(z) = c$; then

$$[\![M]\!]^n_\rho = a_{n+2}(c_{n+1})(b_{n+1}(c_n)).$$

(b) Let $M \equiv xx$ and $\rho(x) = a$; then

$$[\![M]\!]^n_\rho = a_{n+1}(a_n).$$

REMARK 12.46. We shall see below that as n increases, $[\![M]\!]^n_\rho$ approximates closer and closer to $[\![M]\!]_\rho$, in the sense that

$$[\![M]\!]_\rho = \bigsqcup_{n\geq 0} \phi_{n,\infty}([\![M]\!]^n_\rho).$$

The idea of 'approximate' interpretation can be extended to λ-terms in general, and it leads to the type-structure mentioned in the introduction. But that is beyond the scope of this book. (Barendregt 1981, Chapter 14.) All we shall need here are the following two lemmas for combinations of variables.

LEMMA 12.47. *For any combination* M *of variables, any* $\rho : \underline{\text{Vars}} \to D_\infty$, *and any* $n, r \geq 0$:

(a) $\psi_n([\![M]\!]_\rho^{n+1}) \sqsupseteq [\![M]\!]_\rho^n$;

(b) $\phi_{n+r,n}([\![M]\!]_\rho^{n+r}) \sqsupseteq [\![M]\!]_\rho^n$;

(c) $\phi_{n+r,\infty}([\![M]\!]_\rho^{n+r}) \sqsupseteq \phi_{n,\infty}([\![M]\!]_\rho^n)$.

Proof.

(a) Induction on M; for the induction step, use 12.29(a), and for the basis, use the fact that $\rho(x) \in D_\infty$.

(b) Iterate (a).

(c) It is enough to prove the case $r = 1$. For this case,

$$\begin{aligned}\phi_{n+1,\infty}([\![M]\!]_\rho^{n+1}) &\sqsupseteq \phi_{n+1,\infty}(\phi_n(\psi_n([\![M]\!]_\rho^{n+1}))) && \text{by 12.28(c)}\\ &\sqsupseteq \phi_{n+1,\infty}(\phi_n([\![M]\!]_\rho^n)) && \text{by (a)}\\ &= \phi_{n,\infty}([\![M]\!]_\rho^n) && \text{by 12.38(d).}\end{aligned}$$ □

LEMMA 12.48. *For any combination* M *of variables, any* $\rho : \underline{\text{Vars}} \to D_\infty$, *and any* $n, r \geq 0$:

(a) $[\![M]\!]_\rho = \bigsqcup_{n \geq r} \phi_{n,\infty}([\![M]\!]_\rho^n)$;

(b) $([\![M]\!]_\rho)_r = \bigsqcup_{n \geq r} \phi_{n,r}([\![M]\!]_\rho^n)$.

Proof.

(a) By (b) and 12.47(c) and the definition of $\phi_{n,\infty}$.

(b) By induction on M, as follows.

<u>If $M \equiv x$</u>: Let $\rho(x) = d \in D_\infty$. Then $\phi_{n,r}(d_n) = d_r$ when $n \geq r$, by 12.39(a), so

$$\bigsqcup_{n\geq r} \phi_{n,r}([\![x]\!]^n_\rho) = \bigsqcup_{n\geq r} \phi_{n,r}(d_n) = \bigsqcup_{n\geq r} d_r = d_r.$$

<u>If $M \equiv PQ$</u>: First,

$$\begin{aligned}
([\![PQ]\!]_\rho)_r &= \bigsqcup_{n\geq r} \phi_{n,r}(([\![P]\!]_\rho)_{n+1}(([\![Q]\!]_\rho)_n)) && \text{by 12.42.1(b)}\\
&= \bigsqcup_{n\geq r} \phi_{n,r}((\bigsqcup_{p\geq n+1} \phi_{p,n+1}([\![P]\!]^p_\rho))(\bigsqcup_{q\geq n} \phi_{q,n}([\![Q]\!]^q_\rho))) && \text{by ind. hyp.}\\
&= \bigsqcup_{n\geq r} \bigsqcup_{p\geq n+1} \bigsqcup_{q\geq n} a_{n,p,q} && \text{by continuity}
\end{aligned}$$

where

$$a_{n,p,q} = \phi_{n,r}(\phi_{p,n+1}([\![P]\!]^p_\rho)(\phi_{q,n}([\![Q]\!]^q_\rho))).$$

Now by 12.47(b) applied to P and Q,

$$a_{n,p,q} \sqsupseteq \phi_{n,r}([\![P]\!]^{n+1}_\rho([\![Q]\!]^n_\rho)) = \phi_{n,r}([\![PQ]\!]^n_\rho).$$

This gives us half of (b), namely

$$([\![PQ]\!]_\rho)_r \sqsupseteq \bigsqcup_{n\geq r} \phi_{n,r}([\![PQ]\!]^n_\rho).$$

To complete (b) we must prove that the left side $\sqsubseteq$ the right. For this, it is enough to prove that for each $a_{n,p,q}$ there is an $m \geq r$ such that

$$a_{n,p,q} \sqsubseteq \phi_{m,r}([\![PQ]\!]^m_\rho).$$

Now define $m = \max\{p-1,q\}$. Then $m+1 \geq p \geq n+1$ and $m \geq r$, and

$$\begin{aligned}
\phi_{p,n+1}([\![P]\!]^p_\rho) &\sqsubseteq (\phi_{p,n+1} \circ \phi_{m+1,p})([\![P]\!]^{m+1}_\rho) && \text{by 12.47(b)}\\
&= (\psi_{n+1} \circ \ldots \circ \psi_m)([\![P]\!]^{m+1}_\rho).
\end{aligned}$$

Similarly

$$\begin{aligned}
\phi_{q,n}([\![Q]\!]^q_\rho) &\sqsubseteq (\phi_{q,n} \circ \phi_{m,q})([\![Q]\!]^m_\rho)\\
&= (\psi_n \circ \ldots \circ \psi_{m-1})([\![Q]\!]^m_\rho).
\end{aligned}$$

Hence

$$a_{n,p,q} \sqsubseteq \phi_{n,r}((\psi_{n+1} \circ \ldots \circ \psi_m)(\llbracket P \rrbracket_\rho^{m+1})((\psi_n \circ \ldots \circ \psi_{m-1})(\llbracket Q \rrbracket_\rho^m)))$$

$$\sqsubseteq (\phi_{n,r} \circ \psi_n \circ \ldots \circ \psi_{m-1})(\llbracket P \rrbracket_\rho^{m+1}(\llbracket Q \rrbracket_\rho^m)) \qquad \text{by 12.29(a) iterated}$$

$$= \phi_{m,r}(\llbracket PQ \rrbracket_\rho^m) \qquad \text{as required.} \quad \square$$

EXAMPLE 12.49. Let $M \equiv xz(yz)$, and $\rho(x) = a$, $\rho(y) = b$, $\rho(z) = c$; then Lemma 12.48(a) implies that

$$a \cdot c \cdot (b \cdot c) = \bigsqcup_{n \geq 0} \phi_{n,\infty}(a_{n+2}(c_{n+1})(b_{n+1}(c_n))).$$

12E D_∞ IS A λ-MODEL

To prove that D_∞ is a λ-model, it is quickest to show that it is an extensional combinatory algebra and then use Theorem 11.30.

DEFINITION 12.50. Using the k_n from Exercise 12.30, define

$$k = \langle \bot_0, I_{D_0}, k_2, k_3, k_4, \ldots \rangle.$$

LEMMA 12.51. $k \in D_\infty$, *and for all* $a, b \in D_\infty$,

$$k \cdot a \cdot b = a.$$

Proof. First, $k \in D_\infty$ by 12.30. Next, by 12.48(b) applied to $M \equiv uxy$ and $\rho(u) = k$, $\rho(x) = a$, $\rho(y) = b$:

$$(k \cdot a \cdot b)_r = \bigsqcup_{n \geq r} \phi_{n,r}(k_{n+2}(a_{n+1})(b_n))$$

$$= \bigsqcup_{n \geq r} \phi_{n,r}(\psi_n(a_{n+1})) \qquad \text{by 12.30}$$

$$= \bigsqcup \{a_r\} \qquad \text{by 12.39(a)}$$

$$= a_r. \qquad \square$$

DEFINITION 12.52. Using the s_n from Exercise 12.31, define

$$s = \langle \bot_0, I_{D_0}, \psi_2(s_3), s_3, s_4, s_5, \ldots \rangle.$$

LEMMA 12.53. $s \in D_\infty$, *and for all* $a, b, c \in D_\infty$,

$$s \cdot a \cdot b \cdot c = a \cdot c \cdot (b \cdot c).$$

Proof. First, $s \in D_\infty$ by 12.31. Next, apply 12.48(b) to $M \equiv uxyz$, $\rho(u) \equiv s$, $\rho(x) = a$, $\rho(y) = b$, $\rho(z) = c$; this gives

$$(s \cdot a \cdot b \cdot c)_r = \bigsqcup_{n \geq r} \phi_{n,r}(s_{n+3}(a_{n+2})(b_{n+1})(c_n)) \qquad \text{by 12.48(b)}$$

$$= \bigsqcup_{n \geq r} \phi_{n,r}(a_{n+2}(\phi_n(c_n))(b_{n+1}(c_n))) \qquad \text{by 12.31.}$$

Now $\phi_n(c_n) = \phi_n(\psi_n(c_{n+1})) \sqsubseteq c_{n+1}$ by 12.28(c), so

$$(s \cdot a \cdot b \cdot c)_r \sqsubseteq \bigsqcup_{n \geq r} \phi_{n,r}(a_{n+2}(c_{n+1})(b_{n+1}(c_n)))$$

$$= (a \cdot c \cdot (b \cdot c))_r \qquad \text{by 12.49.}$$

To complete the lemma, we must prove that $(s \cdot a \cdot b \cdot c)_r \sqsupseteq (a \cdot c \cdot (b \cdot c))_r$. By above, but taking the l.u.b. for $n \geq r+1$ not $n \geq r$, we have

$$(s \cdot a \cdot b \cdot c)_r = \bigsqcup_{n \geq r+1} \phi_{n,r}(a_{n+2}(\phi_n(c_n))(b_{n+1}(c_n)))$$

$$= \bigsqcup_{n \geq r+1} \phi_{n-1,r}(\psi_{n-1}(a_{n+2}(\phi_n(c_n))(b_{n+1}(c_n)))) \qquad \text{by definition of } \phi_{n,r}$$

$$\sqsupseteq \bigsqcup_{n \geq r+1} \phi_{n-1,r}(\psi_{n+1}(a_{n+2})(\psi_n(\phi_n(c_n)))(\psi_n(b_{n+1})(\psi_{n-1}(c_n)))) \qquad \text{by 12.29}$$

$$= \bigsqcup_{n \geq r+1} \phi_{n-1,r}(a_{n+2}(c_n)(b_n(c_{n-1}))) \qquad \text{by 12.28(b) and definition of } D_\infty$$

$$= (a \cdot c \cdot (b \cdot c))_r \qquad \text{by 12.49.} \qquad \square$$

THEOREM 12.54. D_∞ *is extensional; i.e. if* $a \cdot c = b \cdot c$ *for all* c, *then* $a = b$.

Proof. To show $a = b$, it is enough to show that $a_{r+1} = b_{r+1}$ for all $r \geq 0$. (This implies that $a_0 = b_0$ too, because $a_0 = \psi_0(a_1) = \psi_0(b_1) = b_0$.) Now a_{r+1} and b_{r+1} are functions, so to prove them equal it is enough to prove

$$(\forall d \in D_r)\ a_{r+1}(d) = b_{r+1}(d).$$

Let $d \in D_r$; define $c = \phi_{r,\infty}(d)$, so $c_n = \phi_{r,n}(d)$ for $n \geq 0$. Then

$$
\begin{aligned}
(a \cdot c)_r &= \bigsqcup_{n\geq r} \phi_{n,r}(a_{n+1}(\phi_{r,n}(d))) \quad \text{by 12.42.1(b)} \\
&= \bigsqcup_{n\geq r} (\psi_r \circ \ldots \circ \psi_{n-2} \circ \psi_{n-1} \circ a_{n+1} \circ \phi_{n-1} \circ \phi_{n-2} \circ \ldots \circ \phi_r)(d) \\
&= \bigsqcup_{n\geq r} (\psi_r \circ \ldots \circ \psi_{n-2} \circ (\psi_n(a_{n+1})) \circ \phi_{n-2} \circ \ldots \circ \phi_r)(d) \\
&\qquad \text{by definition of } \psi_n \\
&= \bigsqcup_{n\geq r} (\psi_r \circ \ldots \circ \psi_{n-2} \circ a_n \circ \phi_{n-2} \circ \ldots \circ \phi_r)(d) \quad \text{since } a \in D_\infty \\
&= \bigsqcup_{n\geq r} a_{r+1}(d) \quad \text{by repeating the above} \\
&= a_{r+1}(d).
\end{aligned}
$$

Similarly $(b \cdot c)_r = b_{r+1}(d)$. So if $a \cdot c = b \cdot c$, then $a_{r+1}(d) = b_{r+1}(d)$ as required. □

Theorem 12.55. D_∞ *is an extensional* λ*-model.*

Proof. By 12.51, 12.53, 12.54 and 11.30. □

Now that D_∞ has been proved to be a λ-model, a few interesting properties will be given without proof.

LEMMA 12.56. *Application in* D_∞ *is continuous in both variables; i.e.*

(a) $a \cdot (\bigsqcup X) = \bigsqcup\{a \cdot b : b \in X\}$,

(b) $(\bigsqcup X) \cdot b = \bigsqcup\{a \cdot b : a \in X\}$.

Proof. Straightforward. (Barendregt 1981 Lemma 1.2.12, and 1981(I) Proposition 18.3.11 or 1981(II) Proposition 18.2.11.) □

THEOREM 12.57.

(a) *The continuous functions from* D_∞ *to* D_∞ *are exactly the representable ones.*

(b) $[D_\infty \to D_\infty]$ *is a c.p.o. and is isomorphic to* D_∞.

Proof. Barendregt 1981(I) Theorems 18.3.15, 18.3.16, or 1981(II) Theorems 18.2.15, 18.2.16. □

THEOREM 12.58. *For any c.p.o.* D; *every* $\phi \in [D \to D]$ *has a fixed-point* p *(i.e. such that* (p) = p)*, and the least such* p *is*

$$p_\phi = \bigsqcup_{n \geq 0} \phi^n(\perp).$$

Proof. Easy. □

THEOREM 12.59. *For* D_∞, *the operation of finding the least fixed-point is 'representable' in* D_∞; *in fact, if* Y *is any combinator such that* $Yx =_\beta x(Yx)$, *then for all* $\phi \in [D_\infty \to D_\infty]$ *and all* $f \in D_\infty$ *representing* ϕ,

$$[\![Y]\!] \cdot f = p_\phi.$$

Proof. Barendregt 1981, §19.3. □

REMARK 12.60. It is worth noting that the semantic relation '$D_\infty \models M = N$' can be characterized also in terms of pure syntax. The syntactical structures needed to do this are 'Böhm trees', which are well beyond the scope of this book (Barendregt 1981, Chapter 10), but here is the characterization theorem anyway. It was first proved in Hyland 1976 and Wadsworth 1976, 1978, and the version below is due to Barendregt (1981(I), Corollary 19.2.13 or 1981(II), Corollary 19.2.10).

THEOREM 12.61. *For λ-terms* M *and* N, $D_\infty \models M = N$ *iff the Böhm trees of* M *and* N *have the same 'infinite η-normal form'.*

REMARK 12.62. The D_∞-construction in this book differs slightly from Scott's original one, which used complete lattices not c.p.o's, and thus avoided all problems of proving that l.u.b's exist. (A complete lattice is a c.p.o. in which every subset has a l.u.b.) The reason that c.p.o's are used here, and in Barendregt 1981, is that later work on other models has focussed attention on c.p.o's rather than on lattices, and so any introduction to models that omitted them would be deficient. Also c.p.o's are easier to motivate as models of computability theory (cf. Plotkin 1978b, 1978a §1).

In fact the only difference between using c.p.o's and using lattices is in the starting-set D_0. D_0 was taken to be $\mathbb{N}^+$ here, but any other c.p.o. or complete lattice would have done, and the rest of the construction would not have been affected. Furthermore, the proof of Theorem 12.61 turns out to be independent of D_0, so the set of equations $M = N$ satisfied by D_∞ is independent of D_0.

REMARK 12.63 (Further reading on D_∞). The above account of D_∞ has been very basic indeed. For further insight, see the following works.

Barendregt 1981(I) §§1.2, 18.1, 18.3, 19.2, 19.3 and parts of Chapter 20, give a clear and comprehensive account of D_∞'s construction and main properties. (Or 1981(II) §§1.2, 5.4, 18.2, 19.2, 19.3, 20.)

Scott 1970a, 1972, 1973 are the original publications on D_∞, written with minimum formal details and maximum intuitive motivation. (The informally-circulated manuscripts Scott wrote at this period would also be very helpful if the reader has access to any.) Hyland 1976 and Wadsworth 1976, 1978 are analyses of the structure of D_∞ and $P\omega$, including the characterization theorem. Scott 1976 is also partly about D_∞.

Sanchis 1980 presents a D_∞-construction in a general setting.

Other results on D_∞ are contained in Coppo, Dezani and Ronchi 1978, Ronchi 1981, and Koymans 1984. A category view is in Smyth et al. 1982.

A modified D_∞ obtained by changing the initial projection $\langle \phi_0, \psi_0 \rangle$ is described in Coppo, Dezani and Zacchi 198-.

12F SOME OTHER MODELS

Since D_∞, about a dozen other ways of building λ-models have been found. Here are some of the main ones, with references. (Two λ-models will be called equationally equivalent iff they satisfy the same set of equations $M = N$ (M, N λ-terms).)

12.64. Term models of various theories extending $\lambda\beta$. An important example is the theory $\mathcal{H}$: Barendregt 1981 Definition 4.1.6 and Chapter 16.

12.65. <u>The model</u> D_A (<u>Engeler, Plotkin</u>). For each non-empty set A, define G(A) to be the smallest set such that

(i) $A \subseteq G(A)$,

(ii) if α finite $\subseteq G(A)$ and $m \in G(A)$, then $(\alpha \to m) \in G(A)$, where '$(\alpha \to m)$' denotes any ordered-pair construction such that

$$(\forall \alpha, m) \quad (\alpha \to m) \notin A.$$

Define $D_A = \mathcal{P}(G(A))$, the set of all subsets of G(A). Then for $a, b \in D_A$, define

$$a \cdot b = \{m \in G(A) : (\exists \text{ finite } \beta \subseteq b)\ (\beta \to m) \in a\},$$
$$\Lambda(a) = \{(\beta \to m) : \beta \text{ finite } \subseteq G(A) \text{ and } m \in a \cdot \beta\}.$$

Then $\langle D_A, \cdot, \Lambda \rangle$ is a λ-model. (Longo 1983, Theorem 2.3.) The terms K, $\overline{\text{I}}$ and S are interpreted in D_A thus:

$$k = \{(\alpha \to (\beta \to m)) : \quad \alpha, \beta \text{ finite } \subseteq G(A) \text{ and } m \in \alpha\},$$
$$e = \{(\alpha \to (\beta \to m)) : \quad \alpha, \beta \text{ finite } \subseteq G(A) \text{ and } m \in \alpha \cdot \beta\},$$
$$s = \{(\alpha \to (\beta \to (\gamma \to m))) : \alpha, \beta, \gamma \text{ finite } \subseteq G(A) \text{ and } m \in \alpha \cdot \gamma \cdot (\beta \cdot \gamma)\}.$$

D_A is the shortest known model-construction, apart from term-models. (Contrast D_∞, the longest!) The basic idea was circulated informally by Plotkin in his 1972, and re-discovered by Engeler. The construction is given in Engeler 1981 and Meyer 1982 §5. A motivation for D_A is discussed by Scott in his 1980a §§1-3, and some of its key properties are proved in Longo 1983. There is a special case of D_A in Coppo et al. 1981 §5.

<u>Sample properties</u>:

(a) For no A is D_A extensional. (Engeler 1981 §2.)

(b) D_A is equationally equivalent to the Böhm tree model, 12.68 below. (Longo 1983, Proposition 2.8.)

(c) There are other definitions of Λ which make $\langle D_A, \cdot, \Lambda \rangle$ satisfy different sets of equations. (Longo 1983, Theorem 4.1.)

(d) Every applicative structure $\langle B, \cdot \rangle$ can be isomorphically embedded into $\langle D_B, \cdot \rangle$. (Engeler 1981 §1.)

12.66. The model $P\omega$ (Plotkin, Scott). This model will look at first sight like a special case of D_A, with some trivial differences. But these differences will not be as trivial as they seem (see (d) below).

Let $P\omega$ be the set of all subsets of $\mathbb{N}$. Let $\subseteq$ be set-inclusion as usual. For all $i, j \in \mathbb{N}$, let $\ulcorner i,j \urcorner$ be the number corresponding to $\langle i,j \rangle$ in some given recursive one-to-one coding of ordered pairs in $\mathbb{N}$. Let $\alpha_0, \alpha_1, \alpha_2, \ldots$ be some given effective enumeration of the finite sets of natural numbers. For each α_i and each $m \in \mathbb{N}$, the notation '$(\alpha_i \rightarrow m)$' will be used for $\ulcorner i,m \urcorner$.

Define, for $a, b \in P\omega$,

$$a \cdot b = \{m \in \mathbb{N} : \ (\exists \alpha_i \subseteq b) \ (\alpha_i \rightarrow m) \in a\},$$

$$\Lambda(a) = \{(\alpha_i \rightarrow m) : m \in a \cdot \alpha_i\}.$$

(For the construction-details see Barendregt 1981(I) §18.2 or (II) §18.1.) For the basic properties, including the proof that $\langle P\omega, \cdot, \Lambda \rangle$ is a λ-model, see Barendregt 1981 §§19.1, 19.3, 20.1 - 20.3, and Scott 1976, which has plenty of discussion and motivation. (Note that some of the properties depend on the codings above.) Other properties are in Sanchis 1979, Bruce and Longo 1984 §2, Longo and Martini 1984, and Koymans 1984. Also $P\omega$ is the basis of the textbook Stoy 1977 on denotational semantics.

Sample properties:

(a) $P\omega$ is not extensional. (Scott 1976, Theorem 1.2(iii).)

(b) $P\omega$ is equationally equivalent to the Böhm tree model, 12.68 below. (Barendregt 1981, Corollary 19.1.19(ii).)

(c) $P\omega$ is a complete lattice, and $[P\omega \rightarrow P\omega] = (P\omega \rightarrow P\omega)_{\underline{rep}}$. (Barendregt 1981, Corollary 18.2.8.)

(d) Each of the combinatory algebras $\langle P\omega, \cdot \rangle$ and $\langle D_A, \cdot \rangle$ can be isomorphically embedded in the other (if A is countable), but they are not isomorphic. (Longo 1983, Propositions 4.7, 4.10.)

12.67. The model T^ω (Plotkin). For the construction, see Plotkin 1978a. The model's properties are similar to $P\omega$, but it is not a lattice. It is equationally equivalent to the Böhm tree model, and hence to $P\omega$ and D_A (Barendregt and Longo 1980).

12.68. The Böhm tree model, $\mathcal{B}$ (Barendregt). For the construction and basic properties, see Barendregt 1981(I) §18.4, or (II) §18.3. By 1981(I) Theorem 18.4.14 or (II) Theorem 18.3.10,

$$\mathcal{B} \models M = N \text{ iff } M \text{ has the same Böhm tree as } N.$$

12.69. Filter models (Barendregt, Coppo, Dezani, Venneri). These were developed from type-systems, though they are models of type-free λ-calculus. For the construction and context, see Barendregt et al. 1983. For a generalization and embedding into a D_∞-model which uses $P\omega$ as its starting-point, see Coppo, Dezani, Honsell and Longo 1984.

12.70. The hypergraph model (Sanchis). This is a development of the $P\omega$-construction, and is a combinatory algebra, but is not a λ-model. For the construction and some properties, see Sanchis 1979. For the proof that it is not a λ-model, see Koymans 1984, Chapter 4.

12.71. Information systems (Scott). For the construction and motivation, see Scott 1982. Some of the details are worked out and presented in the thesis Roussou 1983.

REMARK 12.72. (Two other approaches to model-building). The main problem in building a model is to create a structure $\langle D, \cdot \rangle$ such that the members of D behave like functions, and yet $a \cdot b$ is defined for all $a, b \in D$. As we have seen, this problem can be solved in several ways. But there are two other possible approaches.

(I) We could change the set-theory in which we build the models (von Rimscha 1980); see the end of Discussion 3.24. In von Rimscha's model the members are genuine functions and the mapping $\cdot$ is function-application.

(II) We could abandon the requirement that $a \cdot b$ be defined for all a,b. The resulting structures can be called partial models, and the literature contains two main examples.

(a) The uniformly reflexive structures (u.r.s's) of Strong 1968 and Wagner 1969, which are models of an axiomatized abstract theory of partial recursive functions. Some later results on u.r.s's are in Friedman 1971 and Byerly 1982a, 1982b, and the references therein. The simplest such structure is the set $\mathbb{N}$, with $a \cdot b$ defined as $\{a\}(b)$, where $\{a\}$ is the partial recursive function with Gödel number a. If $\{a\}(b)$ has no value, then $a \cdot b$ is not defined, so the model is not 'total'.

(b) Models of typed λ-calculus. We shall meet typed λ-calculus later, but for the moment it is enough to know that application of typed term to typed term is not always defined, so a model of the typed system is only a partial model of the untyped system. The concept of model for logical type-theories based on λ-terms is defined in Henkin 1950.

Some results relating partial and 'total' models are given in Klop 1982 (summarized in Barendregt 1981(II), p. 582), and Shabunin 1983.

CHAPTER THIRTEEN

TYPED TERMS

13A TYPED λ-TERMS

As has been repeatedly emphasized, λ-terms do not represent the mathematician's usual set-theoretic notion of function, with domain and range given as part of the function's definition. Nevertheless, they can easily be modified to fit this notion, and this is what we shall do in this chapter. We shall set up systems of λ and CL whose terms have expressions for domain and range built into their structure, and can be interpreted in ordinary set theory.

We begin by defining expressions called <u>types</u>.

DEFINITION 13.1 (<u>Types</u>). Assume that we have been given some symbols called <u>atomic types</u>; then we define <u>types</u> as follows:

(a) each atomic type is a type;

(b) if α and β are types, then $(\alpha\to\beta)$ is a type.

Each atomic type is intended to denote a particular set; for example we may have N or 0 for the set of natural numbers, and H for the set of truth-values. In general, the atomic types will depend on the intended use of the system we wish to build.

Each type $(\alpha\to\beta)$ (called a <u>compound type</u>) is intended to denote some given set of <u>functions from</u> α <u>to</u> β. More precisely, these are functions whose domain is the set denoted by α, and whose range is a subset of the set denoted by β. The exact set of functions will depend on the intended use of the formal type-theory we shall develop. For example, $(\alpha\to\beta)$ might denote the set of <u>all</u> functions from α to β, or just the <u>continuous</u> functions (cf. Chapter 12), or just the functions <u>representable</u> in some given model (cf. Chapter 10). When this set of functions is specified, then every type comes to denote a set of individuals or functions.

NOTATION 13.2. Lower case Greek letters will denote types. We shall sometimes use the abbreviations

$$\alpha \to \beta \quad \text{for} \quad (\alpha \to \beta)$$
$$\alpha \to \beta \to \gamma \quad \text{for} \quad \alpha \to (\beta \to \gamma)$$
$$\alpha_1 \to \dots \to \alpha_n \to \beta \quad \text{for} \quad \alpha_1 \to (\dots(\alpha_n \to \beta)\dots).$$

Warning: the notation '$\alpha \to \beta$' is becoming increasingly common, but there is no standard notation. Some other notations are $F\alpha\beta$ (used especially by Curry and, in earlier works, by the present authors), β^α, ${}^\alpha\beta$, $(\alpha\beta)$, and $(\beta\alpha)$ (used in Church 1940 and other works depending on it).

DEFINITION 13.3 (Typed λ-terms). For each type α, assume that we have infinitely many variables v^α of type α, and perhaps some constants c^α of type α; then we define typed λ-terms as follows:

(a) each v^α and c^α is a typed λ-term of type α;

(b) if $M^{\alpha \to \beta}$ and N^α are typed λ-terms of types $\alpha \to \beta$ and α respectively, then $(M^{\alpha \to \beta} N^\alpha)^\beta$ is a typed λ-term of type β;

(c) if x^α is a variable of type α and M^β is a typed λ-term of type β, then $(\lambda x^\alpha . M^\beta)^{\alpha \to \beta}$ is a typed λ-term of type $\alpha \to \beta$.

Type superscripts will often be omitted. Other notation-conventions for terms will be the same as in Chapter 1.

Each typed term M^α is intended to denote a member of the set denoted by α. The clauses of Definition 13.3 have been chosen with this intention in mind; for example in (b), if $M^{\alpha \to \beta}$ denotes a function $\phi: \alpha \to \beta$ and N^α denotes a member a of α, then $(M^{\alpha \to \beta} N^\alpha)^\beta$ denotes $\phi(a)$, which is in β.

The atomic constants might include $\bar{0}^N$ and $\bar{\sigma}^{N \to N}$ to denote zero and the successor function.

EXAMPLE 13.4. For each type α, the following is a typed term:

$$I_\alpha \equiv (\lambda x^\alpha . x^\alpha)^{\alpha \to \alpha}.$$

EXAMPLE 13.5. For all types α and β, the following is a typed term:

$$K_{\alpha,\beta} \equiv (\lambda x^\alpha y^\beta . x^\alpha)^{\alpha \to \beta \to \alpha}.$$

Proof. When $K_{\alpha,\beta}$ is written without abbreviations, it becomes

$$(\lambda x^{\alpha}.(\lambda y^{\beta}.x^{\alpha})^{(\beta\to\alpha)})^{(\alpha\to(\beta\to\alpha))}.$$

By working outwards from the inside, it is easy to check that this expression satisfies the conditions of Definition 13.3. □

EXAMPLE 13.6. For all types α, β and γ, the following is a typed term:

$$S_{\alpha,\beta,\gamma} \equiv (\lambda x^{\alpha\to\beta\to\gamma} y^{\alpha\to\beta} z^{\alpha}.xz(yz))^{(\alpha\to\beta\to\gamma)\to(\alpha\to\beta)\to\alpha\to\gamma}.$$

Proof. When $S_{\alpha,\beta,\gamma}$ is written with fewer abbreviations, it becomes

$$(\lambda x^{\alpha\to\beta\to\gamma}.(\lambda y^{\alpha\to\beta}.(\lambda z^{\alpha}.((x^{\alpha\to\beta\to\gamma}z^{\alpha})^{\beta\to\gamma}(y^{\alpha\to\beta}z^{\alpha})^{\beta})^{\gamma})^{\alpha\to\gamma})^{\delta})^{\theta},$$

where

$$\delta \equiv (\alpha\to\beta)\to(\alpha\to\gamma),$$

$$\theta \equiv (\alpha\to\beta\to\gamma)\to(\alpha\to\beta)\to\alpha\to\gamma.$$

To see that this satisfies the definition of typed term, it is best to look at its construction as a tree. This tree is given in Figure 13.1 with some type-superscripts omitted.

$$\dfrac{\dfrac{\dfrac{\dfrac{\dfrac{x^{\alpha\to\beta\to\gamma} \qquad z^{\alpha}}{(xz)^{\beta\to\gamma}} \qquad \dfrac{y^{\alpha\to\beta} \qquad z^{\alpha}}{(yz)^{\beta}}}{(xz(yz))^{\gamma}}}{(\lambda z^{\alpha}.(xz(yz))^{\gamma})^{\alpha\to\gamma}}}{(\lambda y^{\alpha\to\beta}.(\lambda z.xz(yz))^{\alpha\to\gamma})^{\delta}}}{(\lambda x^{\alpha\to\beta\to\gamma}.(\lambda yz.xz(yz))^{\delta})^{\theta}.}$$

Figure 13.1.

The checking process in this proof will crop up again in a different form in the next chapter. □

DEFINITION 13.7 (Substitution). $[N^{\alpha}/x^{\alpha}](M^{\beta})$ is defined in the same way as the substitution of untyped terms in Definition 1.11. It is routine to check that it is a typed term of type β.

If the type of N differs from the type of x, then $[N/x]M$ is not defined.

LEMMA 13.8 (<u>Replacement</u>). *If an occurrence of a typed term* P^{α} *in a typed term* M^{β} *is replaced by another term with type* α, *then the result is a typed term of type* β.

Before coming to the definition of equality and reduction, note that if $((\lambda x^{\alpha}.M^{\beta})^{\alpha\to\beta}N^{\alpha})^{\beta}$ is a typed term (of type β), then $[N^{\alpha}/x^{\alpha}]M^{\beta}$ is also a typed term of type β. Similarly, if $\lambda x^{\alpha}.M^{\beta}$ is a typed term of type $\alpha\to\beta$, then so is $\lambda y^{\alpha}.[y^{\alpha}/x^{\alpha}]M^{\beta}$.
Also all the lemmas on substitution and congruence in Chapter 1 hold for typed terms, with unchanged proofs. (See 1.14 - 1.19.)

DEFINITION 13.9 (<u>Typed</u> β-<u>equality</u>). The theory of typed β-equality will be called $\lambda\beta_t$. It has equations $M^{\alpha} = N^{\beta}$ as its formulas, and the following axiom schemes and rules:

(α) $\lambda x^{\alpha}.M^{\beta} = \lambda y^{\alpha}.[y^{\alpha}/x^{\alpha}]M^{\beta}$ if $y^{\alpha} \notin FV(M^{\beta})$;

(β) $(\lambda x^{\alpha}.M^{\beta})^{\alpha\to\beta}N^{\alpha} = [N^{\alpha}/x^{\alpha}]M^{\beta}$;

(ρ) $M^{\alpha} = M^{\alpha}$;

(μ) $M^{\alpha} = P^{\alpha} \vdash N^{\alpha\to\beta}M^{\alpha} = N^{\alpha\to\beta}P^{\alpha}$;

(ν) $M^{\alpha\to\beta} = P^{\alpha\to\beta} \vdash M^{\alpha\to\beta}N^{\alpha} = P^{\alpha\to\beta}N^{\alpha}$;

(ξ) $M^{\beta} = P^{\beta} \vdash \lambda x^{\alpha}.M^{\beta} = \lambda x^{\alpha}.P^{\beta}$;

(τ) $M^{\alpha} = N^{\alpha},\ N^{\alpha} = P^{\alpha} \vdash M^{\alpha} = P^{\alpha}$;

(σ) $M^{\alpha} = N^{\alpha} \vdash N^{\alpha} = M^{\alpha}$.

For provability in this theory we shall write

$$\lambda\beta_t \vdash M^{\alpha} = N^{\alpha}.$$

<u>Note</u>. By 13.7 and 13.8, if α and β are different types, then no equation $M^{\alpha} = N^{\beta}$ can be proved. In particular, the terms

$$(\lambda x^{\alpha}.x^{\alpha})^{\alpha\to\alpha}, \qquad (\lambda y^{\beta}.y^{\beta})^{\beta\to\beta}$$

are not congruent when $\beta \not\equiv \alpha$.

DEFINITION 13.10 (<u>Typed</u> β-<u>reduction</u>). The theory of typed β-reduction will be called $\lambda\beta_t$. It is obtained from Definition 13.9 by replacing '=' by '▷' and omitting rule (σ). For provability in this theory we shall write

$$\lambda\beta_t \quad \vdash \quad M^{\alpha} \rhd N^{\alpha}.$$

REMARK 13.11. <u>Redexes</u>, <u>contraction</u>, <u>reduction</u> $\rhd_\beta$, and β-<u>normal form</u> are defined exactly as in 1.22 and 1.24. It is routine to prove that

(a) $M^{\alpha} \rhd_\beta N^{\beta} \implies \beta \equiv \alpha$,

(b) $M^{\alpha} \rhd_\beta N^{\alpha} \iff \lambda\beta_t \quad \vdash \quad M^{\alpha} \rhd N^{\alpha}$.

Also, β-<u>equality</u>, $=_\beta$, is defined as in 1.32, and it is routine to prove that

(c) $M^{\alpha} =_\beta N^{\alpha} \iff \lambda\beta_t \quad \vdash \quad M^{\alpha} = N^{\alpha}$.

All the other properties of reduction and equality in Chapter 1 hold for the typed system with the same proofs as in Chapter 1. Note in particular the <u>substitution lemmas</u> (1.28 and 1.34), the <u>Church-Rosser theorems</u> (1.29, 1.35 and Remark A1.13), and the <u>uniqueness of normal forms</u> (1.35.4).

The typed system has one extra property which the untyped system does not have, and which plays a key role in all of its applications. To understand the statement of this property, recall the definition of <u>reductions</u> in 3.14 and <u>reductions with maximal length</u> in 3.15.

THEOREM 13.12 (<u>Strong normalization theorem for typed</u> λ-<u>terms</u>). *In the typed system there are no infinite β-reductions.*

Proof. See Appendix 2, Theorem A2.3. □

COROLLARY 13.12.1 (Normalization theorem). *Every typed term* M^{α} *has a normal form* $M^{*\alpha}$, *and all reductions of* M^{α} *having maximal length end at* $M^{*\alpha}$.

COROLLARY 13.12.2. *β-equality of typed terms is decidable.*

Proof. By 13.12.1 and 1.35.2. □

This completes the discussion of the typed λβ-calculus; we turn next to βη (cf. Chapter 7).

DEFINITION 13.13 (Typed βη-equality and reduction). The equality theory $\lambda\beta\eta_t$ is defined by adding to $\lambda\beta_t$ the following axiom scheme:

$$(\eta) \qquad (\lambda x^{\alpha}.M^{\alpha\to\beta}x^{\alpha})^{\alpha\to\beta} = M^{\alpha\to\beta}, \quad \textit{if} \quad x^{\alpha} \notin FV(M^{\alpha\to\beta}).$$

The reduction theory is defined similarly (using '$\rhd$' instead of '='). The terms η-redex, η-contraction, η-reduction($\rhd_{\eta}$), βη-reduction($\rhd_{\beta\eta}$) and βη-normal form are defined as in 7.6 - 7.8.

REMARK 13.14. The main properties of the untyped βη-system were given in 7.10 - 7.15; these hold for the typed system as well, and the proofs are the same. The strong normalization theorem also holds for typed βη-reductions; see Appendix 2, Theorem A2.4. Hence, Corollaries 13.12.1 and 13.12.2 also extend to the βη-system.

REMARK 13.15 (Definability of recursive functions). By analogy with Chapter 4, it is natural to ask which of the recursive functions can be defined by typed λ-terms, in the sense of Definition 4.4. (By Corollary 13.12.1, only total functions can be so defined.)

If the natural numbers are represented by the Church numerals of Definition 4.2, the complete answer to this question is given by Helmut Schwichtenberg in his 1976, where it is shown that these functions are the polynomials, and functions defined from polynomials by definition-by-cases. (This class of functions is sometimes called the extended polynomials.)

If the natural numbers are represented by the abstract numerals of Remark 4.19, the class of definable functions is even more restricted (though the present authors know of no neat description of it). To get a more useful set of functions in this case, one must add extra constant terms with extra reduction properties, such as recursion combinators; see Chapter 18.

13B TYPED CL-TERMS

The types for typed CL-terms will be the same as those for typed λ-terms.

DEFINITION 13.16 (_Typed_ CL-_terms_). For each type α, assume that we have infinitely many variables v^{α} of type α, and perhaps some constants c^{α} of type α. For all types α, β, assume that there is a constant $K_{\alpha,\beta}$ of type

$$\alpha \to \beta \to \alpha$$

For all α, β, γ, assume that there is a constant $S_{\alpha,\beta,\gamma}$ of type

$$(\alpha \to \beta \to \gamma) \to (\alpha \to \beta) \to \alpha \to \gamma.$$

Then we define _typed_ CL-_terms_ as follows:

(a) each v^{α}, c^{α}, $K_{\alpha,\beta}$, $S_{\alpha,\beta,\gamma}$ is a typed CL-term;

(b) if $X^{\alpha\to\beta}$ and Y^{α} are typed CL-terms of types $\alpha\to\beta$ and α respectively, then the expression $(X^{\alpha\to\beta}Y^{\alpha})^{\beta}$ is a typed CL-term of type β.

Type-superscripts will often be omitted. Other notation-conventions will be the same as in Chapter 2.

The types of $K_{\alpha,\beta}$ and $S_{\alpha,\beta,\gamma}$ are motivated by the types of their λ-counterparts in Examples 13.5 and 13.6. An alternative motivation independent of the λ-system is given in Curry and Feys 1958 §8C.

Substitution of a term U for a variable x is defined as in Definition 2.5, provided that U and x have the same type. As in the case of λ-terms, $[U/x]X$ will have the same type as X.

DEFINITION 13.17 (Typed weak equality). The theory of typed weak equality will be called CLw_t. It has equations $X^\alpha = Y^\beta$ as its formulas, and the following axiom schemes and rules:

(K) $K_{\alpha,\beta}X^\alpha Y^\beta = X^\alpha$;

(S) $S_{\alpha,\beta,\gamma}X^{\alpha\to\beta\to\gamma}Y^{\alpha\to\beta}Z^\alpha = X^{\alpha\to\beta\to\gamma}Z^\alpha(Y^{\alpha\to\beta}Z^\alpha)$;

(ρ) $X^\alpha = X^\alpha$;

(μ) $X^\alpha = Y^\alpha \quad \vdash \quad Z^{\alpha\to\beta}X^\alpha = Z^{\alpha\to\beta}Y^\alpha$;

(ν) $X^{\alpha\to\beta} = Y^{\alpha\to\beta} \quad \vdash \quad X^{\alpha\to\beta}Z^\alpha = Y^{\alpha\to\beta}Z^\alpha$;

(τ) $X^\alpha = Y^\alpha,\ Y^\alpha = Z^\alpha \quad \vdash \quad X^\alpha = Z^\alpha$;

(σ) $X^\alpha = Y^\alpha \quad \vdash \quad Y^\alpha = X^\alpha$.

For provability in this theory we shall write

$$CLw_t \quad \vdash \quad X^\alpha = Y^\alpha.$$

It should be clear that equations $X^\alpha = Y^\beta$ where α and β are different types cannot be proved.

DEFINITION 13.18 (Typed weak reduction). The theory of typed weak reduction will be called CLw_t. It is obtained from Definition 13.17 by replacing '=' by '$\rhd$' and omitting (σ). For provability in this theory we shall write

$$CLw_t \quad \vdash \quad X^\alpha \rhd Y^\alpha.$$

REMARK 13.19. Redexes, contraction, reduction $\rhd_w$, and weak normal form are defined exactly as in 2.7 and 2.8. It is routine to prove that

(a) $X^\alpha \rhd_w Y^\alpha \;\Rightarrow\; \beta \equiv \alpha$,

(b) $X^\alpha \rhd_w Y^\alpha \;\Leftrightarrow\; CLw_t \quad \vdash \quad X^\alpha \rhd Y^\alpha$.

Also weak equality, $=_w$, is defined as in 2.22, and it is routine to prove that

(c) $X^\alpha =_w Y^\alpha \;\Leftrightarrow\; CLw_t \quad \vdash \quad X^\alpha = Y^\alpha$.

All the other properties of reduction and equality in Chapter 2 hold for the typed system with the same proofs as in Chapter 2. Note in particular the substitution lemmas (2.12 and 2.23), the Church-Rosser theorems (2.13, 2.24 and Remark A1.13), and the uniqueness of normal forms (2.24.4).

DEFINITION 13.20 (Abstraction). For each typed term X^β and variable x^α, a term called $\lambda^* x^\alpha . X^\beta$ is defined by induction on the length of X^β as follows (cf. Definition 2.14):

(a) $\lambda^* x^\alpha . X^\beta \equiv K_{\beta,\alpha} X^\beta$ if $x^\alpha \notin FV(X^\beta)$;

(b) $\lambda^* x^\alpha . x^\alpha \equiv I_\alpha \equiv S_{\alpha,\alpha\to\alpha,\alpha} K_{\alpha,\alpha\to\alpha} K_{\alpha,\alpha}$;

(c) $\lambda^* x^\alpha . U^{\alpha\to\beta} x^\alpha \equiv U^{\alpha\to\beta}$ if $x^\alpha \notin FV(U^{\alpha\to\beta})$;

(f) $\lambda^* x^\alpha . (U^{\gamma\to\beta} V^\gamma) \equiv S_{\alpha,\gamma,\beta}(\lambda^* x^\alpha . U^{\gamma\to\beta})(\lambda^* x^\alpha . V^\gamma)$ if (a)-(c) do not apply.

EXERCISE 13.21. Verify that $\lambda^* x^\alpha . X^\beta$ has type $\alpha\to\beta$.

Abstraction has its usual properties, which can be proved by the same proofs as in Chapter 2; note in particular Theorems 2.15, 2.20 and 2.21.

THEOREM 13.22 (Strong normalization theorem for typed CL-terms). *In the typed system there are no infinite weak reductions.*

Proof. Appendix 2, Theorem A2.5. □

COROLLARY 13.22.1. *Every typed* CL-*term has a weak normal form.*

COROLLARY 13.22.2. *Weak equality of typed* CL-*terms is decidable.*

FURTHER READING 13.23. Typed λ and CL have been included in a multitude of type-systems, both in logic and programming-theory, and the literature is extensive. On the logic side, here is a small selection: Church 1940, Schütte 1960, 1968, Takahashi 1967, Prawitz 1968, Henkin 1950, 1963, Andrews 1965, 1971, 1974a, 1974b, Friedman 1973, Gandy 1977, Milner 1977, Statman 1979, 1980, 1982a, 1982b. Some general results are summarized in Barendregt 1981, Appendix A. A particular application will be described in Chapter 18 below, and extended systems will be introduced in Chapter 16.

On the programming side, see the end of Chapter 15.

CHAPTER FOURTEEN

TYPE ASSIGNMENT TO CL-TERMS

14A INTRODUCTION

The typed terms presented in Chapter 13 correspond to functions as they are defined in set theory. On the other hand, if one takes the 'functions-as-operators' approach of Discussion 3.24, one comes away from Chapter 13 with the feeling that something essential to the interpretation of untyped λ-terms has been lost.

For example, for different types α and β, the terms $K_{\alpha,\beta} \equiv \lambda x^{\alpha}y^{\beta}.x^{\alpha}$ are all distinct. But from the viewpoint of discussion 3.24, they all represent special cases of the same operation, that of forming constant-functions. An even more obvious example is given by the terms $I_{\alpha} \equiv \lambda x^{\alpha}.x^{\alpha}$, which are distinct for different types α but which all represent special cases of the one operation of doing nothing. If we admit that these operations are single intuitive concepts, even though perhaps imprecise, then a theory that tries to formalize them by splitting them into an infinity of special cases seems to be heading in the wrong direction.

It seems better to aim for a formalism in which each operation is represented by a single term, and the distinctions that types make are introduced in some other way than by having the types built into the terms' structure. This is what will be done in this chapter.

The idea is to take the untyped terms as given and adopt a set of axioms and rules that will assign certain types to certain terms. Most terms that receive types will receive infinitely many of them, corresponding to the idea that they represent operations with an infinite number of special cases. This infinite number of types will be viewed as special cases of a small number of 'type-schemes', which will be expressions containing variables standing for arbitrary types.

This approach originated in Curry 1934 and 1936, and was developed in Curry et al. 1958 and 1972. It was also developed independently by

J.H. Morris and Robin Milner for use in programming languages; see Milner 1978.

Although at first this approach may seem entirely different from that of the typed terms of Chapter 13, it will turn out that if we take the same types as were used in Chapter 13 and adopt the simplest rules of type-assignment, then a term X is assigned a type α by these rules if and only if there is a corresponding typed term X^{α} in the sense of Chapter 13.

The system below will not be a mere notational variant of Chapter 13, however; it will have more expressive power and more flexibility. For example, a natural question to ask about an untyped term X is whether it has any typed analogues (just as K has the typed analogues $\mathsf{K}_{\alpha,\beta}$); but although this question is about Chapter 13, the easiest way to answer it is to re-state it in the language of the system below. Then its answer can be given for each X by an easy algorithm, as we shall see. Furthermore, the system below can be generalized in ways that the system of Chapter 13 cannot; some of these ways will be sketched later in this book, in Chapter 16.

For technical reasons, the approach below turns out to be significantly simpler for CL-terms than for λ-terms. For this reason this chapter is devoted only to CL-terms. Type assignment to λ-terms and some extensions of both systems, including the extension to 'second-order' types (types involving abstraction and universal quantification) will be taken up in Chapters 15 and 16.

DEFINITION 14.1 (Types and type-schemes). Assume that we have some type-constants and infinitely many type-variables. Then we define type-schemes as follows:

(a) all type-constants and type-variables are type-schemes;

(b) if α and β are type-schemes, then so is $(\alpha\to\beta)$.

A type is a type-scheme containing no variables.

In effect, this definition is obtained by adding variables to Definition 13.1 and calling the atomic types of that definition 'type-constants'.

NOTATION 14.2. Lower-case Greek letters will denote type-schemes, and the same abbreviations will be used as in Notation 13.2. In discussing the types of the Church numerals we shall use the abbreviation

$$N_\alpha \equiv (\alpha \to \alpha) \to \alpha \to \alpha.$$

Letters 'a', 'b', 'c', 'd', 'e' will be used for type-variables in this chapter. The variables x, y, z, ... in terms will be called here _term-variables_. (Type-variables and term-variables can be the same; indeed, see Chapter 16 for an example in which they must be the same.)

The result of simultaneously substituting type-schemes $\alpha_1,\ldots,\alpha_n$ respectively for type-variables $a_1,\ldots,a_n$ in a type-scheme β will be called

$$[\alpha_1/a_1,\ldots,\alpha_n/a_n]\beta.$$

Terms in this chapter will be CL-terms (i.e. the untyped terms from Chapter 2, not the typed terms from Chapter 13). As usual, a _non-redex atom_ is an atom other than K and S. A _non-redex constant_ is a constant other than K and S. A _pure term_ is a term whose only atoms are K, S, and variables. A _combinator_ contains only K and S.

Warning: An expression '$\alpha \to \beta$' containing Greek letters is not a type-scheme, it is only a name in the meta-language for an unspecified type-scheme. Examples of type-schemes are '$(a \to b)$', '$(a \to a)$', 'a', etc.

REMARK 14.3. In this chapter and the next, $\alpha \to \beta$ is intended to denote some set of operators ϕ such that

$$x \in \alpha \Rightarrow \phi(x) \text{ is defined and } \phi(x) \in \beta.$$

This contrasts with Chapter 13; there, a statement $\phi \in \alpha \to \beta$ implied that the domain of ϕ was exactly α, but here it only implies that the domain $\supseteq \alpha$. This comes from our intention that one operator ϕ may have many types, i.e. that $\phi \in \alpha \to \beta$ may be true for many different α and β.

DEFINITION 14.4. _A type-assignment_ (_TA_) _formula_ is any expression

$$X \in \alpha,$$

where X is a CL-term and α is a type-scheme. Its _subject_ is X and its _predicate_ is α.

Such a formula is read informally as 'X is a member of α', or more formally as 'type-scheme α is assigned to term X'. In works by Curry and previous works by the authors, the notations

$$`\vdash \alpha X', \quad `\alpha X'$$

have been used instead (thinking of α as a propositional function rather than a set).

14B THE SYSTEM TA_C

DEFINITION 14.5 (The type-assignment system TA_C). TA_C is a formal theory in the sense of Notation 6.1. Its formulas are arbitrary expressions $X \varepsilon \alpha$. It has two axiom-schemes, motivated by types of $K_{\alpha,\beta}$ and $S_{\alpha,\beta,\gamma}$ in Definition 13.16; they are

(→K) $\qquad K \varepsilon \alpha\to\beta\to\alpha,$

(→S) $\qquad S \varepsilon (\alpha\to\beta\to\gamma)\to(\alpha\to\beta)\to\alpha\to\gamma.$

Its only rule is called the →-elimination rule or (→e); it is motivated by Definition 13.16(b), and says

$$(\to e) \qquad \frac{X \varepsilon \alpha\to\beta \qquad Y \varepsilon \alpha}{XY \varepsilon \beta}.$$

A basis $\mathcal{B}$ is any finite or infinite set of formulas. Iff there is a deduction of $X \varepsilon \alpha$ whose assumptions are all in $\mathcal{B}$, we write

$$\mathcal{B} \vdash_{TA_C} X \varepsilon \alpha,$$

or just $\mathcal{B} \vdash X \varepsilon \alpha$ when there can be no confusion. Iff $\mathcal{B}$ is empty, we write

$$\vdash_{TA_C} X \varepsilon \alpha.$$

EXAMPLE 14.6. Recall that $I \equiv SKK$. Then for all α,

$$\vdash_{TA_C} I \varepsilon \alpha\to\alpha.$$

Proof.

$$\dfrac{\dfrac{\overset{(\to S)}{S\ \varepsilon\ (\alpha\to(\alpha\to\alpha)\to\alpha)\to(\alpha\to\alpha\to\alpha)\to\alpha\to\alpha} \qquad \overset{(\to K)}{K\ \varepsilon\ \alpha\to(\alpha\to\alpha)\to\alpha}}{SK\ \varepsilon\ (\alpha\to\alpha\to\alpha)\to\alpha\to\alpha}\,(\to e) \qquad \overset{(\to K)}{K\ \varepsilon\ \alpha\to\alpha\to\alpha}}{SKK\ \varepsilon\ \alpha\to\alpha.}\,(\to e)$$

□

EXAMPLE 14.7. Recall that $B \equiv S(KS)K$. Then for all α, β, γ,

$$\vdash_{TA_C} B\ \varepsilon\ (\beta\to\gamma)\to(\alpha\to\beta)\to\alpha\to\gamma\ .$$

Proof.

$$\dfrac{\dfrac{\overset{(\to S)}{S\ \varepsilon\ (\beta_1\to\alpha_1)\to(\alpha_2\to\beta_2)\to\alpha_2\to\gamma_2} \qquad \dfrac{\overset{(\to K)}{K\ \varepsilon\ \alpha_1\to\beta_1\to\alpha_1} \qquad \overset{(\to S)}{S\ \varepsilon\ \alpha_1}}{KS\ \varepsilon\ \beta_1\to\alpha_1}\,(\to e)}{S(KS)\ \varepsilon\ (\alpha_2\to\beta_2)\to\alpha_2\to\gamma_2}\,(\to e) \qquad K\ \varepsilon\ \alpha_2\to\beta_2}{S(KS)K\ \varepsilon\ \alpha_2\to\gamma_2}\,(\to e)$$

Here

$$\alpha_1 \equiv (\alpha\to\beta\to\gamma)\to(\alpha\to\beta)\to\alpha\to\gamma \equiv \beta_2\to\gamma_2,$$

$$\beta_1 \equiv \beta\to\gamma,$$

$$\alpha_2 \equiv \beta\to\gamma \quad \equiv \beta_1,$$

$$\beta_2 \equiv \alpha\to\beta\to\gamma \equiv \alpha\to\beta_1 \equiv \alpha\to\alpha_2,$$

$$\gamma_2 \equiv (\alpha\to\beta)\to\alpha\to\gamma.$$

□

EXERCISE 14.8. For each of the terms on the left below, give a TA_C-deduction to show it has all the type-schemes on the right (one scheme for each α, β, γ).

Term			Type-schemes
(a)*	$Z_o \equiv KI$	(cf. 4.2)	$\beta \to \alpha \to \alpha$
(b)*	$\bar{\sigma} \equiv SB$	(cf. 4.2)	$((\beta \to \gamma) \to \alpha \to \beta) \to (\beta \to \gamma) \to \alpha \to \gamma$,
(c)*	$W \equiv SS(KI)$	(cf. 2.11)	$(\alpha \to \alpha \to \beta) \to \alpha \to \beta$,
(d)	KK		$\alpha \to \beta \to \gamma \to \beta$,
(e)	$Z_o \equiv KI$		$N_\alpha \qquad (N_\alpha \equiv (\alpha \to \alpha) \to \alpha \to \alpha)$,
(f)	$\bar{\sigma} \equiv SB$		$N_\alpha \to N_\alpha$,
(g)	$Z_n \equiv (SB)^n(KI)$		N_α.

EXERCISE 14.9. Give TA_C-deductions to show that for all α, β, γ,

(a) $U \varepsilon \alpha \to \beta \to \gamma, \quad V \varepsilon \alpha \to \beta, \quad W \varepsilon \alpha \quad \vdash \quad SUVW \varepsilon \gamma$,

(b) $U \varepsilon \alpha \to \beta \to \gamma, \quad V \varepsilon \alpha \to \beta, \quad W \varepsilon \alpha \quad \vdash \quad UW(VW) \varepsilon \gamma$,

(c) $U \varepsilon \alpha, \quad V \varepsilon \beta \quad \vdash \quad KUV \varepsilon \alpha$,

(d) $x \varepsilon \alpha \to \beta, \quad x \varepsilon \alpha \quad \vdash \quad xx \varepsilon \beta$.

These exercises raise some obvious points of interest, which the next few notes and lemmas will discuss, before we move on to the main metatheorems about TA_C.

NOTE 14.10 (<u>Type-variables</u>). The type-variables play very little formal role in TA_C. In particular, TA_C has no substitution rule, so an assumption like

$$x \varepsilon a$$

does not imply that x has every type. The purpose of type-variables is to make certain metatheorems easier to state, and to prepare the way for more expressive systems in which the types may contain quantifiers or even λ (Chapter 16).

However, although substitution is not derivable, it is admissible, as the lemma below shows.

LEMMA 14.11 (<u>Substitution lemma</u>). *Let* $\mathcal{B}$ *be any basis and let*

$$\mathcal{B} \vdash_{TA_C} X \varepsilon \alpha.$$

Let $[\beta_1/a_1,\ldots,\beta_k/a_k]\mathcal{B}$ *be the basis obtained by substituting* $\beta_1,\ldots,\beta_k$ *for* $a_1,\ldots,a_k$ *simultaneously in all the predicates in* $\mathcal{B}$. *Then*

$$[\beta_1/a_1,\ldots,\beta_k/a_k]\mathcal{B} \vdash_{TA_C} X \varepsilon [\beta_1/a_1,\ldots,\beta_k/a_k]\alpha.$$

Proof. Substitute β's for a's throughout the given deduction. (You probably obtained your answer to 14.8(e) from that to 14.8(a) in this way.) Substitution creates from an (→K)- or (→S)-axiom a new (→K)- or (→S)-axiom, and from an instance of (→e) a new instance of (→e). □

NOTE 14.12. What makes this proof work is the fact that we have assumed in (→K) and (→S), not just two formulas, but an infinite set. (The axiom-schemes (→K) and (→S) themselves are not formulas of TA_C at all, but just meta-language expressions which conveniently summarize the infinite set of axioms.) Since TA_C has no substitution-rule, the lemma above would fail if we assumed just the two formulas

$$K \varepsilon a \to b \to a, \qquad S \varepsilon (a \to b \to c) \to (a \to b) \to a \to c.$$

But because all the other axioms are substitution-instances of these two, they have an obvious interest; let us call them the <u>principal axioms</u>.

NOTE 14.13 (<u>Deduction from bases</u>). Another point of interest is seen in Exercise 14.9, which shows that in TA_C we can already answer questions of a kind that could not even have been asked in Chapter 13, namely, 'What type would X have, if certain parts of X had certain types?'. The possibility of making deductions from bases is an important advantage of the present approach over that in Chapter 13.

Because of this, let us look at some different kinds of bases that a deduction may use as assumptions. All the kinds in the following definition will be used later; some are taken from the literature, but with slight modifications.

DEFINITION 14.14 (Kinds of bases). As before, a basis $\mathcal{B}$ is any finite, infinite or empty set of formulas, say

$$U_1 \varepsilon \delta_1, U_2 \varepsilon \delta_2, \ldots .$$

$\mathcal{B}$ is a weakly- [strongly-] normal-subjects basis iff each U_i is weakly [strongly] irreducible and begins with a non-redex atom. (Each such U_i must have form

$$U_i \equiv q_i V_{i,1} \ldots V_{i,k_i}$$

where q_i is a non-redex atom and each $V_{i,j}$ is irreducible.)

$\mathcal{B}$ is a monoschematic basis iff each U_i is a non-redex constant, and the formulas are exactly the set of all substitution-instances of some set of 'principal formulas', one for each constant in $\mathcal{B}$. (Compare the set of all the axioms for K and S.) More precisely, $\mathcal{B}$ is monoschematic iff each U_i is a non-redex constant and, for each U occurring as a subject in $\mathcal{B}$ (say $U \equiv U_{i_1} \equiv U_{i_2} \equiv \ldots$), there is one i_j such that $\{\delta_{i_1}, \delta_{i_2}, \ldots\}$ is exactly the set of all substitution-instances of δ_{i_j}. The formula $U \varepsilon \delta_{i_j}$ is called the principal formula for U in $\mathcal{B}$.

EXAMPLE 14.15 (A monoschematic basis). Suppose we include atoms $\bar{0}$, $\bar{\sigma}$, Z in the definition of term, as suggested in Discussion 4.19, and include a type-constant N for the set of all natural numbers. Let the definition of reduction be modified to include

$$Z\bar{n} \rhd Z_n,$$

as suggested in (4.22). Then a natural set of assumptions is

(a) $$\bar{0} \varepsilon N, \quad \bar{\sigma} \varepsilon N \to N, \quad Z \varepsilon N \to N_\alpha,$$

where $N_\alpha \equiv (\alpha \to \alpha) \to \alpha \to \alpha$, cf. Exercise 14.8(g). This basis contains an infinite number of formulas $Z \varepsilon N \to N_\alpha$, namely one for each α. But all the type-schemes $N \to N_\alpha$ are substitution-instances of the one type-scheme

$$N \to N_a,$$

so the basis is monoschematic.

If a basis is monoschematic, its members can be used just like the axioms for K and S. Deductions from monoschematic bases have turned out to be nice and tidy and easy to study. For example, in Lemma 14.11, if $\mathcal{B}$ is monoschematic, then we can replace

$$[\beta_1/a_1,\ldots,\beta_n/a_n]\mathcal{B}$$

by just '$\mathcal{B}$'. Also, we shall see that for any term X, the type-schemes deducible for X from a monoschematic basis are all substitution-instances of one 'principal type-scheme'. (Theorem 14.40.)

But other kinds of bases have considerable interest too, and are worth more study than they have been given in the literature up to now. Here are two examples.

EXAMPLE 14.16. (<u>Two bases for number theory</u>). Suppose we take pure terms, and a type-constant N, and look at the Church numerals (see Exercise 14.8(e) - (g)). Then a natural set of assumptions is

(a) $$\mathsf{KI} \in \mathsf{N} \qquad \mathsf{SB} \in \mathsf{N}\to\mathsf{N}.$$

This basis is not monoschematic, since its subjects are not atoms. Nor is it weakly-normal-subjects, since its subjects begin with combinators.

Alternatively, we could assume the infinity of formulas

(b) $$(\mathsf{SB})^n(\mathsf{KI}) \in \mathsf{N} \qquad (n = 0,1,2,\ldots).$$

This is also not monoschematic. It is also not a weakly-normal-subjects basis, but it has some of the properties of such bases, as we shall see.

EXAMPLE 14.17 (<u>Bases with proper inclusions</u>). An assumption of the form

$$\mathsf{I} \in \alpha\to\beta \qquad (\alpha \not\equiv \beta)$$

says intuitively that the identity operator maps α into β, i.e. that α is a subset of β. Such a formula is called a <u>proper inclusion</u>. By rule ($\to$e),

$$X \in \alpha, \quad \mathsf{I} \in \alpha\to\beta \ \vdash\ \mathsf{I}X \in \beta.$$

To use proper inclusions effectively, one needs also to be able to deduce $X \in \beta$ from $\mathsf{I}X \in \beta$. There is no way to do this in TA_C. So, when proper inclusions and other extra assumptions are of interest, a rule of equality-invariance of types has to be added to TA_C; see §E below.

NOTE 14.18 (The subject-construction property). Returning to Exercises 14.8 and 9, notice that the deduction of a formula $X \in \alpha$ closely follows the construction of X. In rule ($\to$e), the subject of the conclusion is built from the subjects of the premises, so as we move down the deduction the subject grows in length, and contains all earlier subjects.

In more detail: let $\mathcal{D}$ be a tree-form deduction of $X \in \alpha$ from assumptions

(a) $$U_1 \in \delta_1, \ldots, U_n \in \delta_n \qquad (n \geq 0),$$

such that each of these formulas actually occurs in $\mathcal{D}$.

Suppose, for the moment, that each U_i is a non-redex atom. Then $U_1, \ldots, U_n$ will be exactly the non-redex atoms occurring in X, and if we strip all type-schemes from $\mathcal{D}$, we shall get the construction-tree for X (i.e. the tree which shows how X is built up from $U_1, \ldots, U_n$ and perhaps also some occurrences of K and S.) To each occurrence of U_i in X there will correspond an assumption $U_i \in \delta_i$ in $\mathcal{D}$, and to each occurrence of S or K in X there will correspond an axiom in $\mathcal{D}$. (Hence $n = 0$ iff X is a combinator.) In general, if Z is any subterm of X, then to each occurrence of Z in X there will correspond a formula in $\mathcal{D}$ with Z as subject.

For example, look at the deduction for Exercise 14.9(b):

(b) $$\dfrac{\dfrac{U \in \alpha\to\beta\to\gamma \qquad W \in \alpha}{UW \in \beta\to\gamma}(\to e) \qquad \dfrac{V \in \alpha\to\beta \qquad W \in \alpha}{VW \in \beta}(\to e)}{UW(VW) \in \gamma.}(\to e)$$

Here $X \equiv UW(VW)$, and the assumptions are $U \in \alpha\to\beta\to\gamma$, $V \in \alpha\to\beta$, $W \in \alpha$ $(n = 3)$. Stripping the type-schemes away, we get this tree:

(c) $$\dfrac{\dfrac{U \quad W}{UW} \qquad \dfrac{V \quad W}{VW}}{UW(VW).}$$

If U, V, W are atoms, then this is the construction-tree of X.

Return now to the general case (a); suppose some of the U's are composite. Then X will be an applicative combination of $U_1, \ldots, U_n$ and S, K, and the stripped tree will show how X is built up from these terms; though it will not be the whole construction-tree of X, but only a lower part. Each term-occurrence Z in X will either (i) be inside an occurrence of a U_i corresponding to an assumption $U_i \varepsilon \delta_i$, or (ii) be an applicative combination of U's and S, K, and have a corresponding formula in $\mathcal{D}$ with Z as subject. (For example, in (b) above, if

$$X \equiv V(VW)W(VW), \quad U \equiv V(VW), \quad Z \equiv VW,$$

then X contains two occurrences of Z, one in U and one at the right-hand end of X. The one in U has no corresponding formula in (b), but the second corresponds to the formula $VW \varepsilon \beta$ in (b).)

The term-construction/deduction correspondence is described formally in Curry et al. 1958 §9B (the Subject-construction theorem). It is the key to the study of TA_C, and will be used repeatedly throughout this chapter.

One of its corollaries is that if each subject in a basis $\mathcal{B}$ is an atom, and

$$\mathcal{B} \vdash_{TA_C} X \varepsilon \alpha,$$

then $\mathcal{B}$ contains a formula $q_i \varepsilon \delta_i$ for each non-redex atom q_i in X.

It is also used in the proof of the following theorem. This is the first main metatheorem about TA_C, and shows that the type of $\lambda^*x.X$ is exactly what one would expect from the informal interpretation of $\lambda^*x.X$ as a function. It will help us build complex deductions fast, just as the deduction-theorem for propositional logic helps to build propositional deductions. Curry called it the Stratification theorem (Curry et al. 1958 §9D, Corollary 1.1).

THEOREM 14.19 (Abstraction and types). *Let* $\mathcal{B}$ *be any basis. If* x *does not occur in any subject in* $\mathcal{B}$, *and if*

$$\mathcal{B},\ x \varepsilon \alpha \vdash_{TA_C} X \varepsilon \beta,$$

then

$$\mathcal{B} \vdash_{TA_C} \lambda^*x.X \varepsilon \alpha \to \beta,$$

where λ^* *is either* λ^η *(i.e. the* λ^* *of Definition* 2.14*) or* λ^w *(Definition* 9.20*) or* λ^β *(Definition* 9.34*).*

Proof. Induction on X, with cases corresponding to Definition 2.14, which includes all the cases in 9.20 and 9.34. Let $\mathcal{D}$ be the given deduction of $X \varepsilon \beta$. The restriction that x not occur in $\mathcal{B}$ implies that whenever x occurs in $\mathcal{D}$, its type-scheme must be α.

Case 1: $x \notin FV(X)$ and $\lambda^*x.X \equiv KX$. Then the assumption $x \varepsilon \alpha$ is not used in $\mathcal{D}$, so $\mathcal{D}$ is a deduction of

$$\mathcal{B} \vdash X \varepsilon \beta.$$

Now the formula $K \varepsilon \beta\to\alpha\to\beta$ is an instance of axiom-scheme $(\to K)$, and hence by $(\to e)$,

$$\mathcal{B} \vdash KX \varepsilon \alpha\to\beta.$$

Case 2: $X \equiv x$ and $\lambda^*x.X \equiv I$. Then $\beta \equiv \alpha$. The desired result is

$$\mathcal{B} \vdash I \varepsilon \alpha\to\alpha,$$

which follows by Example 14.6.

Case 3: $X \equiv Ux$, where $x \notin FV(U)$, and $\lambda^*x.X \equiv U$. Then by the correspondence between deductions and constructions, $\mathcal{D}$ must have form

$$\dfrac{\begin{matrix}\mathcal{D}_1\\ U \varepsilon \alpha\to\beta\end{matrix} \qquad x \varepsilon \alpha}{Ux \varepsilon \beta}\,(\to e),$$

where x does not occur in $\mathcal{D}_1$. (The above notation means that $\mathcal{D}_1$ is a deduction of the formula $U \varepsilon \alpha\to\beta$, and $\mathcal{D}$ is the result of applying $(\to e)$ to $\mathcal{D}_1$ and an assumption $x \varepsilon \alpha$.) But $\lambda^*x.X \equiv U$. Hence $\mathcal{D}_1$ is a deduction of the desired result.

Case 4: $X \equiv X_1X_2$ and $\lambda^*x.X \equiv S(\lambda^{*\prime}x.X_1)(\lambda^{*\prime}x.X_2)$, where $\lambda^{*\prime}$ is λ^* if λ^* is λ^η or λ^w, but $\lambda^{*\prime}$ is λ^η if λ^* is λ^β. Then $\mathcal{D}$ must have form

$$\dfrac{\begin{matrix}\mathcal{D}_1\\ X_1 \varepsilon \gamma\to\beta\end{matrix} \qquad \begin{matrix}\mathcal{D}_2\\ X_2 \varepsilon \gamma\end{matrix}}{X_1X_2 \varepsilon \beta}\,(\to e).$$

We are proving the theorem for all three forms of λ^* at once, so by the induction-hypothesis,

$$\mathcal{B} \vdash (\lambda^{*\prime}x.X_1) \; \varepsilon \; \alpha \to \gamma \to \beta,$$

$$\mathcal{B} \vdash (\lambda^{*\prime}x.X_2) \; \varepsilon \; \alpha \to \gamma.$$

Now the formula

$$S \; \varepsilon \; (\alpha \to \gamma \to \beta) \to (\alpha \to \gamma) \to (\alpha \to \beta)$$

is an instance of axiom-scheme $(\to S)$, so the result follows by $(\to e)$. This completes the proof. □

Notice how the type-schemes for S and K fit the cases in this proof; if they had not already been motivated by analogy with $S_{\alpha,\beta,\gamma}$ and $K_{\alpha,\beta}$ in Chapter 13 (which were motivated by a λ-analogy), the above proof would have been their principal motivation.

COROLLARY 14.19.1. *If no subject in* $\mathcal{B}$ *contains the (distinct) variables* $x_1,\ldots,x_n$, *and*

$$\mathcal{B}, x_1 \; \varepsilon \; \alpha_1,\ldots,x_n \; \varepsilon \; \alpha_n \vdash_{TA_C} X \; \varepsilon \; \beta,$$

and λ^* *is defined by any of Definitions* 2.4, 9.20, 9.34, *then*

$$\mathcal{B} \vdash_{TA_C} \lambda^* x_1 \ldots x_n.X \; \varepsilon \; \alpha_1 \to \ldots \to \alpha_n \to \beta.$$

EXERCISE 14.20. Let $C \equiv \lambda^* xyz.xzy$. Let D and $R_{Bernays}$ be as defined in (4.7) and (4.10). For each α let $\xi \equiv N_{N_\alpha} \to N_\alpha$. Prove in TA_C:

(a) $C \; \varepsilon \; (\beta \to \alpha \to \gamma) \to (\alpha \to \beta \to \gamma)$,

(b) $D \; \varepsilon \; \alpha \to \alpha \to N_\alpha \to \alpha$,

(c) $R_{Bernays} \; \varepsilon \; N_\alpha \to (N_\alpha \to N_\alpha \to N_\alpha) \to N_\xi \to N_\alpha$.

EXERCISE 14.21. To see the advantage of using Theorem 14.19, write out C in terms of S and K and prove (a) of Exercise 14.20 without using Theorem 14.19!

EXERCISE 14.22. Let R be as defined in (4.23), and $\mathcal{B}$ be the basis in Example 14.15(a). Show that, in TA_C,

$$\mathcal{B} \vdash R \; \varepsilon \; N \to (N \to N \to N) \to N \to N.$$

The next theorem will be the second main meta-theorem about TA_C, and will show that type-schemes are preserved by both weak and strong reduction. Its proof will need the following lemma.

LEMMA 14.23 (Replacement). *Let* $\mathcal{B}_1, \mathcal{B}_2$ *be any bases and* $\mathcal{D}$ *be a deduction giving*

$$\mathcal{B}_1 \vdash_{TA_C} X \in \alpha.$$

Let V *be any term-occurrence in* X, *such that* $\mathcal{D}$ *contains a formula* $V \in \gamma$ *in the same position as* V *has in the construction-tree of* X. *Let* X^* *be the result of replacing* V *by a term* W *such that*

$$\mathcal{B}_2 \vdash_{TA_C} W \in \gamma.$$

Then

$$\mathcal{B}_1, \mathcal{B}_2 \vdash_{TA_C} X^* \in \alpha.$$

Proof. First cut off from $\mathcal{D}$ the subtree above the formula $V \in \gamma$. The result is a deduction $\mathcal{D}_1$ with form

$$\begin{array}{c} V \in \gamma \\ \mathcal{D}_1 \\ X \in \alpha. \end{array}$$

(This notation means that $\mathcal{D}_1$ has conclusion $X \in \alpha$ and that one of its assumptions is the formula $V \in \gamma$.) Then replace V by W in the assumption $V \in \gamma$, and in all formulas below it in $\mathcal{D}_1$. The result is a deduction $\mathcal{D}_1^*$ with form

$$\begin{array}{c} W \in \gamma \\ \mathcal{D}_1^* \\ X^* \in \alpha. \end{array}$$

Then take the given deduction $\mathcal{D}_2$ of $W \in \gamma$ and place it over the assumption $W \in \gamma$ in $\mathcal{D}_1^*$. The result is a deduction

$$\begin{array}{c} \mathcal{D}_2 \\ W \in \gamma \\ \mathcal{D}_1^* \\ X^* \in \alpha \end{array}$$

as desired. □

THEOREM 14.24 (<u>Subject-reduction theorem</u>). *Let* $\mathcal{B}$ *be a weakly-[strongly-] normal-subjects basis; if*

$$\mathcal{B} \vdash_{TA_C} X \varepsilon \alpha$$

and $X \vartriangleright_w X'$ $[X \succ_{\beta\eta} X']$, *then*

$$\mathcal{B} \vdash_{TA_C} X' \varepsilon \alpha.$$

Proof. The proofs for weak and strong reduction are in Curry et al. 1958 §§9C2, 9C6 and Curry et al. 1972 §14B2. We shall keep to weak reduction here for simplicity. By the replacement lemma, it is enough to take care of the case that X is a redex and X′ is its contractum.

<u>Case</u> 1: $X \equiv KX'Y$. Let $\mathcal{D}$ be a deduction of $X \varepsilon \alpha$ from $\mathcal{B}$. By the nature of $\mathcal{B}$, the first K in X cannot be part of a subject in $\mathcal{B}$, so it must correspond to an instance of axiom-scheme $(\rightarrow K)$. Hence, using the correspondence between deductions and term-constructions, $\mathcal{D}$ must have form

$$\dfrac{\dfrac{K \varepsilon \alpha\rightarrow\beta\rightarrow\alpha \qquad \overset{\mathcal{D}_1}{X' \varepsilon \alpha}}{KX' \varepsilon \beta\rightarrow\alpha}(\rightarrow e) \qquad \overset{\mathcal{D}_2}{Y \varepsilon \beta}}{KX'Y \varepsilon \alpha.}(\rightarrow e)$$

Then $\mathcal{D}_1$ is the desired deduction.

<u>Case</u> 2: $X \equiv SUVW$ and $X' \equiv UW(VW)$. By the argument of the previous case, the deduction of $X \varepsilon \alpha$ must have the following form:

$$\dfrac{\dfrac{\dfrac{S \varepsilon (\gamma\rightarrow\beta\rightarrow\alpha)\rightarrow(\gamma\rightarrow\beta)\rightarrow(\gamma\rightarrow\alpha) \qquad \overset{\mathcal{D}_1}{U \varepsilon \gamma\rightarrow\beta\rightarrow\alpha}}{SU \varepsilon (\gamma\rightarrow\beta)\rightarrow(\gamma\rightarrow\alpha)}(\rightarrow e) \qquad \overset{\mathcal{D}_2}{V \varepsilon \gamma\rightarrow\beta}}{SUV \varepsilon \gamma\rightarrow\alpha}(\ e) \qquad \overset{\mathcal{D}_3}{W \varepsilon \gamma}}{SUVW \varepsilon \alpha.}(\rightarrow e)$$

From this we can construct a deduction of $X' \varepsilon \alpha$ as follows:

$$\dfrac{\dfrac{\begin{matrix}\mathcal{D}_1 & \mathcal{D}_3\\ U \in \gamma\to\beta\to\alpha & W \in \gamma\end{matrix}}{UW \in \beta\to\alpha}\,(\to e) \qquad \dfrac{\begin{matrix}\mathcal{D}_2 & \mathcal{D}_3\\ V \in \gamma\to\beta & W \in \gamma\end{matrix}}{VW \in \beta}\,(\to e)}{UW(VW) \in \alpha.}\,(\to e)$$

□

REMARK 14.25.

(a) The above proof will still work if the condition on the basis $\mathcal{B}$ is relaxed slightly, to say that every subject in $\mathcal{B}$ is in nf, and if a subject in $\mathcal{B}$ begins with S or K, then every type-scheme it receives in $\mathcal{B}$ is atomic. An example of such a basis is Example 14.16(b); so the subject-reduction theorem holds for that basis, even though it is not a weakly-normal-subjects basis.

(b) In contrast, an example of an interesting basis for which the theorem's conclusion fails is

$$Z_0 \in N, \qquad BW(BB) \in N\to N,$$

where $Z_0 \equiv KI$ as usual. (The significance of this is that BW(BB) behaves in many ways like $\bar{\sigma}$, since if $\bar{n}$ is Z_n, then

$$\begin{aligned} BW(BB)\bar{n}xy &\rhd_w W(BB\bar{n})xy\\ &\rhd_w BB\bar{n}xxy\\ &\rhd_w B(\bar{n}x)xy\\ &\rhd_w \bar{n}x(xy)\\ &\rhd_w x^n(xy) \equiv x^{n+1}y.) \end{aligned}$$

From this basis, we can easily deduce

$$BW(BB)\bar{0} \in N;$$

but the theorem's conclusion fails, because

$$BW(BB)\bar{0} \rhd_w W(BB\bar{0}),$$

and we cannot deduce $W(BB\bar{0}) \in N$.

(c) For $\succ$, Curry's proof of 14.24 assumed that I was an atom, with an axiom-scheme $I \in \alpha\to\alpha$ in TA_C. (See 8.18 for the reason.) But it works also when I is SKK, as here.

REMARK 14.26 (Subject-expansion). It is natural to ask whether Theorem 14.24 can be reversed; that is, whether

$$\mathcal{B} \vdash X' \varepsilon \alpha$$

implies

$$\mathcal{B} \vdash X \varepsilon \alpha,$$

if X reduces to X′. The answer is that reversal is possible only under certain very restrictive conditions (Curry et al. 1958 §9C3).

For an example where reversal is not possible, take

(a) $$X \equiv \mathsf{SKSI}, \qquad X' \equiv \mathsf{KI(SI)}.$$

Here $X \rhd_w X'$, and in TA_C we can prove $X' \varepsilon \beta\to\beta$ for all type-schemes β; but by Exercise 14.38 later, we can only prove $X \varepsilon \beta\to\beta$ for composite β.

An even stronger example is

(b) $$X \equiv \mathsf{SIII}, \qquad X' \equiv \mathsf{II(II)}.$$

In this case we can prove $X' \varepsilon \beta\to\beta$ for every β, but X has no types at all (by Example 14.36).

The non-reversibility of Theorem 14.24 means that the set of type-schemes assigned to a term is not invariant under conversion. This is a serious drawback to the system TA_C. The obvious answer is to add an equality-invariance rule (which will of course be non-recursive), and this is what will be done in §E. We shall see that despite the rule's non-recursiveness, the extended system is still fairly manageable.

14C STRATIFIED CL-TERMS AND THE LINK WITH CHAPTER 13

Some untyped CL-terms have typed analogues in the system of Chapter 13, for example K, S and B, but others have not. In this section the definition of 'has a typed analogue' will be made precise, and the set of untyped terms that have typed analogues will be characterized and shown to be decidable.

We shall assume that the type-constants in Chapter 14 are the same as the atomic types in Chapter 13, and that there is a one-to-one correspondence between the typed term-variables of Chapter 13, and the untyped variables of Chapter 14 (so that v^α, u^β correspond to the same untyped variable iff $v \equiv u$ and $\alpha \equiv \beta$).

For simplicity, the emphasis will be on pure closed terms (combinations of K and S only). Other terms will be mentioned for completeness' sake, but should be ignored at a first reading.

DEFINITION 14.27. For each pure typed CL-term Y^α, define $|Y^\alpha|$ to be the untyped term obtained by replacing each typed variable in Y^α by the corresponding untyped variable, and striking out all other type-superscripts from Y^α. In particular, define

$$|S_{\alpha,\beta,\gamma}| \equiv S, \qquad |K_{\alpha,\beta}| \equiv K.$$

Clearly the natural meaning of 'X has a typed analogue' is that $X \equiv |Y^\alpha|$ for some Y^α. The following definition is another approach to the same concept.

DEFINITION 14.28 (Stratified CL-terms). Let X be any CL-term, with $FV(X) = \{x_1,\ldots,x_n\}$ $(n \geq 0)$.

(a) For a closed pure X: we say that X is stratified iff there is a type-scheme α such that

$$\vdash_{TA_C} X \in \alpha.$$

(b) For any pure X: we say that X is stratified iff there exist $\delta_1,\ldots,\delta_n$, α such that

$$x_1 \in \delta_1,\ldots,x_n \in \delta_n \vdash_{TA_C} X \in \alpha.$$

(c) For any X: we say that X is stratified relative to a basis $\mathcal{B}$ iff $x_1,\ldots,x_n$ do not occur in any of the subjects of $\mathcal{B}$ and there exist $\delta_1,\ldots,\delta_n$, α such that

$$\mathcal{B},\ x_1 \in \delta_1,\ldots,x_n \in \delta_n \vdash_{TA_C} X \in \alpha.$$

EXAMPLES 14.29. By 14.6 - 8 and 14.20, the following closed terms are stratified:

$$K,\ S,\ I,\ B,\ C,\ W,\ KK,\ SB,\ Z_n,\ D,\ R_{Bernays}.$$

By 14.22, the R defined in (4.23) is stratified relative to the basis in Example 14.15(a).

In contrast, look at Exercise 14.9(d); this is not a proof that xx is stratified, because x occurs twice with two different type-schemes. (In the definition of 'stratified', $x_1,\ldots,x_n$ are assumed by the usual convention to be distinct.)

LEMMA 14.30. *Let* X *be a pure* CL-*term*:

(a) X *is stratified iff each subterm of* X *is stratified.*

(b) X *is stratified iff there exist types (not just type-schemes)* $\delta_1,\ldots,\delta_n,\alpha$ *satisfying Definition* 14.28(b).

(c) X *is stratified iff* $X \equiv |Y^\alpha|$ *for some pure typed term* Y^α.

(d) *The set of all stratified pure* CL-*terms is closed under weak and strong reduction, but not expansion.*

(e) *The set of all stratified pure* CL-*terms is closed under abstraction* (λ^*), *but not application.*

Proof.

(a) By the subject-construction property, 14.18.

(b) By the substitution lemma, 14.11.

(c) By (b) and comparing Definitions 14.27 and 28.

(d) By the subject-reduction theorem, 14.24.

(e) By the theorem on abstraction and types, 14.19, and the fact that certain terms are not stratified (Examples 14.35 and 36). □

DEFINITION 14.31 (<u>Principal type-scheme</u>, <u>p.t.s.</u>). Let X be any CL-term, with $FV(X) = \{x_1,\ldots,x_n\}$ $(n \geq 0)$.

(a) For a closed pure X: α is a <u>principal type-scheme</u> (<u>p.t.s.</u>) of X iff

$$\vdash_{TA_C} X \in \alpha'$$

holds for a type-scheme α' when and only when α' is a substitution-instance of α. (Since α is a substitution-instance of itself, α is a p.t.s. of X only if $\vdash$ $X \in \alpha$.)

(b) For any pure X: first, an FV(X)-basis is any basis $\mathcal{B}$ with form

$$\mathcal{B} = \{x_1 \in \delta_1,\dots,x_n \in \delta_n\}.$$

A pair $\langle\mathcal{B},\alpha\rangle$ is a principal pair (p.p.) of X, and α is a p.t.s. of X, iff $\mathcal{B}$ is an FV(X)-basis and the relationship

$$\mathcal{B}' \quad \vdash_{TA_C} \quad X \in \alpha'$$

holds for an FV(X)-basis $\mathcal{B}'$ and a type-scheme α' when and only when $\langle\mathcal{B}',\alpha'\rangle$ is a substitution-instance of $\langle\mathcal{B},\alpha\rangle$.

(c) For any X: let $\mathcal{B}_0$ be any basis whose subjects do not contain any of $x_1,\dots,x_n$; a pair $\langle\mathcal{B},\alpha\rangle$ is a p.p., and α is a p.t.s., of X relative to $\mathcal{B}_0$, iff $\mathcal{B}$ is an FV(X)-basis and

$$\mathcal{B}_0 \cup \mathcal{B}' \quad \vdash_{TA_C} \quad X \in \alpha'$$

holds for an FV(X)-basis $\mathcal{B}'$ and a type-scheme α' when and only when $\langle\mathcal{B}',\alpha'\rangle$ is a substitution-instance of $\langle\mathcal{B},\alpha\rangle$.

NOTE 14.32. A substitution-instance of $\langle\mathcal{B},\alpha\rangle$ is a pair $\langle\mathcal{B}',\alpha'\rangle$ that is the result of a simultaneous substitution

$$[\beta_1/a_1,\dots,\beta_k/a_k]$$

of type-schemes for type-variables in α and the predicates in $\mathcal{B}$. The subjects in $\mathcal{B}$ are unchanged.

DISCUSSION 14.33. In Exercise 14.8, the type-schemes assigned to Z_0 and $\bar{\sigma}$ in (e) and (f) are substitution-instances of those assigned to the same terms in (a) and (b) respectively. Furthermore, the type-schemes in (a) and (b) have substitution-instances which those in (e) and (f) do not, so the type-schemes in (e) and (f) are non-principal. As it happens, the type-schemes in (a) and (b) are principal. But how can we establish this fact?

The answer is to consider the correspondence between deduction-trees and construction-trees described in Note 14.18. A deduction of a formula $X \in \alpha$ in TA_C must follow the construction-tree of X. So to determine whether X is stratified, it is sufficient to write down the construction-tree of X, and try to fill in a suitable type-scheme at each stage, conforming to the patterns demanded by the axiom-schemes and deduction-rule of TA_C.

This process leads to an algorithm for deciding whether or not X is stratified, and for computing a p.t.s. for X if X is stratified. The algorithm is a little complicated to define formally. (See Curry 1969a, or see Hindley 1969a for a different and slightly later definition). But it is relatively simple to operate in practice, as the examples below will show.

If there is no way to fill in the type-schemes that is consistent with TA_C, the attempt will lead to a contradiction, and we can conclude that X is not stratified. But if suitable type-schemes can be filled in throughout the tree, the process of filling them in will indicate the most general possible type-scheme at each stage, and the final one will be the principal type-scheme for X.

EXAMPLE 14.34. I has p.t.s. $a \to a$.

Proof. By definition, $I \equiv SKK$, and its construction-tree is

$$\dfrac{\dfrac{S \qquad K}{SK} \qquad K}{SKK.}$$

If we can deduce

$$SKK \ \varepsilon \ \eta$$

for some η, then this must be the conclusion of an inference by rule ($\to$e), for which the premises are

$$SK \ \varepsilon \ \xi \to \eta, \qquad K \ \varepsilon \ \xi,$$

for some ξ. The second of these must be an instance of axiom-scheme ($\to$K), while the first must be the conclusion of an inference for which the premises are

$$S \ \varepsilon \ \zeta \to \xi \to \eta, \qquad K \ \varepsilon \ \zeta,$$

for some ζ. These last two formulas must be instances of the axiom-schemes ($\to$S) and ($\to$K) respectively. To get an instance of ($\to$S), we need

(a) $$\zeta \equiv \alpha \to \beta \to \gamma,$$

(b) $$\xi \equiv \alpha \to \beta,$$

(c) $$\eta \equiv \alpha \to \gamma,$$

for suitable type-schemes α, β, and γ. To make the formulas $\mathsf{K} \in \xi$ and $\mathsf{K} \in \zeta$ to be instances of $(\to \mathsf{K})$, we need

(d) $$\beta \equiv \delta \to \alpha,$$

(e) $$\gamma \equiv \alpha,$$

for some δ. Now $\eta \equiv \alpha \to \alpha$ follows immediately from (c) and (e). Thus every type-scheme η for I must be a substitution-instance of $a \to a$.

Conversely, every substitution-instance of $a \to a$ is a type-scheme for I, by Exercise 14.6. Thus $a \to a$ is a p.t.s. of I. □

REMARK 14.35 (Pseudo-uniqueness of p.t.s.). Clearly $b \to b$ is also a p.t.s. of I, where b is a type-variable distinct from a, so the p.t.s. of a term is not unique. However, it is easy to see that the p.t.s's of a term X differ only by the substitution of distinct variables for distinct variables, and we shall often refer to 'the p.t.s. of X', as if it were unique.

REMARK 14.36. From the above example, it is not hard to see that I is stratified if and only if the equations (a) - (e), in the unknowns α, β, γ, δ, ξ, ζ, η, possess a solution. And the p.t.s. is the solution for η whose discovery demands the fewest possible substitutions. This equation-solving process is the core of the algorithm of Curry 1969a.

EXAMPLE 14.37. SII is unstratified.

Proof. If we could deduce

$$\mathsf{SII} \in \eta$$

for some type-scheme η, then this must be the conclusion of an inference by $(\to e)$, whose premises are

$$\mathsf{SI} \in \xi \to \eta, \qquad \mathsf{I} \in \xi,$$

for some ξ. By the second of these and Example 14.34, we must have

(a) $$\xi \equiv \zeta \to \zeta$$

for some ζ. Hence, the first of these premises is

$$\mathsf{SI}\ \varepsilon\ (\zeta \to \zeta) \to \eta.$$

This, in turn, must be the conclusion of an inference whose premises are

$$\mathsf{S}\ \varepsilon\ \theta \to (\zeta \to \zeta) \to \eta, \qquad \mathsf{I}\ \varepsilon\ \theta,$$

for some θ. By the second of these and Example 14.34, we must have

$$\theta \equiv \delta \to \delta$$

for some δ. Thus, the first of these is

$$\mathsf{S}\ \varepsilon\ (\delta \to \delta) \to (\zeta \to \zeta) \to \eta.$$

But this must be an instance of axiom-scheme $(\to \mathsf{S})$. Hence there must be type-schemes α, β and γ such that

$$\delta \to \delta \equiv \alpha \to \beta \to \gamma,$$
$$\zeta \to \zeta \equiv \alpha \to \beta,$$
$$\eta \equiv \alpha \to \gamma.$$

It follows that

$$\beta \equiv \zeta \equiv \alpha \equiv \delta \equiv \beta \to \gamma,$$

and $\beta \equiv \beta \to \gamma$ is impossible. □

EXAMPLE 14.38. xx is unstratified.

Proof. Since the only free variable in xx is x, it follows from the definition that xx is stratified iff there are type-schemes α and β such that

$$x\ \varepsilon\ \alpha \quad \vdash_{TA_C} \quad xx\ \varepsilon\ \beta.$$

If there is a deduction of this, the last inference must be by $(\to e)$, and the premises must be

$$x\ \varepsilon\ \gamma \to \beta, \qquad x\ \varepsilon\ \gamma,$$

for some type-scheme γ. But this implies

$$\gamma \to \beta \equiv \alpha \equiv \gamma,$$

which is impossible. □

EXERCISE 14.39. Prove that SKSI is stratified, and has principal type-scheme $(a \to b) \to a \to b$.

These examples and exercise should make it clear that the following two theorems hold; their proofs, although simple in principle, are complicated to describe formally and will not be given here; see Curry 1969a Theorem 1, Hindley 1969a Theorem 1.

THEOREM 14.40 (P.t.s. theorem).

(a) *Every stratified pure* CL-*term has a p.t.s. and a p.p.*

(b) *If* $\mathcal{B}_0$ *is a monoschematic basis, then every term stratified relative to* $\mathcal{B}_0$ *has a p.t.s. and a p.p. relative to* $\mathcal{B}_0$.

THEOREM 14.41. *The set of all stratified pure* CL-*terms is recursively decidable.*

Some terms and their p.t.s's

Term		Principal type-scheme	
I		$a \to a$	
K		$a \to b \to a$	
S		$(a \to b \to c) \to (a \to b) \to a \to c$	
B		$(b \to c) \to (a \to b) \to a \to c$	
C		$(b \to a \to c) \to (a \to b \to c)$	
W		$(a \to a \to b) \to a \to b$	
CI		$a \to (a \to b) \to b$	
CB		$(a \to b) \to (b \to c) \to (a \to c)$	
$\bar{\sigma}$	(≡ SB)	$((b \to c) \to (a \to b)) \to (b \to c) \to a \to c$	
Z_0	(≡ KI)	$b \to a \to a$	
Z_1	(≡ SB(KI))	$(a \to b) \to a \to b$	
Z_n	$(n > 1)$	$(a \to a) \to a \to a$	(≡ N_a)
D	(4.7)	$a_1 \to a_2 \to ((d \to a_2) \to (a_1 \to c)) \to c$	

Warning. The type-scheme for R_{Bernays} in 14.20(c) is not principal.

A term X is called <u>strongly normalizable</u> (SN), with respect to $\rhd_w$, iff all weak reductions starting at X are finite.

THEOREM 14.42 (<u>SN theorem for stratified terms and $\rhd_w$</u>).

(a) *Every stratified pure* CL-*term is* SN *with respect to* $\rhd_w$.

(b) *Every* CL-*term stratified with respect to a monoschematic or weakly-normal-subjects basis is* SN *with respect to* $\rhd_w$.

Proof. (a): by (b) for the empty basis. (b): Corollary 14.52.1. □

COROLLARY 14.42.1 (<u>Normalization theorem for stratified terms</u>).

(a) *Every stratified pure* CL-*term has a weak normal form.*

(b) *If* $\mathcal{B}$ *is a monoschematic or weakly-normal-subjects basis, then every* CL-*term stratified relative to* $\mathcal{B}$ *has a weak normal form.*

REMARK 14.43 (<u>Strong reduction and</u> I). In the literature on $\succ_{\beta\eta}$, I is always assumed to be an atom, with reduction-schemes

$$\text{I}X \rhd_w X, \qquad \text{I}X \succ_{\beta\eta} X;$$

also 'non-redex atom' means 'atom not I, K or S', and when TA_C is discussed, it is assumed to include an axiom-scheme

$$(\to \text{I}) \qquad\qquad \text{I} \in \alpha \to \alpha \,.$$

These changes are to make 'strongly irreducible' equivalent to 'in strong nf'; see 8.18 and 3.7. (They thus affect the definition of strongly-normal-subjects basis, 14.14.) All this chapter's results stay valid, as is easily checked. Also, although 14.42 does not easily extend to $\succ$ (see 14.56), its corollary does, as follows.

THEOREM 14.44. *Let* I *be an atom as in* 14.43. *Corollary* 14.42.1 *stays true if we replace 'weak', 'weakly' by 'strong', 'strongly'.*

Proof. Hindley, Lercher and Seldin 1975, Theorems 9.19 - 21. (In 9.19, the second 'M' should be 'X'.) By the way, 9.19 - 20 are corrected versions of the 9.19 - 20 in Hindley et al. 1972, which contain errors. (In 9.19, the Y in the conclusion should be a Y′ such that $Y \succ Y'$, and 9.20 needs a similar change. But 9.21 is correct.)

COROLLARY 14.44.1. *In* CL, *the* Y-*combinators in* 3.4 *are not stratified.*

Proof. Theorem 14.44 and Exercise 9.16(f). □

REMARK 14.45 (<u>Correspondence with Chapter 13</u>). Although it should be clear by now that stratified terms correspond to typed terms in the sense of Chapter 13, that correspondence is not one-to-one. For example the one stratified term K corresponds to an infinite number of typed terms $\mathsf{K}_{\alpha,\beta}$.

To get a natural one-to-one correspondence, we must pass from stratified terms to type-assignment deductions. Each pure typed term Y^α, with

$$FV(Y^\alpha) = \{v_1^{\delta_1},\ldots,v_n^{\delta_n}\},$$

determines, in an obvious way, a deduction giving

$$x_1 \in \delta_1,\ldots,x_n \in \delta_n \quad \vdash_{TA_C} \quad |Y^\alpha| \in \alpha,$$

where the x's correspond to the v's in the typed/untyped variables correspondence that was assumed at the start of this section. And conversely, from such a deduction we can construct a typed term.

Note that not every deduction corresponds to a typed term, however. A deduction may contain type-schemes that are not types, or assumptions $U \in \delta$ where U is not a variable; or even assumptions $x \in \delta$ where x corresponds, in the typed/untyped variables correspondence, to a typed variable v^θ with $\theta \not\equiv \delta$.

14D FORMULAS-AS-TYPES AND NORMALIZATION

DISCUSSION 14.46. In Curry and Feys 1958 §9E it is pointed out that if all subjects are removed from a deduction in TA_C, and $\to$ is interpreted as implication, then the result is a deduction in the pure implicational fragment of the intuitionistic propositional calculus. This is because the above transformation takes rule ($\to$e) to <u>modus ponens</u>,

$$\frac{\alpha\to\beta \qquad \alpha}{\beta},$$

and the axiom-schemes ($\to\mathsf{K}$) and ($\to\mathsf{S}$) to well-known theorem-schemes of the intuitionistic propositional calculus, namely

$$\alpha\to\beta\to\alpha, \qquad (\alpha\to\beta\to\gamma)\to(\alpha\to\beta)\to\alpha\to\gamma.$$

For example, if this transformation is carried out on the deduction in Example 14.6, the result is the following standard proof of $\alpha\to\alpha$:

$$\dfrac{\dfrac{(\alpha\to(\alpha\to\alpha)\to\alpha)\to(\alpha\to\alpha\to\alpha)\to\alpha\to\alpha \qquad \alpha\to(\alpha\to\alpha)\to\alpha}{(\alpha\to\alpha\to\alpha)} \qquad \alpha\to\alpha\to\alpha}{\alpha\to\alpha.}$$

Furthermore, the term SKK which has been deleted from the conclusion by the transformation determines the tree-structure of the deduction. Even better, if α is a type, then the whole propositional deduction can be coded as a single typed term

$$\mathsf{S}_{\alpha,\alpha\to\alpha,\alpha}\,\mathsf{K}_{\alpha,\alpha\to\alpha}\,\mathsf{K}_{\alpha,\alpha}\,.$$

Thus, roughly speaking, type-schemes correspond to propositional formulas, and typed terms to propositional deductions. (It is interesting to note that Theorem 14.19 corresponds to the deduction theorem, Mendelson 1964 Proposition 1.8.)

One corollary is that type-schemes which correspond to formulas unprovable in intuitionistic propositional logic, cannot be assigned to any closed pure term. Examples are

$$((a\to b)\to a)\to a, \quad a\to b, \quad a.$$

Since the publication of Curry and Feys 1958, a number of people have taken up this idea and extended it to other connectives and quantifiers. These include H. Läuchli (see his 1965 and 1970), N.G. de Bruijn (see his 1970 and 1980), D.S. Scott (his 1970b), and W.A. Howard, whose 1980 (written in 1969) is probably the best source for readers of this book. More on the idea can be found in Martin-Löf 1972.

DISCUSSION 14.47. The deductions/terms correspondence suggests that we should be able to reduce or simplify deductions just like terms, and perhaps define a concept of irreducible or 'simplest' deduction of a formula, corresponding to terms in normal form, and even prove Church-Rosser and strong normalization theorems for deductions, corresponding to those theorems for typed terms. And indeed all this can be done. (Troelstra 1973, Chapter IV.)

In fact, the theory of proof-reductions was begun independently of the correspondence with terms, by Dag Prawitz in his 1965, and it is now a standard tool in proof-theory. The correspondence with terms has helped illuminate parts of this theory, and it, in turn, has illuminated parts of the theory of typed terms, for example the strong normalization theorem, 13.22.

To get a little of its flavour, let us sketch the basic definitions for a reduction-theory of deductions. Since this book is not about propositional logic, we shall do it for TA_C-deductions, not propositional deductions.

DEFINITION 14.48 (Deduction-reductions for TA_C). A reduction of one deduction to another consists of a sequence of replacements by the following two reduction-rules:

K-reductions for deductions. A deduction of the form

$$\dfrac{\dfrac{K\ \varepsilon\ \alpha\to\beta\to\alpha \qquad \begin{array}{c}\mathcal{D}_1\\ X\ \varepsilon\ \alpha\end{array}}{KX\ \varepsilon\ \beta\to\alpha}\,(\to e) \qquad \begin{array}{c}\mathcal{D}_2\\ Y\ \varepsilon\ \beta\end{array}}{\begin{array}{c}KXY\ \varepsilon\ \alpha\\ \mathcal{D}_3\end{array}}\,(\to e)$$

may be reduced to

$$\begin{array}{c}\mathcal{D}_1\\ X\ \varepsilon\ \alpha\\ \mathcal{D}_3',\end{array}$$

where $\mathcal{D}_3'$ is obtained from $\mathcal{D}_3$ by replacing appropriate occurrences of KXY by X.

S-reductions for deductions. A deduction of the form

$$\dfrac{\dfrac{\dfrac{S\ \varepsilon\ (\alpha\to\beta\to\gamma)\to(\alpha\to\beta)\to\alpha\to\gamma \qquad \begin{array}{c}\mathcal{D}_1\\ X\ \varepsilon\ \alpha\to\beta\to\gamma\end{array}}{SX\ \varepsilon\ (\alpha\to\beta)\to\alpha\to\gamma}\,(\to e) \qquad \begin{array}{c}\mathcal{D}_2\\ Y\ \varepsilon\ \alpha\to\beta\end{array}}{SXY\ \varepsilon\ \alpha\to\gamma}\,(\to e) \qquad \begin{array}{c}\mathcal{D}_3\\ Z\ \varepsilon\ \alpha\end{array}}{\begin{array}{c}SXYZ\ \varepsilon\ \gamma\\ \mathcal{D}_4\end{array}}\,(\to e)$$

may be reduced to

$$\dfrac{\dfrac{\begin{matrix}\mathcal{D}_1 & \mathcal{D}_3\\ X\ \varepsilon\ \alpha\to\beta\to\gamma & Z\ \varepsilon\ \alpha\end{matrix}}{XZ\ \varepsilon\ \beta\to\gamma}(\to e) \qquad \dfrac{\begin{matrix}\mathcal{D}_2 & \mathcal{D}_3\\ Y\ \varepsilon\ \alpha\to\beta & Z\ \varepsilon\ \alpha\end{matrix}}{YZ\ \varepsilon\ \beta}(\to e)}{\begin{matrix}XZ(YZ)\ \varepsilon\ \gamma\\ \mathcal{D}'_4\end{matrix}}(\to e),$$

where $\mathcal{D}'_4$ is obtained from $\mathcal{D}_4$ by replacing appropriate occurrences of SXYZ by XZ(YZ).

REMARK 14.49. Let $\mathcal{D}$ be a deduction giving

$$\mathcal{B} \vdash_{TA_C} X\ \varepsilon\ \alpha.$$

(a) From the above definition it is clear that if a reduction of $\mathcal{D}$ is possible, then X must contain a weak redex. Also, if $\mathcal{D}$ reduces to $\mathcal{D}'$, the X will be weakly reduced, so $\mathcal{D}'$ will give

$$\mathcal{B} \vdash_{TA_C} X'\ \varepsilon\ \alpha \qquad (X \rhd_w X').$$

(b) Conversely, if X contains a weak redex, and $\mathcal{B}$ is a weakly-normal-subjects basis, then by the proof of the subject-reduction theorem (14.24), a reduction of $\mathcal{D}$ is possible. And if X weakly reduces to Y, then $\mathcal{D}$ can be reduced to a deduction giving

$$\mathcal{B} \vdash_{TA_C} Y\ \varepsilon\ \alpha.$$

(c) A deduction which cannot be reduced is called <u>normal</u>.

The most important property of deduction-reductions is that they cannot go on for ever. This can be proved directly, but here it will be deduced from the corresponding property for typed terms, via the following definition and lemma.

DEFINITION 14.50 (<u>Assignment of typed terms to deductions</u>). To each deduction $\mathcal{D}$ in TA_C, assign a typed term $T(\mathcal{D})$ as follows. $T(\mathcal{D})$ will encode just enough of the structure of $\mathcal{D}$ to serve in the proof of the SN theorem, and no more.

First, choose any atomic type from the definition of types in 13.1 (call it c), and change all predicates in $\mathcal{D}$ from type-schemes to types, by substituting c for all their type-variables. Call the result $\mathcal{D}'$.

Then, for each type α, choose one typed term-variable, call it v^{α}. Assign a typed term to each part of $\mathcal{D}'$, by induction, thus:

(a) to an assumption $x \in \delta$, assign v^{δ};

(b) to an assumption $U \in \delta$ where U is not a variable, assign v^{δ};

(c) to an axiom $K \in \alpha \to \beta \to \alpha$, assign $K_{\alpha,\beta}$;

(d) to an axiom $S \in (\alpha \to \beta \to \gamma) \to (\alpha \to \beta) \to \alpha \to \gamma$, assign $S_{\alpha,\beta,\gamma}$;

(e) to the conclusion of an application of Rule $(\to e)$, say

$$\frac{U \in \alpha \to \beta \qquad V \in \alpha}{UV \in \beta,}(\to e)$$

assign $(X^{\alpha \to \beta} Y^{\alpha})^{\beta}$, where $X^{\alpha \to \beta}$ has been assigned to the premise $U \in \alpha \to \beta$, and Y^{α} to the premise $V \in \alpha$.

The typed term $T(\mathcal{D})$ contains only one variable of each type (though that variable may occur many times), and no non-redex constants. But it contains all the occurrences of S and K that have been introduced into $\mathcal{D}$ by axioms, and this is enough to give us the following key lemma.

LEMMA 14.51. *Let* $\mathcal{D}, \mathcal{E}$ *be any* TA_C*-deductions, and let* $\mathcal{D}$ *reduce to* $\mathcal{E}$ *by one of the replacements in Definition* 14.48. *Then* $T(\mathcal{D})$ *reduces to* $T(\mathcal{E})$ *by one weak contraction.*

Proof. Exercise. □

THEOREM 14.52 (Strong normalization theorem for deductions). *Every reduction of a* TA_C*-deduction terminates in a normal deduction.*

Proof. Suppose we have an infinite reduction of deductions. Then, by Lemma 14.51, we shall get an infinite weak reduction of typed terms. But this contradicts the strong normalization theorem for these terms, Theorem 13.22 or A2.5. □

COROLLARY 14.52.1. *If* $\mathcal{B}$ *is a weakly-normal-subjects basis, and*

$$\mathcal{B} \vdash_{TA_C} X \in \alpha,$$

then X *is strongly normalizable w.r. to weak reduction.*

Proof. By the theorem, and Remark 14.49(b). □

REMARK 14.53. If the condition of the Corollary that $\mathcal{B}$ be a weakly-normal-subjects basis (Definition 14.14) be abandoned, then it is possible to have a weak redex in a term X for which there is a normal deduction of $\mathcal{B} \vdash X \in \alpha$, so the Corollary may fail. For example, if $\mathcal{B} = \{YK \in \alpha\}$ for some α, then there would be a one-step normal deduction of

$$\mathcal{B} \vdash_{TA_C} YK \in \alpha,$$

yet YK has no weak normal form. However, there are cases in which the hypothesis of the corollary is not satisfied but its conclusion is true or partly true. Two such cases are discussed in the following two remarks.

REMARK 14.54 (Bases with a universal type). Suppose a basis $\mathcal{B}$ contains a formula $X \in \omega$ for each term X, where ω is a type-constant. Then $\mathcal{B}$ is clearly not a weakly-normal-subjects basis. However, if the part of $\mathcal{B}$ left over after all of the formulas $X \in \omega$ are removed is a weakly-normal-subjects basis, then in any normal deduction, the weak redexes in the term in the conclusion will all be in components assigned type ω in the deduction. (See Seldin 1977b, pp. 26-27, where, as in the work of Curry, ω is called E.) This fact has been used by Coppo et al. 1979 in a generalization of this system in which all terms have types, and the positions of the ω's in a term's type indicate important aspects of the term's reduction-behaviour, for example, whether it has a normal form. (See 15.41(e).)

REMARK 14.55 (Bases with proper inclusions). A basis $\mathcal{B}$ which contains a proper inclusion $I \in \alpha \to \beta$ (see Example 14.17) is clearly not a weakly-normal-subjects basis. Furthermore, it is clear that if such an assumption is used in a normal deduction of a formula $X \in \gamma$, there can easily be a redex in X. For example, let a, b, c, and d be type-variables, let G and H be non-redex atoms, and let $\mathcal{B}$ be the basis

$$\mathcal{B} = \{I \in (a \to b) \to (c \to d),\ G \in b,\ H \in c\}.$$

Then, using the axiom $K \in b \to a \to b$, we can easily deduce

$$\mathcal{B} \vdash I(KG)H \in d$$

by a normal deduction. Yet, I(KG) is a redex; furthermore, if this redex is contracted to KG, then the term becomes KGH, which is also a redex. This shows that contracting all redexes of the form IU need not lead to a term in normal form. However, if that part of $\mathcal{B}$ left over when the proper inclusions are removed is a weakly-normal-subjects basis, then $\mathcal{B} \vdash X \varepsilon \gamma$ implies that X has a normal form, provided that each proper inclusion $I \varepsilon \alpha \to \beta$ in $\mathcal{B}$ satisfies either of the following two conditions: (1) β is a type-constant; or (2) for each term U such that $\mathcal{B} \vdash U \varepsilon \alpha$ and for each natural number n, each reduction of $Ux_1 \ldots x_n$ proceeds entirely inside U. (See Seldin 1977b Remark 2, p. 23.) Proper inclusions satisfying (1) can occur in a system of transfinite type theory, where a type-constant T is made into a transfinite type by postulating $I \varepsilon \alpha \to T$ for each finite type α. To see an application of condition (2), it is necessary to consider a type theory that is to include statements about types. In this case, each type must be a term. This can be accomplished by making each type-constant a non-redex constant, each type-variable an ordinary term-variable, and taking another non-redex constant, say F (Curry's notation), so that $\gamma \to \delta$ will be regarded as an abbreviation for $F\gamma\delta$. Then if H is the type-constant of propositions, we can turn each type into a propositional function by making a new type-constant L, postulating $\alpha \varepsilon L$ for each type-scheme α in which L does not occur, and postulating the proper inclusion

$$I \varepsilon L \to \alpha \to H$$

for each type-scheme α in which L does not occur. Type theories with both of these properties occur in Andrews 1965 and in Curry et al. 1972 Chapter 17.

REMARK 14.56 (Extension to $\beta\eta$-strong reduction). Since the notion of SN has not yet been defined for $\beta\eta$-strong reduction in the literature, (see Remark A2.8 in Appendix 2), we do not have SN results for $\beta\eta$-strong reduction. But we can obtain normalization results for deductions. (Seldin 1977b Remark 3, p. 24.) First, as usual when considering strong reduction, we change the system to one in which I is an atom (Remark 14.43). Then the key step is to add to the definition of deduction-reduction (Definition 14.48) the following additional reduction rule:

<u>$\beta\eta$-strong reductions for deductions</u>. If

$$\begin{array}{ccc} x \in \alpha & & x \in \alpha \\ \mathcal{D}_1 & \text{reduces to} & \mathcal{D}_2 \\ X \in \beta & & Y \in \beta \end{array}$$

and if

$$\begin{array}{ccc} \mathcal{D}'_1 & \text{and} & \mathcal{D}'_2 \\ \lambda^{\eta}x.X \in \alpha\to\beta & & \lambda^{\eta}x.Y \in \alpha\to\beta \end{array}$$

are the corresponding deductions obtained by Theorem 14.19, then

$$\begin{array}{ccc} \mathcal{D}'_1 & & \mathcal{D}'_2 \\ \lambda^{\eta}x.X \in \alpha\to\beta & \text{reduces to} & \lambda^{\eta}x.Y \in \alpha\to\beta \\ \mathcal{D}_3 & & \mathcal{D}'_3, \end{array}$$

where $\mathcal{D}'_3$ is obtained from $\mathcal{D}_3$ by replacing appropriate occurrences of $\lambda^{\eta}x.X$ by $\lambda^{\eta}x.Y$.

It is also necessary, of course, to refer to strongly-normal-subjects bases instead of weakly-normal-subjects bases.

14E THE EQUALITY-RULE Eq′

As we have seen in Remark 14.26, TA_C is not invariant under combinatory equality. For example, SKSI does not have all of the type schemes that KI(SI) has; furthermore, although II(II) has p.t.s. $a\to a$, SIII is unstratified. This is a defect in a system using combinators, for the usefulness of combinators comes from the transformations that can be carried out using their reduction properties. To remedy this defect we can add the following rule:

<u>Rule</u> Eq′
$$\frac{X \in \alpha \qquad X =_{*} Y}{Y \in \alpha.}$$

Of course, this is really three rules: one when $=_{*}$ is weak equality, a second when it is $=_{c\beta}$, and a third when it is $=_{c\beta\eta}$. When we need to distinguish them, they will be called, respectively,

Eqw′, Eqβ′, Eq$\beta\eta$′.

DEFINITION 14.57 (The systems $TA_{C=}$). These systems are obtained from TA_C (Definition 14.5) as follows:

$TA_{C=w}$: add rule Eqw′ (and the axioms and rules for $=_w$);

$TA_{C=\beta}$: add I as a new atom (see Remark 14.43), and add rule Eqβ′;

$TA_{C=\beta\eta}$: add I as a new atom, and add rule Eqβη′.

The name $TA_{C=}$ will mean any or all of these systems, according to context.

REMARK 14.58. Note that $TA_{C=}$ is undecidable, unlike TA_C which is decidable. The reason for this is that in systems with rule Eq′, deductions need not follow the constructions of the terms as they do in TA_C (see Note 14.18), because it is possible for a deduction in $TA_{C=}$ to consist of a deduction in TA_C followed by an inference by rule Eq′. Since $=_*$ is undecidable, so is $TA_{C=}$.

DISCUSSION 14.59. It might seem that, since Rule Eq′ can occur anywhere in a deduction, $TA_{C=}$ is a much richer system than TA_C. But this is not the case. Every deduction in $TA_{C=}$ can be replaced by one in which Rule Eq′ occurs only at the end. For, suppose an inference by Eq′ occurs before the end of a deduction. Then its conclusion is a premise for an inference by $(\to e)$, or else for another inference by Eq′. Now since $=_*$ is transitive, successive inferences by Rule Eq′ can always be combined into one; so we may assume that the next rule is $(\to e)$, and our deduction has one of the forms

$$\dfrac{\dfrac{\begin{array}{c}\mathcal{D}_1\\ X \in \alpha\to\beta\end{array} \qquad X =_* Y}{Y \in \alpha\to\beta}\,\text{Eq}' \qquad \begin{array}{c}\mathcal{D}_2\\ Z \in \alpha\end{array}}{\begin{array}{c}YZ \in \beta\\ \mathcal{D}_3\end{array}}\,(\to e)$$

or

$$\dfrac{\begin{array}{c}\mathcal{D}_2\\ Z\ \varepsilon\ \alpha\to\beta\end{array}\qquad \dfrac{\begin{array}{c}\mathcal{D}_1\\ X\ \varepsilon\ \alpha\end{array}\qquad X =_* Y}{Y\ \varepsilon\ \alpha}\,\mathrm{Eq}'}{\begin{array}{c}ZY\ \varepsilon\ \beta\\ \mathcal{D}_3\,.\end{array}}\,(\to e)$$

These forms can be replaced, respectively, by

$$\dfrac{\dfrac{\begin{array}{c}\mathcal{D}_1\\ X\ \varepsilon\ \alpha\to\beta\end{array}\qquad \begin{array}{c}\mathcal{D}_2\\ Z\ \varepsilon\ \alpha\end{array}}{XZ\ \varepsilon\ \beta}\,(\to e)\qquad X =_* Y}{\begin{array}{c}YZ\ \varepsilon\ \beta\\ \mathcal{D}_3\end{array}}\,\mathrm{Eq}'$$

or

$$\dfrac{\dfrac{\begin{array}{c}\mathcal{D}_2\\ Z\ \varepsilon\ \alpha\to\beta\end{array}\qquad \begin{array}{c}\mathcal{D}_1\\ X\ \varepsilon\ \alpha\end{array}}{ZX\ \varepsilon\ \beta}\,(\to e)\qquad X =_* Y}{\begin{array}{c}ZY\ \varepsilon\ \beta\\ \mathcal{D}_3\,.\end{array}}\,\mathrm{Eq}'$$

If these replacements are made systematically in any deduction beginning at the top, and if consecutive inferences by Rule Eq′ are combined whenever they occur in this process, we eventually wind up with a new deduction of the same formula in which there is at most one inference by Rule Eq′, and that inference occurs at the end. This proves the following theorem.

THEOREM 14.60 (Eq′-postponement theorem). *If* $=_*$ *is* $=_w$ *or* $=_{c\beta}$ *or* $=_{c\beta\eta}$, *and*

$$\mathcal{B} \vdash_{TA_{C=}} X\ \varepsilon\ \alpha,$$

where $\mathcal{B}$ *is any basis, then there is a term* Y *such that* $Y =_* X$ *and*

$$\mathcal{B} \vdash_{TA_C} Y\ \varepsilon\ \alpha.$$

COROLLARY 14.60.1 (Normalization theorem). *If* $\mathcal{B}$ *is a weakly-* [*strongly-*] *normal-subjects basis, and if*

$$\mathcal{B} \vdash X\ \varepsilon\ \alpha$$

in $TA_{C=w}$ [$TA_{C=\beta\eta}$], *then* X *has a weak* [*strong*] *normal form.*

REMARK 14.61.

(a) At present there is no definition of β-normal form in CL, so Corollary 14.60.1 does not mention β-equality.

(b) Corollary 14.60.1 cannot be extended to conclude that X is SN, because there are terms with normal forms which are not SN. An example is $X \equiv Y(KI)$. Now X has a normal form since

$$X \rhd KIX \rhd I,$$

but it also has an infinite reduction:

$$X \rhd KIX \rhd KI(KIX) \rhd KI(KI(KIX)) \rhd \ldots .$$

We clearly have

$$\vdash_{TA_{C=}} X \, \varepsilon \, \alpha \to \alpha$$

for each type-scheme α, but X is not SN.

The definitions of <u>stratified</u>, <u>p.t.s.</u>, <u>p.p.</u> are exactly the same for the present system as for TA_C (Definitions 14.28, 14.31), but with $TA_{C=}$-deducibility instead of TA_C-deducibility. The following theorem then comes easily from the p.t.s. theorem for TA_C.

THEOREM 14.62 (<u>P.t.s. theorem for</u> $TA_{C=}$). *Let* $=_*$ *be* $=_w$ *or* $=_{c\beta}$ *or* $=_{c\beta\eta}$.

(a) *Every stratified pure* CL-*term has a p.p. and a p.t.s. in* $TA_{C=}$.

(b) *If* $\mathcal{B}_0$ *is a monoschematic basis, then every term stratified relative to* $\mathcal{B}_0$ *has a p.t.s. and a p.p. relative to* $\mathcal{B}_0$.

REMARK 14.63. Although this theorem may appear to be the same as the p.t.s. theorem for TA_C (14.40), it is not quite. For terms may have different p.t.s's in $TA_{C=}$ than in TA_C. For example, the p.t.s. of SKSI in TA_C is

$$(a \to b) \to a \to b,$$

whereas in $TA_{C=}$ it is $a \to a$. (See Exercise 14.34).

Also, the class of stratified terms in $TA_{C=}$ differs from that in TA_C. In fact, the classes are related by the following theorem.

THEOREM 14.64. *Let* $=_*$ *be* $=_w$ *or* $=_{c\beta\eta}$. *Then a pure* CL-*term* X *is stratified in* $TA_{C=}$ *iff* X *has a normal form* X^* *which is stratified in* TA_C. *Further, the type-schemes that* $TA_{C=}$ *assigns to* X *are exactly those that* TA_C *assigns to* X^*.

This theorem and the Eq′-postponement theorem show that the study of $TA_{C=}$ reduces to the study of TA_C, which is much easier. This is why so much of the present chapter was devoted to TA_C, despite the fact that TA_C is incomplete with respect to equality.

FURTHER READING. See the end of the next chapter.

CHAPTER FIFTEEN

TYPE ASSIGNMENT TO λ-TERMS

15A INTRODUCTION

This chapter will do for λ-terms what Chapter 14 does for CL-terms. There is not a major difference in either the basic ideas or the main results between the two chapters, but there is a major technical complication in the proofs, caused by the fact that λ-terms have bound variables while CL-terms do not.

In this chapter the terms are λ-terms exactly as defined in Chapter 1, not the typed terms of Chapter 13. Recall the abbreviations

$$\mathsf{I} \equiv \lambda x.x, \qquad \mathsf{K} \equiv \lambda xy.x, \qquad \mathsf{S} \equiv \lambda xyz.xz(yz).$$

Types and type-schemes are defined exactly as in Definition 14.1. The basic conventions for type-schemes are those of Notation 14.2. Type-schemes are interpreted exactly as in Remark 14.3. Finally, type-assignment formulas are as defined in Definition 14.4, except that now their subjects are λ-terms, not CL-terms.

15B THE SYSTEM TA_λ

DISCUSSION 15.1 (<u>The →-introduction rule</u>). A comparison of Definition 14.5 and Definition 13.16 shows that the axiom-schemes $(\rightarrow \mathsf{K})$ and $(\rightarrow \mathsf{S})$ of the former correspond to the constants $\mathsf{K}_{\alpha,\beta}$ and $\mathsf{S}_{\alpha,\beta,\gamma}$ of the latter, and that rule $(\rightarrow e)$ of the former corresponds to the construction of $(X^{\alpha\rightarrow\beta}Y^{\alpha})^{\beta}$ in the latter. To assign types to λ-terms, we will not need $(\rightarrow \mathsf{K})$ and $(\rightarrow \mathsf{S})$, but instead we will need a new rule corresponding to the construction of

$$(\lambda x^{\alpha}.M^{\beta})^{\alpha\rightarrow\beta}$$

in the definition of typed λ-terms (13.3). This new rule will not be a straightforward rule like $(\rightarrow e)$, but an introduction-rule in the style of Gerhard Gentzen's 'Natural Deduction' systems; see Prawitz 1965. This rule is called →-introduction or $(\rightarrow i)$, and says

$$(\rightarrow i)\qquad \begin{cases} \text{If } x \notin FV(L_1 \ldots L_n) \text{ and} \\ L_1 \in \delta_1, \ldots, L_n \in \delta_n, x \in \alpha \ \vdash\ M \in \beta, \\ \text{then} \\ L_1 \in \delta_1, \ldots, L_n \in \delta_n \ \vdash\ (\lambda x.M) \in (\alpha \rightarrow \beta). \end{cases}$$

It is usually written thus:

$$(\rightarrow i)\qquad \frac{\begin{matrix}[x \in \alpha] \\ M \in \beta\end{matrix}}{\lambda x.M \ \in \alpha \rightarrow \beta}\,.$$

This needs some explanation. Gentzen's Natural Deduction systems are very like formal theories as defined in Chapter 6, but they are not quite the same. A deduction is a tree of formulas just like deductions in Notation 6.1, but Gentzen allowed that some of the assumptions at the branch-tops may be only temporary assumptions, to be employed at an early stage in the deduction and then discharged (or cancelled) at a later stage. After an assumption has been discharged, it is marked in some way; we shall enclose it in brackets.

In such a system, rule $(\rightarrow i)$ is read as 'If $x \notin FV(L_1 \ldots L_n)$, and $M \in \beta$ is the conclusion of a deduction whose not-yet-discharged assumptions are $x \in \alpha, L_1 \in \delta_1, \ldots, L_n \in \delta_n$, then you may deduce

$$(\lambda x.M) \in (\alpha \rightarrow \beta),$$

and whenever the assumption $x \in \alpha$ occurs undischarged at a branch-top above $M \in \beta$, you must enclose it in brackets to show that it has now been discharged.'

When an assumption has been discharged, it ceases to count as an assumption. More formally, in a completed deduction-tree in a natural-deduction system, some formulas at branch-tops may be enclosed in brackets; and if $\mathcal{B}$ is any set of formulas, the notation

$$\mathcal{B} \ \vdash\ M \in \alpha$$

is defined to mean that there is a deduction-tree whose bottom formula is $M \varepsilon \alpha$, and whose <u>unbracketed</u> top-formulas are members of $\mathcal{B}$. As usual, when $\mathcal{B}$ is empty we say

$$\vdash \quad M \varepsilon \alpha.$$

Here are three examples. Further details on Natural Deduction are in Prawitz 1965, Chapter 1.

EXAMPLE 15.2. In any system containing rules $(\to e)$ and $(\to i)$, we have, for all type-schemes α, β, γ,

$$\vdash \quad S \varepsilon (\alpha\to\beta\to\gamma)\to(\alpha\to\beta)\to\alpha\to\gamma.$$

Proof. The following is a deduction of the required formula. For ease of reading, each assumption is numbered, and whenever $(\to i)$ is used, the number of the assumption it discharges is shown; e.g. '$(\to i - 2)$'.

$$\dfrac{\dfrac{\dfrac{\dfrac{\dfrac{\overset{3}{[x \varepsilon \alpha\to\beta\to\gamma]} \quad \overset{1}{[z \varepsilon \alpha]}}{xz \varepsilon \beta\to\gamma}(\to e) \qquad \dfrac{\overset{2}{[y \varepsilon \alpha\to\beta]} \quad \overset{1}{[z \varepsilon \alpha]}}{yz \varepsilon \beta}(\to e)}{xz(yz) \varepsilon \gamma}(\to e)}{\lambda z.xz(yz) \varepsilon (\alpha\to\gamma)}(\to i - 1)}{\lambda yz.xz(yz) \varepsilon (\alpha\to\beta)\to\alpha\to\gamma}(\to i - 2)}{\lambda xyz.xz(yz) \varepsilon (\alpha\to\beta\to\gamma)\to(\alpha\to\beta)\to\alpha\to\gamma}(\to i - 3). \qquad \square$$

<u>Comments</u>.

(a) Although all the branch-top formulas have brackets in the completed deduction above, each one starts life without brackets around it, and only receives brackets when it is discharged after a use of $(\to i)$.

(b) Only formulas at branch-tops can be discharged, never those in the body of the deduction.

(c) The rule-names and the numbers in the above deduction are included merely for the reader's convenience; if they were omitted, the form of the rule would still show which rule it was, and which assumptions (if any) it discharged.

(d) The conditions in Rule $(\to i)$ prevent the use of $(\to i)$ when x occurs free in two distinct undischarged assumptions, e.g. $x \varepsilon \alpha$, $x \varepsilon \alpha\to\beta$.

EXAMPLE 15.3. In any system containing $(\to e)$ and $(\to i)$,

$$\vdash \ \mathsf{K} \ \varepsilon \ \alpha \to \beta \to \alpha.$$

Proof. Here is a deduction of the required formula. In it, the first application of $(\to i)$ discharges all assumptions $y \ \varepsilon \ \beta$ that occur. But none in fact occur, so nothing is discharged. This is perfectly legitimate; it is called 'vacuous discharge', and is shown by '$(\to i - v)$'.

$$\dfrac{\dfrac{\overset{1}{[x \ \varepsilon \ \alpha]}}{\lambda y.x \ \varepsilon \ \beta \to \alpha}(\to i - v)}{\lambda xy.x \ \varepsilon \ \alpha \to \beta \to \alpha}(\to i - 1). \qquad \square$$

EXAMPLE 15.4. In any system containing $(\to e)$ and $(\to i)$,

$$\vdash \ \mathsf{I} \ \varepsilon \ \alpha \to \alpha.$$

Proof.

$$\dfrac{\overset{1}{[x \ \varepsilon \ \alpha]}}{\lambda x.x \ \varepsilon \ \alpha \to \alpha}(\to i - 1). \qquad \square$$

DISCUSSION 15.5. Before we come to define the type-assignment system, we need to consider one further rule. Since two terms that are α-convertible (Definition 1.16) are intended to represent the same operation, any two such terms should be assigned exactly the same types. That is, we want an α-invariance property:

(a) $$\mathcal{B} \vdash M \ \varepsilon \ \beta, \ M \equiv_{\alpha} N \quad \Rightarrow \quad \mathcal{B} \vdash N \ \varepsilon \ \beta$$

(where $\mathcal{B}$ is any set of formulas). If all the terms in $\mathcal{B}$ are atoms, then α-invariance will turn out to be provable by induction on the lengths of deductions. But sometimes it will be interesting to consider more general sets $\mathcal{B}$. For example, if there is an atomic type N for the natural numbers, we will want to assume

$$\bar{0} \ \varepsilon \ \mathsf{N} \qquad\qquad (\bar{0} \equiv \lambda xy.y).$$

The α-invariance property will then fail unless we also assume

$$\lambda xz.z \ \varepsilon \ \mathsf{N}, \qquad \lambda uv.v \ \varepsilon \ \mathsf{N},$$

etc. But this makes our set of assumptions infinite, in a rather boring way. To avoid this, we shall postulate a formal rule which, in effect, closes every set of assumptions under α-conversion. (It is tempting to simply postulate (a) as an unrestricted rule, but this would make the subject-construction property harder to state and use.)

DEFINITION 15.6 (The type-assignment system TA_λ). TA_λ is a Natural Deduction system. Its formulas, called type-assignment (TA) formulas, are expressions

$$M \in \alpha$$

for all λ-terms M and type-schemes α. (M is called the formula's subject, and α its predicate.) TA_λ has no axioms. Its rules are

(→e) $$\frac{M \in \alpha \to \beta \qquad N \in \alpha}{MN \in \beta}$$

(→i) $$\frac{\begin{matrix}[x \in \alpha] \\ M \in \beta\end{matrix}}{\lambda x.M \in \alpha \to \beta}$$ Condition: $x \in \alpha$ is the only undischarged assumption in whose subject x occurs free. After (→i) is used, every occurrence of $x \in \alpha$ above $M \in \beta$ is called 'discharged'.

($\equiv'_\alpha$) $$\frac{M \in \beta \qquad M \equiv_\alpha N}{N \in \beta}$$ Condition: $M \in \beta$ is not the conclusion of a rule.

(Strictly speaking, TA_λ also contains axioms and rules to define α-conversion, but we shall leave these to the imagination.) For any finite or infinite set $\mathcal{B}$ of formulas, the notation

$$\mathcal{B} \vdash_{TA_\lambda} M \in \alpha$$

means that there is a deduction of $M \in \alpha$ whose undischarged assumptions are members of $\mathcal{B}$. If $\mathcal{B}$ is empty, we say simply

$$\vdash_{TA_\lambda} M \in \alpha.$$

EXAMPLE 15.7. Recall that $B \equiv \lambda xyz.x(yz)$. Then for all type-schemes α, β, γ,

$$\vdash_{TA_\lambda} \quad B \ \varepsilon\ (\beta\to\gamma)\to(\alpha\to\beta)\to\alpha\to\gamma.$$

Proof.

$$\dfrac{\dfrac{\dfrac{\dfrac{\overset{3}{[x\ \varepsilon\ \beta\to\gamma]} \qquad \dfrac{\overset{2}{[y\ \varepsilon\ \alpha\to\beta]} \quad \overset{1}{[z\ \varepsilon\ \alpha]}}{yz\ \varepsilon\ \beta}(\to e)}{x(yz)\ \varepsilon\ \gamma}(\to e)}{\lambda z.x(yz)\ \varepsilon\ \alpha\to\gamma}(\to i-1)}{\lambda yz.x(yz)\ \varepsilon\ (\alpha\to\beta)\to\alpha\to\gamma}(\to i-2)}{\lambda xyz.x(yz)\ \varepsilon\ (\beta\to\gamma)\to(\alpha\to\beta)\to\alpha\to\gamma}(\to i-3). \qquad \square$$

EXERCISE 15.8. For each of the terms on the left below, give a TA_λ-deduction to show that it has all the type-schemes on the right (one scheme for each α, β, γ). Cf. Exercise 14.8.

	Term		Type-schemes
(a)	$Z_0 \equiv \lambda xy.y$	(cf. 4.2)	$\beta\to\alpha\to\alpha$;
(b)	$\bar{\sigma} \equiv \lambda uxy.x(uxy)$	(cf. 4.2)	$((\beta\to\gamma)\to\alpha\to\beta)\to\ (\beta\to\gamma)\to\alpha\to\gamma$;
(c)	$W \equiv \lambda xy.xyy$	(cf. 2.11)	$(\alpha\to\alpha\to\beta)\to\alpha\to\beta$;
(d)	$\lambda xyz.y$		$\alpha\to\beta\to\gamma\to\beta$;
(e)	$Z_0 \equiv \lambda xy.y$		$N_\alpha \quad (N_\alpha \equiv (\alpha\to\alpha)\to\alpha\to\alpha)$;
(f)	$\bar{\sigma} \equiv \lambda uxy.x(uxy)$		$N_\alpha\to N_\alpha$;
(g)	$Z_n \equiv \lambda xy.x^n y$		N_α.

DEFINITION 15.9 (<u>Kinds of bases</u>). <u>Monoschematic</u> bases are defined as in Definition 14.14. A β- [$\beta\eta$-] <u>normal-subjects basis</u> is one in which each subject is in β- [$\beta\eta$-] normal form, and does not begin with a λ.

Every monoschematic basis is both a β- and $\beta\eta$-normal-subjects basis.

LEMMA 15.10 (Substitution). *Let* $\mathcal{B}$ *be any basis, and let*

$$\mathcal{B} \vdash_{TA_\lambda} X \varepsilon \alpha.$$

Then

$$[\beta_1/a_1,\ldots,\beta_n/a_n]\mathcal{B} \vdash_{TA_\lambda} X \varepsilon [\beta_1/a_1,\ldots,\beta_n/a_n]\alpha.$$

Proof. Like the proof of Lemma 14.11, with the additional remark that substitution creates from an instance of $(\to i)$ or $(\equiv'_\alpha)$ a new instance of $(\to i)$ or $(\equiv'_\alpha)$. □

LEMMA 15.11 (α-invariance). *Let* $\mathcal{B}$ *be any basis, and let*

$$\mathcal{B} \vdash_{TA_\lambda} M \varepsilon \beta,$$

and $M \equiv_\alpha N$. *Then*

$$\mathcal{B} \vdash_{TA_\lambda} N \varepsilon \beta.$$

Proof. Curry et al 1972, §14D3, Case 1, replacing assumption (ii) in that proof by rule $(\equiv'_\alpha)$. □

NOTE 15.12 (The subject-construction property). As with TA_C, the deduction of a formula $X \varepsilon \alpha$ in TA_λ closely follows the construction of X. (See Examples 15.2, 3, 4, 7.) The only extra complication here is rule $(\equiv'_\alpha)$, and this can only be used at the top of a branch in a deduction-tree. Indeed, if all the subjects of the assumptions are atoms, it cannot be used at all. Just as with TA_C, we shall not state the property in full formal detail here, but merely give some simple examples of its use. (The property for a system without the α – conversion rule is expressed formally in Curry et al. 1972, §14D2, subject-construction theorem.)

EXAMPLE 15.13. Every type-scheme assigned to I $(\equiv \lambda x.x)$ by TA_λ has the form $\alpha \to \alpha$.

Proof. By the construction of I, every type assigned to it must be compound, and the last inference in the deduction must be by $(\to i)$. In other words, if there is a deduction of

$$\vdash_{TA_\lambda} I \varepsilon \alpha \to \beta,$$

then removing the last inference from the deduction leaves a deduction of

$$x \varepsilon \alpha \vdash x \varepsilon \beta.$$

But this latter deduction can only be a deduction with one step, and it follows that $\beta \equiv \alpha$. □

EXERCISE 15.14. Prove that every type-scheme assigned to B ($\equiv \lambda xyz.x(yz)$) by TA_λ has the form $(\beta \to \gamma) \to (\alpha \to \beta) \to \alpha \to \gamma$.

EXERCISE 15.15. Prove that there is no type-scheme assigned by TA_λ to $Y_{Curry} \equiv \lambda x.VV$, where $V \equiv \lambda y.x(yy)$ (Definition 3.4).

Like TA_C, TA_λ preserves type-schemes under reductions, though here, of course, the reductions are β- and $\beta\eta$-reductions. And, as in TA_C, we need a replacement lemma to prove this, though here the presence of bound variables will complicate the lemma's statement and proof.

LEMMA 15.16 (Replacement). *Let* $\mathcal{B}_1$ *be any basis, and* $\mathcal{D}$ *be a deduction giving*

$$\mathcal{B}_1 \vdash_{TA_\lambda} X \varepsilon \alpha.$$

Let V *be a term-occurrence in* X, *and let* $\lambda x_1, \ldots, \lambda x_n$ *be those* λ's *in* X *whose scope contains* V. *Let* $\mathcal{D}$ *contain a formula* $V \varepsilon \gamma$ *in the same position as* V *has in the construction-tree for* X, *and let*

$$x_1 \varepsilon \delta_1, \ldots, x_n \varepsilon \delta_n$$

be the assumptions above $V \varepsilon \gamma$ *that are discharged by applications of* $(\to i)$ *below it. Assume that* $V \varepsilon \gamma$ *is not in* $\mathcal{B}_1$. *Let* W *be a term such that* $FV(W) \subseteq FV(V)$, *and let* $\mathcal{B}_2$ *be a basis not containing* $x_1, \ldots, x_n$ *free. Let*

$$\mathcal{B}_2, x_1 \varepsilon \delta_1, \ldots, x_n \varepsilon \delta_n \vdash_{TA_\lambda} W \varepsilon \gamma.$$

Let X^* *be the result of replacing* V *by* W *in* X. *Then*

$$\mathcal{B}_1 \cup \mathcal{B}_2 \vdash_{TA_\lambda} X^* \varepsilon \alpha.$$

Proof. A full proof requires a careful induction on X. Here is an outline. First cut off from $\mathcal{D}$ the subtree above the formula $V \varepsilon \gamma$. The result is a tree $\mathcal{D}_1$ with form

$$\begin{array}{c} V \varepsilon \gamma \\ \mathcal{D}_1 \\ X \varepsilon \alpha. \end{array}$$

And, since $V \varepsilon \gamma$ is not in $\mathcal{B}_1$, the first step below $V \varepsilon \gamma$ cannot be rule $(\equiv'_\alpha)$. In fact, that rule cannot be used anywhere below $V \varepsilon \gamma$. Replace V by W in $V \varepsilon \gamma$, and in all formulas below it in $\mathcal{D}_1$; then take the given deduction of $W \varepsilon \gamma$ and place it above. The result is a tree, leading from assumptions in $\mathcal{B}_1 \cup \mathcal{B}_2$ to the conclusion $X^* \varepsilon \alpha$. And each $(\rightarrow i)$-step in this tree satisfies the conditions required in this rule, because $x_1, \ldots, x_n$ do not occur in $\mathcal{B}_2$. So the tree is a deduction. □

THEOREM 15.17 (Subject-reduction theorem). *Let* $\mathcal{B}$ *be a* β- $[\beta\eta$-$]$ *normal-subjects basis. If*

$$\mathcal{B} \vdash_{TA_\lambda} X \varepsilon \alpha$$

and $X \rhd_\beta X'$ $[X \rhd_{\beta\eta} X']$, *then*

$$\mathcal{B} \vdash_{TA_\lambda} X' \varepsilon \alpha.$$

Proof. If the step from X to X′ is an α-conversion, use Lemma 15.11. Now suppose X βη-contracts to X′. By the replacement lemma, it is enough to take care of the case that X is a redex and X′ is its contractum. By Lemma 15.11, we can assume that no variable bound in XX′ occurs free in a subject in $\mathcal{B}$.

Case 1: X is a β-redex, say $(\lambda x.M)N$, and X′ is $[N/x]M$. Since $\mathcal{B}$ is a β- [βη-] normal-subjects basis, the formula $X \varepsilon \alpha$ is not in $\mathcal{B}$, and is not the conclusion of the α-rule. Hence, it is the conclusion of an inference by $(\rightarrow e)$, for which the premises are

$$\mathcal{B} \vdash \lambda x.M \varepsilon \beta\rightarrow\alpha, \qquad \mathcal{B} \vdash N \varepsilon \beta.$$

Now the formula $\lambda x.M \varepsilon \beta\rightarrow\alpha$ is also not in $\mathcal{B}$, and is not the conclusion of $(\equiv'_\alpha)$, so the first of these premises must be the conclusion of an inference by $(\rightarrow i)$, the premise for which is

$$\mathcal{B}, x \varepsilon \beta \vdash M \varepsilon \alpha,$$

where x does not occur free in any subject in $\mathcal{B}$. The conclusion, which is

$$\mathcal{B} \vdash [N/x]M \varepsilon \alpha,$$

then follows by the replacement lemma.

Case 2: ($\beta\eta$-reduction only). X is an η-redex. Then $X \equiv \lambda x.Mx$, where x does not occur free in M, and $X' \equiv M$. Since $\mathcal{B}$ is a $\beta\eta$-normal-subjects basis, the formula $X \varepsilon \alpha$ is not in $\mathcal{B}$. Hence, it is the conclusion of an inference by $(\rightarrow i)$, where $\alpha \equiv \beta \rightarrow \gamma$ and the premise is

$$\mathcal{B}, x \varepsilon \beta \vdash Mx \varepsilon \gamma.$$

Since x does not occur free in any subject of $\mathcal{B}$, the formula $Mx \varepsilon \gamma$ does not occur in $\mathcal{B}$, and is not the conclusion of the α-rule; hence it must be the conclusion of an inference by $(\rightarrow e)$ whose premises are

$$\mathcal{B}, x \varepsilon \beta \vdash M \varepsilon \delta \rightarrow \gamma, \qquad \mathcal{B}, x \varepsilon \beta \vdash x \varepsilon \delta.$$

By the second of these, and the fact that x does not occur free in any subject of $\mathcal{B}$, we have that $\delta \equiv \beta$, and hence the first of these is

$$\mathcal{B}, x \varepsilon \beta \vdash M \varepsilon \alpha.$$

Since x does not occur free in M or in any subject of $\mathcal{B}$, the assumption $x \varepsilon \beta$ is not used in this deduction and can thus be omitted from the list of undischarged assumptions. □

REMARK 15.18.

(a) The proof will still work if the condition on $\mathcal{B}$ is relaxed slightly, to say that every subject of $\mathcal{B}$ is in normal form and if a subject in $\mathcal{B}$ begins with a λ, then every type-scheme it receives in $\mathcal{B}$ is a type constant. An example of such a basis is

$$\mathcal{B} = \{Z_n \varepsilon N : n \varepsilon \mathbb{N}\},$$

where $Z_n \equiv \lambda xy.x^n y$, so the subject-reduction theorem holds for this basis although it is not a β- or $\beta\eta$-normal-subjects basis.

(b) An example of a basis for which the theorem fails is

$$Z_0 \varepsilon N, \qquad \bar{\sigma} \varepsilon N \rightarrow N,$$

since $Z_0 \equiv \lambda xy.y$ and $\bar{\sigma} \equiv \lambda uxy.x(uxy)$ both begin with λ, so the hypothesis of the theorem fails. The conclusion also fails, since from this basis it is possible to prove $\bar{\sigma}Z_0 \varepsilon N$, and $\bar{\sigma}Z_0 \rhd_\beta Z_1 \equiv \lambda xy.xy$, but it is impossible to prove $Z_1 \varepsilon N$.

REMARK 15.19. Theorem 15.17 cannot be reversed; it is not, in general, true that if $\vdash X \in \alpha$ and $X' \rhd_\beta X$ then $\vdash X' \in \alpha$. For example, let $X \equiv Z_o$ and let $X' \equiv \lambda xy.Ky(xy)$. Then, as indicated in Exercise 15.8(a), we have $\vdash X \in \beta\to\alpha\to\alpha$ for any type-scheme β, but $\vdash X' \in \beta\to\alpha\to\alpha$ only if $\beta \equiv \alpha\to\gamma$ for some type-scheme γ. An even stronger example is $X \equiv I$ and $X' \equiv (\lambda z.zz)I$, since we have $\vdash X \in \alpha\to\alpha$ whereas TA_λ assigns no type-scheme at all to X' (by Example 15.25 and the fact that TA_λ assigns a type to a term only if it assigns a type to each of its subterms.)

However, reversal is possible under certain very restricted conditions; see Curry et al. 1972 §14D4.

In §E we shall study a system defined by adding a rule of equality-invariance to TA_λ.

15C STRATIFIED λ-TERMS

The notion of stratified term is the same for TA_λ as for TA_C. However, for simplicity we shall only consider pure terms here. For such terms, the α-rule cannot occur in deductions. The α-invariance lemma (15.11) is still valid, however.

DEFINITION 15.20 (Stratified λ-terms). A pure λ-term X, with $FV(X) = \{x_1,\ldots,x_n\}$, is said to be stratified iff there are type-schemes $\alpha_1,\ldots,\alpha_n,\alpha$ such that

$$x_1 \in \alpha_1, \ldots, x_n \in \alpha_n \vdash_{TA_\lambda} X \in \alpha.$$

If X is closed, this means that X is stratified iff there is an α such that

$$\vdash_{TA_\lambda} X \in \alpha.$$

EXAMPLES 15.21. By 15.2, 4, 7, 8, the following are stratified terms:

$$K,\ S,\ I,\ B,\ W,\ Z_o\ (\equiv \lambda xy.y),\ \bar{\sigma}\ (\equiv \lambda uxy.x(uxy)),\ Z_n\ (\equiv \lambda xy.x^n y).$$

It is not hard to show that the following are also stratified:

$$C\ (\equiv \lambda xyz.xzy),\quad D\ (\equiv \lambda xyz.z(Ky)x),\quad R_{\text{Bernays}} \quad \text{(see 4.10)}.$$

But Exercise 14.9(d), although it works in TA_λ as well as in TA_C since no axiom-schemes are involved, fails to prove that xx is stratified in TA_λ for the same reason that it failed to prove xx stratified in TA_C (Examples 14.29).

LEMMA 15.22. *In* TA_λ:

(a) X *is stratified iff each subterm of* X *is stratified.*

(b) X *is stratified iff there exist types (not just type-schemes)* $\alpha_1,\ldots,\alpha_n,\alpha$ *satisfying Definition* 15.20.

(c) *The set of all stratified terms is closed under* β- *and* βη-*reduction, but not expansion.*

(d) *The set of all stratified terms is closed under abstraction but not application.*

Proof. (a) by the subject-construction property, 15.12; (b) by the substitution-lemma, 15.10; (c) by the subject-reduction theorem 15.17; (d) by rule (→ i) and the fact that certain terms are not stratified. □

DEFINITION 15.23 (Principal type-scheme, p.t.s.)

(a) If X is pure and closed: α is called a p.t.s. of X iff

$$\vdash_{TA_\lambda} X \in \alpha'$$

holds for a type-scheme α′ when and only when α′ is a substitution-instance of α.

(b) If X is pure and $FV(X) = \{x_1,\ldots,x_n\}$: a pair $\langle \mathcal{B},\alpha\rangle$ is called a principal pair (p.p.) of X, and α a p.t.s. of X, iff $\mathcal{B}$ is an FV(X)-basis (see 14.31(b)) and the relation

$$\mathcal{B}' \vdash_{TA} X \in \alpha'$$

holds for an FV(X)-basis $\mathcal{B}'$ and a type-scheme α′ when and only when $\langle \mathcal{B}',\alpha'\rangle$ is a substitution-instance of $\langle \mathcal{B},\alpha\rangle$. (Cf. Note 14.32.)

EXAMPLE 15.24. I has p.t.s. $a \to a$.

Proof. See Example 15.13. □

EXAMPLE 15.25. $\lambda x.xx$ is unstratified.

Proof. By the proof of Example 14.38, xx is unstratified. The result then follows by Lemma 15.22(a). □

As in Chapter 14, these examples should make it clear that the following theorems hold; their proofs, although simple in principle, are complicated to write out and will not be given here. (Ben-Yelles 1979, Theorem 2.1.)

THEOREM 15.26 (P.t.s. theorem). *Every stratified pure λ-term has a p.t.s. and a p.p.*

THEOREM 15.27. *The set of all stratified pure λ-terms is recursively decidable.*

If each CL-term in the table after Theorem 14.41 is replaced by the corresponding λ-term, the result is a table of p.t.s's of λ-terms. (It is easy to check that this table is correct.)

The fact that these corresponding terms have the same p.t.s's suggests that TA_C and TA_λ are equivalent. To state the precise form of this equivalence, first define, for each basis $\mathcal{B}$ in TA_C,

$$\mathcal{B}_\lambda = \{\text{formulas } X_\lambda \in \alpha : X \in \alpha \text{ is in } \mathcal{B}\}.$$

Similarly, for a basis $\mathcal{B}$ in TA_λ, and any H-transformation, define

$$\mathcal{B}_H = \{\text{formulas } X_H \in \alpha : X \in \alpha \text{ is in } \mathcal{B}\}.$$

Then an easy proof gives the following result.

THEOREM 15.28 (Equivalence of TA_C and TA_λ). *Let* H *be* H_η, H_w, *or* H_β (9.7, 9.21, 9.34-5); *then*

(a) $\mathcal{B} \vdash_{TA_C} X \in \alpha \quad\Rightarrow\quad \mathcal{B}_\lambda \vdash_{TA_\lambda} X_\lambda \in \alpha$,

(b) $\mathcal{B} \vdash_{TA_\lambda} M \in \alpha \quad\Rightarrow\quad \mathcal{B}_H \vdash_{TA_C} M_H \in \alpha$.

EXERCISE 15.29.

(a) Prove that, for a pure CL-term X, X is stratified in TA_C iff X_λ is stratified in TA_λ; also X and X_λ have the same p.t.s.

(b) Let H be H_η; find a pure λ-term M such that M_H has a different p.t.s. from M.

(c) What are the corresponding results for H_w and H_β?

15D FORMULAS-AS-TYPES AND NORMALIZATION

Deduction-reductions work for TA_λ much as they do for TA_C. Of course the S- and K-reductions in the last chapter must be replaced by β-reductions as defined below, but this is the same sort of replacement one makes in passing from weak CL-reduction to λβ-reduction.

DEFINITION 15.30 (Deduction-reductions for TA_λ). A reduction of one deduction to another consists of a sequence of replacements by the following reduction rule:

β-reductions for deductions. A deduction of the form

$$\dfrac{\dfrac{\begin{array}{c}1\\ [x\ \varepsilon\ \alpha]\\ \mathcal{D}_1(x)\\ M\ \varepsilon\ \beta\end{array}}{\lambda x.M\ \varepsilon\ \alpha\to\beta}(\to i-1) \qquad \begin{array}{c}\mathcal{D}_2\\ N\ \varepsilon\ \alpha\end{array}}{(\lambda x.M)N\ \varepsilon\ \beta}(\to e)$$
$$\mathcal{D}_3$$

reduces to

$$\begin{array}{c}\mathcal{D}_2\\ N\ \varepsilon\ \alpha\\ \mathcal{D}_1(N)\\ [N/x]M\ \varepsilon\ \beta\\ \mathcal{D}_3',\end{array}$$

where $\mathcal{D}_3'$ is obtained from $\mathcal{D}_3$ by replacing appropriate occurrences of $(\lambda x.M)N$ by $[N/x]M$.

Note that carrying out this reduction step has the effect of performing one contraction on the subject of the conclusion.

Notice also that this reduction step corresponds, if we delete all the subjects, to the implication-reduction step in Prawitz 1965. In fact, the proof of Prawitz 1965 Chapter III Theorem 2 can be combined with the formulas-as-types transformation to show that every deduction can be reduced to a normal (irreducible) deduction. See Seldin 1977b Theorem 6, p. 22. This gives us the following result:

THEOREM 15.31 (Normalization theorem for deductions). *Every* TA_λ*-deduction can be reduced to a normal deduction.*

COROLLARY 15.31.1 (Normalization theorem for terms). *Let* $\mathcal{B}$ *be a* β*-normal-subjects basis. If* $\mathcal{B} \vdash_{TA_\lambda} X \in \alpha$, *then* X *has a* β*-normal form.*

Proof. It is easy, in the light of the proof of Theorem 15.17, to see that if a β-redex occurs in X, then any deduction of $X \in \alpha$ can be reduced by a β-reduction. See Seldin 1977b Corollary 6.2, p. 23. This depends, of course, on the fact that the condition on $\mathcal{B}$ in the above corollary is the same as that in Theorem 15.17. □

REMARK 15.32. Note that this theorem and corollary are weaker than Theorem 14.52 and Corollary 14.52.1, in that the latter prove that the deductions and terms involved in TA_C are SN. It might seem possible to obtain these stronger results here by the same method as there, namely by mapping deductions to typed λ-terms and using the strong normalization theorem for the latter (Theorem A2.3). However, the presence of bound variables, and the effect that this has on substitution, makes it more difficult to define the correspondence between deductions and typed terms for TA_λ than for TA_C, and, in fact, we the authors have not yet found a definition of this correspondence and a proof of its required properties that seems completely satisfactory, at least for all of the deductions we are considering. An alternative method of obtaining a strong normalization result is to apply the method of proof of Theorem A2.3 directly to deductions; we have not had time to fully check the details, but have no doubt that such a proof is possible.

Remark 14.54, on bases with a universal type, and Remark 14.55, on bases with proper inclusions, apply to TA_λ as well as to TA_C.

REMARK 15.33. A deduction-reduction analogous to $\beta\eta$-reduction can be defined by adding to Definition 15.30 the following additional reduction rule:

<u>η-reductions for deductions</u>. A deduction of the form

$$\dfrac{\dfrac{\begin{matrix}\mathcal{D}_1\\ M \in \alpha\to\beta\end{matrix} \qquad \begin{matrix}1\\ [x \in \alpha]\end{matrix}}{Mx \in \beta}(\to i)}{\begin{matrix}\lambda x.Mx \in \alpha\to\beta\\ \mathcal{D}_2\end{matrix}}(\to i - 1)$$

where x does not occur free in $\mathcal{D}_1$ (and hence does not occur free in M), reduces to

$$\begin{matrix}\mathcal{D}_1\\ M \in \alpha\to\beta\\ \mathcal{D}_2',\end{matrix}$$

where $\mathcal{D}_2'$ is obtained from $\mathcal{D}_2$ by replacing appropriate occurences of $\lambda x.Mx$ by M.

For the normalization theorem for $\beta\eta$-reductions of deductions, see Seldin 1977b Corollary 6.1, p.22. For $\beta\eta$-reductions of terms, use the fact that a term has a $\beta\eta$-normal form iff it has a β-normal form (7.13).

15E THE EQUALITY-RULE Eq′

The system TA_λ is like TA_C in failing to be invariant under equality (Remark 15.19). Hence, we have the same interest we had in the case of TA_C for adding the following rule:

<u>Rule</u> Eq′. $$\frac{X \in \alpha \qquad X =_* Y}{Y \in \alpha}.$$

In connection with TA_λ, this is really two rules, depending on whether $=_*$ is $=_\beta$ or $=_{\beta\eta}$.

DEFINITION 15.34 (<u>The systems</u> $TA_{\lambda=}$). These systems are obtained from TA_{λ} (Definition 15.6) by adding Rule Eq′. If $=_*$ is $=_{\beta}$, then the rule is called Eqβ′ and the system $TA_{\lambda=\beta}$. If $=_*$ is $=_{\beta\eta}$, then the rule is called Eqβη′ and the system $TA_{\lambda=\beta\eta}$. The names Eq′ and $TA_{\lambda=}$ will mean either and/or both of these rules and systems, according to the context.

(The fact that the names of the varieties of Rule Eq′ are the same for $TA_{C=}$ and TA_{λ} is not a major problem, since which ones is intended will invariably be clear from the context.)

REMARK 15.35. $TA_{\lambda=}$ is undecidable, because $=_{\beta}$ and $=_{\beta\eta}$ are undecidable.

DISCUSSION 15.36. The Eq′-postponement theorem can be proved for $TA_{\lambda=}$ by adding to the proof for $TA_{C=}$, in Discussion 14.59, the extra case in which an inference by Eq′ occurs directly above an inference by $(\to i)$. In this case, the given deduction has the form

$$\begin{array}{c}
\begin{array}{l} \quad 1 \\ [x \in \alpha] \\ \mathcal{D}_1(x) \end{array} \\
\dfrac{\dfrac{X \qquad\qquad X =_* Y}{Y \in \beta}\,\text{Eq}'}{\lambda x.Y \in \alpha\to\beta}\,(\to i - 1) \\
\mathcal{D}_2,
\end{array}$$

and it can be replaced by

$$\begin{array}{c}
\begin{array}{l} \quad 1 \\ [x \in \alpha] \\ \mathcal{D}_1(x) \end{array} \\
\dfrac{\dfrac{X \in \beta}{\lambda x.X \in \alpha\to\beta}\,(\to i - 1) \qquad \lambda x.X =_* \lambda x.Y}{\lambda x.Y \in \alpha\to\beta}\,\text{Eq}' \\
\mathcal{D}_2.
\end{array}$$

If this replacement is used with the others in Discussion 14.59, the result is a proof of the following theorem.

THEOREM 15.37 (Eq′-postponement theorem). *If* $=_*$ *is* $=_{\lambda\beta}$ *or* $=_{\lambda\beta\eta}$, *and*

$$\mathcal{B} \vdash_{TA_{\lambda=}} X \in \alpha,$$

where $\mathcal{B}$ *is any basis, then there is a term* Y *such that* $Y =_* X$ *and*

$$\mathcal{B} \vdash_{TA_{\lambda}} Y \in \alpha.$$

COROLLARY 15.37.1 (Normalization theorem). *If* $\mathcal{B}$ *is a* β- [βη-] *normal-subjects basis, and if*

$$\mathcal{B} \vdash X \in \alpha$$

in $TA_{\lambda=\beta}$ [$TA_{\lambda=\beta\eta}$], *then* X *has a* β- [βη-] *normal form.*

For reasons stated in Remark 14.61(b), this Corollary cannot be extended to conclude that X is SN.

COROLLARY 15.37.2 (P.t.s. theorem for $TA_{\lambda=}$). *Let* $=_*$ *be* $=_{\lambda\beta}$ *or* $=_{\lambda\beta\eta}$; *then every stratified pure* λ*-term has a p.t.s. and a p.p. in* $TA_{\lambda=}$.

REMARK 15.38. The definitions of stratified, p.t.s., and p.p. used in this Corollary are the same as for TA_{λ}, but with $TA_{\lambda=}$-deducibility replacing TA_{λ}-deducibility. Note that a term may have a p.t.s. in $TA_{\lambda=}$ and a different one, or none at all, in TA_{λ}. The following theorem connects the two systems, and, together with the Eq′-postponement theorem, reduces the study of $TA_{\lambda=}$ to that of TA_{λ}.

THEOREM 15.39. *Let* $=_*$ *be* $=_{\lambda\beta}$ *or* $=_{\lambda\beta\eta}$. *Then a pure* λ*-term* X *is stratified in* $TA_{\lambda=}$ *iff* X *has a normal form* X^* *which is stratified in* TA_{λ}. *Further, the type-schemes that* $TA_{\lambda=}$ *assigns to* X *are exactly those that* TA_{λ} *assigns to* X^*.

Finally, the systems $TA_{\lambda=}$ are linked to $TA_{C=}$ by the following theorem. To state it, let us say that systems $TA_{C=}$, $TA_{\lambda=}$, and an H-transformation are compatible iff either they are $TA_{C=\beta}$, $TA_{\lambda=\beta}$, H_β, or they are $TA_{C=\beta\eta}$, $TA_{\lambda=\beta\eta}$, H_η.

THEOREM 15.40 (Equivalence of $TA_{C=}$ and $TA_{\lambda=}$). *If* $TA_{C=}$, $TA_{\lambda=}$, *and* H *are compatible, then*

(a) $$B \vdash_{TA_{C=}} X \in \alpha \iff B_\lambda \vdash_{TA_{\lambda=}} X_\lambda \in \alpha,$$

(b) $$B \vdash_{TA_{\lambda=}} X \in \alpha \iff B_H \vdash_{TA_{C=}} X_H \in \alpha.$$

FURTHER READING 15.41.

(a) On TA_C. Sanchis 1964, Curry 1969a, Hindley 1969a.

(b) On TA_λ. Seldin 1977b, Reynolds 1974, Milner 1978, Pottinger 1978.

(c) On $TA_{C=}$ and $TA_{\lambda=}$. Seldin 1978; and for semantics, Hindley 1983, Coppo 1984.

(d) On types and cartesian closed categories. Lambek 1958-72, 1980, D. Scott 1980b, Seely 1984, Curien 1985, Lambek and P. Scott 198-.

(e) On Coppo-Dezani-Sallé types. These are an extension of the definition of type-scheme to include intersection-types $(\alpha\cap\beta)$ and a universal type ω. (Cf. Remark 14.54.) Origins: Coppo and Dezani 1978, Sallé 1978. Good introduction: Coppo and Dezani 1981. Further results: Barendregt et al. 1983, Coppo, Dezani et al. 1979, 1980, 1984, 198-, Dezani and Margaria 198-, Hindley 1982, Ronchi 198-, Ronchi and Venneri 1984.

(f) On types in programming. Two bibliographies (hundreds of items): Dungan 1979, Meyer 198-. Informal specialist newsletter: Cardelli and MacQueen 1983+. A few papers: Berry 1978, Bruce and Longo 1984, 1985, Constable 1980, Coppo 1983, Damas and Milner 1982, Donahue 1979, Fortune et al. 1983, Gordon et al. 1979, Hennessy and Plotkin 1979, Huet 1973, 1975, Kahn et al. 1984, Leivant 1983, 198-, MacQueen et al. 1982, 1984, Milner 1977, 1978, Milner et al. 1975, Mitchell 1984, Plotkin 1977, Reynolds 1974, 1984, 1985, Robinet 1979. See also the references at the ends of §§16B2 and 16B3 in the next chapter.

CHAPTER SIXTEEN

GENERALIZED TYPE ASSIGNMENT

16A INTRODUCTION

In recent years, there have been several applications of type assignment that require generalizations of the theories we have considered so far. These generalizations are the subject of this chapter.

Of course, it is easy to generalize any theory of type assignment by just adding new type-forming operations. For example, to relate type-assignment to cartesian closed categories, one needs ordered pairs in which the first and second elements may have arbitrarily different types. This is impossible in TA_C or TA_λ, by Barendregt 1974, so it is necessary to introduce a new type-forming operation $\times$ and to postulate

$$D \ \varepsilon \ \alpha \to \beta \to (\alpha \times \beta),$$
$$D_1 \ \varepsilon \ (\alpha \times \beta) \to \alpha,$$
$$D_2 \ \varepsilon \ (\alpha \times \beta) \to \beta,$$

where, as in (4.7),

$$D \equiv \lambda xyz.z(Ky)x,$$
$$D_1 \equiv \lambda x.x\bar{0},$$
$$D_2 \equiv \lambda x.x\bar{1}.$$

Although an extension like this adds new types and assigns new types to terms, it does not represent a major change in the way type-assignment operates. The extensions we will consider in this chapter, however, require major changes in the foundations of the theories of type-assignment.

In Section 16B we shall look briefly at four systems from the literature; then in 16C onwards, we shall go more deeply into a fifth, in which the other four can be interpreted.

NOTATION 16.1. In the parts of the chapter that are valid for both λ- and CL-terms, and for any reduction and equality, the following 'neutral' notation will be used:

Notation	Meaning for λ	Meaning for CL
λ	λ	λ^* as defined in 2.14
$X =_* Y$	$X =_\beta Y$ or $X =_{\beta\eta} Y$	$X =_w Y$ or $X =_{c\beta} Y$ or $X =_{c\beta\eta} Y$
$X \rhd_* Y$	$X \rhd_\beta Y$ or $X \rhd_{\beta\eta} Y$	$X \rhd_w Y$ or $X \succ_{\beta\eta} Y$.

(When a theorem is not valid for all of these systems, then the systems for which it holds will be made clear in its statement.)

As usual, a <u>basis</u> $\mathcal{B}$ will be any finite, infinite, or empty set of formulas.

16B FOUR EXTENSIONS

16B1 INFINITE TERMS

When Spector 1962 extended to analysis Gödel's consistency proof for arithmetic (see Chapter 18), he needed to refer to infinite sequences of functionals. This led Tait 1965 and Martin-Löf 1972 to introduce extensions of typed λ-calculus in which an infinite sequence of typed terms itself is a new term. To take the version of Martin-Löf 1972, suppose that we have $X_n \in \alpha_n$ for each natural number n. Then the infinite sequence $(X_0, X_1, \ldots, X_n, \ldots)$ is a term, which is assigned a type called $\Pi\alpha_n$. This infinite term is intended to satisfy the reduction

$$(X_0, X_1, \ldots, X_n, \ldots)\bar{m} \rhd X_m.$$

To make both sides of this reduction have the same type, we need the type-assignment rule

$$\frac{X \in \Pi\alpha_n}{X\bar{m} \in \alpha_m}.$$

By analogy with the representation of the partial recursive functions in λ and CL, it is natural to think of giving a finite representation to these infinite terms by constructing finite terms X and α such that

$$X\bar{m} \rhd X_m, \qquad \alpha\bar{m} \rhd \alpha_m.$$

Then the type $\Pi\alpha_n$ could be written $\Pi\alpha$, and the above rule for Π would become

$$\frac{X \in \Pi\alpha}{X\bar{m} \in \alpha\bar{m}\,.}$$

Of course, we would need Rule Eq′ and a corresponding rule Eq″ for equality between types:

Eq″ $$\frac{X \in \alpha \qquad \alpha =_* \beta}{X \in \beta,}$$

where $=_*$ is the equivalence generated by the system's reducibility relation. (If the sequences X_n and α_n were recursively enumerable, we could expect X and α to be pure terms; but for general sequences, new constants and reduction-axioms would have to be added before X could be constructed.)

However, the type-assignment machinery we have so far is not sufficient for such a finite representation of infinite terms. For one thing, it would no longer be enough to have just types; we would also need functions whose values are types. Furthermore, the type of $X\bar{n}$ depends not only on the type of X and the type of $\bar{n}$, but also on $\bar{n}$ itself. In other words, the type of the value of a function depends on the argument as well as the type of the argument. This is a significant extension of the notion of type assignment that we have been considering up to now.

16B2 SECOND-ORDER POLYMORPHIC TYPE-ASSIGNMENT

In the type-assignment system TA_λ we took, for example, the identity-operator I and gave it every type of the form $\alpha \to \alpha$, and we expressed this in terms of type-schemes by saying that $a \to a$ is the p.t.s. of I. In contrast, in the system $\lambda\beta_t$ in Chapter 13 there were an infinite number of functions I_α, each with a single type $\alpha \to \alpha$. Second-order polymorphic type-assignment gives a way of combining these two approaches in a single system, with formal notation for I as well as for every I_α, and formal rules relating their type-schemes.

This may seem a trivial thing to do, as we have already described the relations between TA_λ and $\lambda\beta_t$ in Chapter 15. But that description was not a formal amalgamation of the two systems, and in modern programming contexts, such an amalgamation has become important. For example, in programming-languages such as FORTRAN and PASCAL, programs that differ only in the types of their variables need to be duplicated, and compiled separately. To use an example from Fortune et al. 1983 §5.3 p. 166: distinct programs are necessary for sorting integers, sorting character-strings, and sorting other structured objects. But, at the level of using comparisons to determine what swap operations to perform, the sort-algorithms used in these programs may all be identical. It would therefore save both programming and compiling time if a single sort-program could be written once and then run for any of the appropriate types. (Cf. also the discussions in §§3.24, 14A.)

An amalgamated system would require at the very least a substitution-rule for types, allowing a type-scheme to be substituted for any type-variable. A further requirement would be a formal notation for expressing generalization over types.

The design of languages with one or both of these features became a priority as soon as computer hardware could handle sufficiently high-powered languages; some recent languages with this kind of type-structure include ADA (see Barnes 1982 or Wegner 1980), ML (Gordon et al. 1979), HOPE (Burstall et al. 1980), and PLEXP (McCracken 1979).

For theoretical purposes, it is useful to see how to add these features onto a language based on λ or CL. Formal machinery for doing this was originally introduced by Girard in his 1971 and 1972, in connection with proof theory. It was introduced independently in Reynolds 1974 in the context of programming theory, and is now called <u>second-order polymorphic type-assignment</u> in a growing body of literature. The present account follows the notation of Reynolds 1974.

The first step in setting up this formal machinery is to introduce functions from types to values, called <u>polymorphic functions</u> (Reynolds 1974). As an example, consider I_α, the restriction of I to the domain α; following Reynolds, we call it

$$\lambda x \varepsilon \alpha . x.$$

Then we introduce the notation

$$\Lambda a . \lambda x \varepsilon a . x$$

for the function which, when applied to a set α, outputs the function I_α.

(So 'Λ' denotes abstraction with respect to type-variables, and 'λ' abstraction with respect to term-variables.) With this notation there is associated a new application of functions to types or type-schemes:

$$(\Lambda a.\ \lambda x \in a.\ x)\alpha,$$

and a new β-reduction:

$$(\Lambda a.\ \lambda x \in a.\ x)\alpha \rhd_\beta \lambda x \in \alpha.\ x.$$

Finally, we introduce a new type-forming operator Δ to give types for Λ-abstraction terms; for example, the type of $\Lambda a.\ \lambda x \in a.\ x$ is called

$$\Delta a.\ a \to a.$$

This leads to the following formal definitions.

DEFINITION 16.2 (Second-order polymorphic types and type-schemes). Assume that we have some type-constants and infinitely many type-variables, as in Definition 14.1. Then we define second order polymorphic type-schemes as follows:

(a) all type-constants and type-variables are type-schemes;
(b) if α and β are type-schemes, then so is $(\alpha \to \beta)$;
(c) if α is a type-scheme and a is a type-variable, then $(\Delta a.\ \alpha)$ is a type-scheme.

An occurrence of a type-variable a in a type-scheme α is said to be bound if it is inside a subtype-scheme of the form $\Delta a.\ \beta$; otherwise it is free. A second-order polymorphic type is a second-order polymorphic type-scheme in which every occurrence of a type-variable is bound. The set of all type-variables free in α is called

$$FV(\alpha).$$

DEFINITION 16.3 (Second order polymorphic λ-terms). Assume that we have infinitely many term-variables, distinct from the type-variables, and perhaps some constants, each constant having a type-scheme assigned to it. Then we define second order polymorphic λ-terms as follows:

(a) every constant and variable is a term;
(b) if M and N are terms, so is (MN);
(c) if x is a variable, α a type-scheme, and M a term, then $(\lambda x \in \alpha.\ M)$ is a term;
(d) if M is a term and α is a type-scheme, then $(M\alpha)$ is a term;
(e) if a is a type-variable and M is a term, then $(\Lambda a.\ M)$ is a term.

An occurrence of a term-variable x in a term P is said to be <u>bound</u> if it is inside a subterm of the form $\lambda x \in \alpha. M$; otherwise it is <u>free</u>. An occurrence of a type-variable a in a term P is <u>bound</u> if it is inside a subterm of the form $\Lambda a. M$; otherwise it is <u>free</u>. The set of all term- and type-variables free in M is called

$$FV(M).$$

DEFINITION 16.4 (<u>Substitution</u>). Substitution of terms for term-variables and type-schemes for type-variables is defined much as in Definition 1.11; in particular, bound term- and type-variables are automatically changed to avoid conflicts.

DEFINITION 16.5 (<u>Change of bound variables, congruence</u>). A <u>change of bound variable</u> in a type-scheme or term is any of the replacements

(a) $\Delta a. \beta \equiv_\alpha \Delta b. [b/a]\beta$ if $b \notin FV(\beta)$,

(b) $\Lambda a. M \equiv_\alpha \Lambda b. [b/a]M$ if $b \notin FV(M)$,

(c) $\lambda x \in \beta. M \equiv_\alpha \lambda y \in \beta. [y/x]M$ if $y \notin FV(M)$.

For terms, as in Definition 1.16 we say that $P \equiv_\alpha Q$ iff Q has been obtained from P by a finite (perhaps empty) sequence of changes of bound variable, not necessarily all of the same kind. For type-schemes we define $\beta \equiv_\alpha \gamma$ similarly, but by replacements (a) only.

DEFINITION 16.6 (β-<u>reduction</u>). For terms P and Q, we say that P β-<u>reduces to</u> Q ($P \triangleright_\beta Q$) iff Q is obtained from P by a finite (perhaps empty) series of changes of bound variables and the following kinds of contraction:

(β^1) $(\lambda x \in \alpha. M)N \triangleright_\beta [N/x]M$;

(β^2) $(\Lambda a. M)\alpha \triangleright_\beta [\alpha/a]M$.

REMARK 16.7 ($\beta\eta$-<u>reduction</u>). It is possible to define a $\beta\eta$-reduction for this system, using η^1-<u>contractions</u> (replacing $\lambda x \in \alpha. Mx$ by M if $x \notin FV(M)$) and, by analogy, η^2-<u>contractions</u> (replacing $\Lambda a. Ma$ by M if $a \notin FV(M)$). See Bruce and Meyer [1984].

DEFINITION 16.8 (The type-assignment system TAP). TAP (second-order polymorphic type-assignment) is a natural-deduction system. Its formulas are the type-assignment formulas

$$M \, \varepsilon \, \alpha,$$

where M is a second-order polymorphic term (Definition 16.3) and α is a second-order polymorphic type-scheme (Definition 16.2). TAP has no axioms; its rules are as follows:

($\to$e) $$\frac{M \, \varepsilon \, \alpha \to \beta \qquad N \, \varepsilon \, \alpha}{MN \, \varepsilon \, \beta}$$

($\to$i) $$\frac{\begin{array}{c}[x \, \varepsilon \, \alpha] \\ M \, \varepsilon \, \beta\end{array}}{\lambda x \varepsilon \alpha . M \, \varepsilon \, \alpha \to \beta}$$ Condition: x is not free in any undischarged assumption;

(Δe) $$\frac{M \, \varepsilon \, \Delta a . \alpha}{M\beta \, \varepsilon \, [\beta/a]\alpha}$$ Condition: β is a type-scheme;

(Δi) $$\frac{M \, \varepsilon \, \alpha}{\Lambda a . M \, \varepsilon \, \Delta a . \alpha}$$ Condition: a is a type-variable which is not free in any undischarged assumption;

($\equiv'_\alpha$) $$\frac{M \, \varepsilon \, \beta \qquad M \equiv_\alpha N}{N \, \varepsilon \, \beta}$$

($\equiv''_\alpha$) $$\frac{M \, \varepsilon \, \beta \qquad \beta \equiv_\alpha \gamma}{M \, \varepsilon \, \gamma}$$ Condition: $M \, \varepsilon \, \beta$ is not the conclusion of a rule.

NOTES 16.9.

(a) Neither of the equality-rules Eq′ and Eq″ is postulated in TAP.

(b) Rules ($\equiv'_\alpha$) and ($\equiv''_\alpha$) have not been postulated in the literature; however, it is standard to ignore changes of bound variables, and the rules seem necessary to formalize this practice. Note that while rule ($\equiv''_\alpha$) is restricted to the situation in which the left premise is an assumption (like rule ($\equiv'_\alpha$) in TA_λ, Definition 15.6), rule ($\equiv'_\alpha$) cannot be so restricted here. If it were, then the α-invariance lemma (cf. 15.11) would fail to hold: if rule ($\equiv'_\alpha$) could only be applied to assumptions, then in most cases it would be possible to deduce statements of the form $\Lambda a . M \, \varepsilon \, \Delta b . \beta$ only if $a \equiv b$. (Cf. Remark 16.32.) On the other hand, for rule $(\equiv''_{\alpha})$, it is straightforward to prove the corresponding α-invariance lemma: if $\mathcal{B}$ is a basis such that

$$\mathcal{B} \vdash_{TAP} M \varepsilon \beta,$$

and $\beta \equiv_{\alpha} \gamma$, then

$$\mathcal{B} \vdash_{TAP} M \varepsilon \gamma.$$

EXAMPLE 16.10. The informal discussion just before Definition 16.2 corresponds to the following deduction in TAP:

$$\dfrac{\dfrac{\dfrac{\overset{1}{[x \, \varepsilon \, a]}}{\lambda x \varepsilon a. x \; \varepsilon \; a \to a}(\to i - 1)}{\Lambda a. \lambda x \varepsilon a. x \; \varepsilon \; \Delta a. a \to a}(\Delta i)}{(\Lambda a. \lambda x \varepsilon a. x)\alpha \; \varepsilon \; \alpha \to \alpha.}(\Delta e)$$

Note that the term in the conclusion reduces to $\lambda x \varepsilon \alpha. x$.

REMARK 16.11. For the further theory of TAP, including the normalization theorem, see Fortune et al. 1983 §6, pp. 172ff. Some other papers that touch on particular technical aspects of second-order polymorphism in programming are Böhm and Berarducci 1985, Bruce and Longo 1985, Bruce and Meyer 1984, Coppo 1983, Leivant 1983, 198-, Reynolds 1984, 1985. The newsletter Cardelli and MacQueen 1983+ is also relevant.

16B3 AUTOMATH

In 1968, N.G. de Bruijn introduced the language AUTOMATH for automatic theorem proving. The idea behind this language is to use the formulas-as-types notion to construct a language in which the term translating a candidate for a proof will be well formed iff the candidate for a proof is indeed a valid proof. AUTOMATH and the project created around the language at Eindhoven are described in de Bruijn 1980, where there are other references.

The type theory used in AUTOMATH is interesting as a formal structure in its own right, and has been studied in Nederpelt 1980 and van Daalen 1980. The types are simply terms. Abstraction is in the form $\lambda x \varepsilon \alpha. M$, just as it is in second order polymorphic type assignment, but there is no special quantified type like the second order type $\Delta a.\alpha$. Instead, any term $\lambda x\varepsilon\alpha.\beta$ can serve as a type. There are two natural-deduction type-assignment rules; they are stated here in Reynolds' notation for ease of comparison, though the de Bruijn-Nederpelt notation is different.

(t e) $$\frac{M\ \varepsilon\ \lambda x\varepsilon\alpha.\beta \qquad N\ \varepsilon\ \alpha}{MN\ \varepsilon\ (\lambda x\varepsilon\alpha.\beta)N}$$

(t i) $$\frac{\begin{matrix}[x\ \varepsilon\ \alpha]\\ M\ \varepsilon\ \beta\end{matrix}}{\lambda x\varepsilon\alpha.M\ \varepsilon\ \lambda x\varepsilon\alpha.\beta}$$ {Condition: x is not free in any undischarged assumption.

Note that by rule (t e), the type of the value of the application of a function to an argument depends on the argument as well as on the type of the argument; this was the significant feature of the two previous systems too.

For an overview of this system, its role in theorem proving, and such properties of the system as the subject-reduction theorem and the strong normalization theorem, see Nederpelt 1980 and the references given there. For proofs and details, see the theses Nederpelt 1973 and van Daalen 1980. (The rules above are for what Nederpelt calls strongly functional terms.)

16B4 MARTIN-LÖF'S THEORY OF TYPES

Martin-Löf 1975a introduced a variant of typed λ-calculus, one of whose features is a compound type called the cartesian product $(\Pi x\varepsilon\alpha)\beta$, where x is not free in α but may occur free in β, thus making β a type-scheme denoting a family of sets. Type- and term- variables are taken to be the same, and the types are special λ-terms. The natural-deduction rules for this cartesian product are as follows.

(Π e) $$\frac{M\ \varepsilon\ (\Pi x\varepsilon\alpha)\beta \qquad N\ \varepsilon\ \alpha}{MN\ \varepsilon\ [N/x]\beta}$$

(Π i) $$\frac{\begin{matrix}[x\ \varepsilon\ \alpha]\\ M\ \varepsilon\ \beta\end{matrix}}{\lambda x.M\ \varepsilon\ (\Pi x\varepsilon\alpha)\beta}$$ {Condition: x is not free in α or in any undischarged assumption.

NOTE 16.12. If x does not occur free in β, then $(\Pi x \in \alpha)\beta$ is essentially the type $\alpha \to \beta$, and rules $(\Pi\,e)$ and $(\Pi\,i)$ become the same as $(\to e)$ and $(\to i)$. But since we allow types β in which x does occur free, we have made a generalization of TA_λ in which, again, the type of the value of the application of a function to its argument depends on the argument as well as the type of the argument.

NOTE 16.13. A comparison of rules $(\Pi\,e)$ and $(\Pi\,i)$ with rules $(t\,e)$ and $(t\,i)$ of §16B3 indicates that the type $(\Pi x \in \alpha)\beta$ behaves exactly like the AUTOMATH term $\lambda x \in \alpha.\,\beta$ when that term is used as a type. The important difference between AUTOMATH and Martin-Löf's systems is that in the former any term can be a type, whereas in the latter the types are a special class of terms.

NOTE 16.14. Actually, Martin-Löf has formulated not just one system, but a series of systems. One is presented in Martin-Löf 1975a, another in his 1982 (which really dates from 1979). A still more recent version is presented in Beeson 1985 Chapter XI. All of these systems have other type-forming operators, and the more recent ones have added different equality-relations for each type. (Thus, although Martin-Löf has rules for equality of both terms and types, these rules are not our Eq′ and Eq″.) But these features are outside the range of this chapter.

16C CURRY'S GENERALIZED TYPE-ASSIGNMENT: INTRODUCTION

The common thread to the four extensions of ordinary type-assignment in §B is the idea that the type of the result of applying a function to an argument may depend on the argument as well as the type of the argument. Curry had the idea of extending type-assignment this way as long ago as the 1950's. The letter 'G' was reserved for a type-forming operator with this property in Appendix A of Curry and Feys 1958, and the operator was discussed in §15A8 of Curry et al. 1972. (By the way, readers of Curry's writings will have to get used to a notation different from that used here.)

The notation in this section will be the neutral notation of 16.1, applicable equally to λ- and to CL-terms.

DISCUSSION 16.15. To see the way this generalization occurred to Curry, we need first to note that Curry did not use the notation '$\alpha \to \beta$' for compound types in ordinary type-assignment, but '$F\alpha\beta$', thinking of F as a non-redex atom, and of the types as special terms. (Though he originally defined F in terms of logical operators, as in §17.7.) If we used Curry's notation, then the rules ($\to$e) and ($\to$i) would be written as follows:

$$(F\,e) \qquad \frac{X \in F\alpha\beta \qquad Y \in \alpha}{XY \in \beta}$$

$$(F\,i) \qquad \frac{\begin{matrix}[x \in \alpha] \\ X \in \beta\end{matrix}}{\lambda x.X \in F\alpha\beta}$$

Condition: x is not free in any undischarged assumption.

As for axiom-schemes ($\to$K) and ($\to$S), to translate them it is best to first make the following abbreviation:

$$(16.16) \qquad F_n\alpha_1\alpha_2\ldots\alpha_n\beta \equiv F\alpha_1(F\alpha_2(\ldots(F\alpha_n\beta)\ldots)).$$

So $F_n\alpha_1\alpha_2\ldots\alpha_n\beta$ translates $\alpha_1 \to \alpha_2 \to \ldots \to \alpha_n \to \beta$. Then ($\to$K) and ($\to$S) become

$$(F\,K) \qquad K \in F_2\alpha\beta\alpha,$$

$$(F\,S) \qquad S \in F_3(F_2\alpha\beta\gamma)(F\alpha\beta)\alpha\gamma.$$

Now suppose we want a system in which the type of XY depends on Y as well as on the type of Y. If we think in terms of the formal changes needed in rules (F e) and (F i), we will understand what Curry did; he replaced F by a new type-forming operator G, and replaced rules (F e) and (F i) by the following:

$$(G\,e) \qquad \frac{X \in G\alpha\beta \qquad Y \in \alpha}{XY \in \beta Y}$$

$$(G\,i) \qquad \frac{\begin{matrix}[x \in \alpha] \\ X \in \beta\end{matrix}}{\lambda x.X \in G\alpha(\lambda x.\beta)}$$

Condition: x is not free in α or in any undischarged assumption.

In rule (G e), β is a type-scheme, and $\lambda x.\beta$ is a type-function; that is, a function which has types as values, at least for arguments of type α. This means that our type-language, when we define it, will have to include a notation for such functions as well as for the types themselves.

REMARK 16.17. A comparison of the rules for G with those for Π in §B4 indicates an equivalence between G and Martin-Löf's cartesian product. Because

$$(\Pi x \varepsilon \alpha)\beta$$

behaves just like $G\alpha(\lambda x.\beta)$; and conversely,

$$G\alpha\beta$$

behaves just like $(\Pi x \varepsilon \alpha)(\beta x)$, if $x \notin FV(G\alpha\beta)$.

REMARK 16.18 (Rule Eq″). To manipulate type-functions, we shall need to postulate this rule:

(Eq″)
$$\frac{X \varepsilon \alpha \qquad \alpha =_* \beta}{X \varepsilon \beta}$$

One of the most typical uses of Rule Eq″ is that given in the following deduction, in which a type is assigned to $xz(yz)$, assuming that only one type is assigned to z. (This can be done in ordinary →-type-assignment, and hence should be possible in any system we build that claims to be stronger.) Assume $z \varepsilon \alpha$, let β and γ be type-functions with $z \notin FV(\beta\gamma)$, and deduce as follows:

$$\dfrac{\dfrac{\dfrac{x \varepsilon G\alpha(\lambda z.G(\beta z)(\gamma z)) \qquad z \varepsilon \alpha}{xz \varepsilon (\lambda z.G(\beta z)(\gamma z))z}(G e)}{xz \varepsilon G(\beta z)(\gamma z)}Eq'' \qquad \dfrac{y \varepsilon G\alpha\beta \qquad z \varepsilon \alpha}{yz \varepsilon \beta z}(G e)}{xz(yz) \varepsilon \gamma z(yz).}(G e)$$

REMARK 16.19 (The F_n-sequence). Since the types of Chapters 14 and 15 were not terms, rule Eq″ could not have been used with those systems. However, if these chapters had been written in terms of F, then this rule could have been added. Given Eq″, the informal abbreviation (16.16) can be formalized as a definition of a sequence $F_1, F_2, \ldots,$ as follows:

$$F_1 \equiv F,$$
$$F_{n+1} \equiv \lambda x_0 x_1 \ldots x_n y . F x_0 (F_n x_1 \ldots x_n y).$$

From this definition, (16.16) can be proved as an equation in $\lambda\beta$ or CLw.

REMARK 16.20 (Definition of F). Given Eq″, we can define F in terms of G (Curry, unpublished notes, 1966):

$$F \equiv \lambda xy.Gx(Ky).$$

Rules (F i) and (F e) are derivable from (G i), (G e) and Eq″, via this definition. (Exercise).

REMARK 16.21 (The G_n-sequence). There is a G_n-sequence corresponding to the F_n-sequence above. For $n = 2$, $F_2\alpha\beta\gamma$ represents, roughly speaking, a set of 2-argument functions with values in γ; in contrast, a function f is in $G_2\alpha\beta\gamma$ when β and γ are type-functions and for each x in α and y in βx, we have fxy in γxy. Following this line of thought gives us a sequence

$$G_1 \equiv G,$$
$$G_{n+1} \equiv \lambda x_0 x_1 \ldots x_n y.Gx_0(\lambda u.G_n(x_1 u)\ldots(x_n u)(yu)).$$

From these definitions and the above definition of F, we can deduce, in both $\lambda\beta$ and CLw,

$$F_n\alpha_1\ldots\alpha_n\beta =_* G_n\alpha_1(K\alpha_2)\ldots(K^{n-1}\alpha_n)(K^n\beta)$$
$$=_* G_n\alpha_1(\lambda u_1.\alpha_2)\ldots(\lambda u_1\ldots u_{n-1}.\alpha_n)(\lambda u_1\ldots u_n.\beta),$$

where $u_1,\ldots,u_n \notin FV(\alpha_1\ldots\alpha_n\beta)$, and are distinct (the usual convention).

REMARK 16.22. One more point of comparison between F and G is worth mentioning. From $X \varepsilon F_n\alpha_1\alpha_2\ldots\alpha_n\beta$, we can use (F e) and Eq″ to derive

$$x_1 \varepsilon \alpha_1,\ x_2 \varepsilon \alpha_2,\ \ldots,\ x_n \varepsilon \alpha_n \quad \vdash \quad Xx_1\ldots x_n \varepsilon \beta.$$

On the other hand, from $X \varepsilon G_n\alpha_1\alpha_2\ldots\alpha_n\beta$, what follows by (G e) and Eq″ is

$$x_1 \varepsilon \alpha_1,\ x_2 \varepsilon (\alpha_2 x_1),\ \ldots,\ x_n \varepsilon (\alpha_n x_1\ldots x_{n-1}) \quad \vdash \quad (Xx_1\ldots x_n) \varepsilon (\beta x_1\ldots x_n).$$

This corresponds to the comment in Remark 16.21: for $n = 2$, if a function f is in $G\alpha_1\alpha_2\beta$, then α_2 and β are type-functions, and for each x_1 in α_1 and x_2 in $\alpha_2 x_1$, we have fx_1x_2 in $\beta x_1 x_2$.

REMARK 16.23. There are a number of different G-based systems possible. In the next section we shall consider 'basic' generalized type-assignment for λ-terms and for CL-terms. This is the theory that was taken up in Seldin 1979, and frequent reference will be made to that paper. In the following section we shall take a brief look at the way the theory might be generalized further.

(By the way, in Seldin 1979, types were thought of as predicates rather than sets, and so αX was written there instead of $X \varepsilon \alpha$. Furthermore, rule Eq″ was there called Eqp, for 'equality of predicates'.)

16D BASIC GENERALIZED TYPE-ASSIGNMENT

16D1 BASIC GENERALIZED TYPES

We have seen that when $G\alpha\beta$ is a type, although α is also a type, β is not, but is rather a function such that βY is a type when Y has type α. When Y does not have type α, we do not in general demand that βY be a type.

But if we do demand this, i.e. that βY is always a type, whether Y has type α or another type or no type at all, then the theory becomes simpler. This theory is called basic generalized type-assignment.

To build it, we shall first define a set of expressions called type-functions. These are intended to denote functions whose output-values are sets. Type-functions with no argument-places will be called types, and will denote sets as before.

The definition requires a few preliminary technicalities. We begin by taking a finite or infinite sequence $\theta_1, \theta_2, \ldots$ of constant non-redex atoms distinct from G, called type-constants, and we assume that there is associated with each θ_i a non-negative integer $\underline{dg}(\theta_i)$, called its degree. (Each θ_i is intended to denote a set or a set-valued function, and $\underline{dg}(\theta_i)$ is its number of arguments.)

DEFINITION 16.24 (Atomic type-function). The term α is said to be an atomic type-function of degree n iff

$$\alpha \equiv \theta U_1 \ldots U_k,$$

where θ is a type-constant of degree $k+n$ and $U_1,\ldots,U_k$ are any terms.

DEFINITION 16.25 (Proper basic generalized type-function). The term α is a proper basic generalized type-function (or, when the context is clear, a proper type-function) of rank m and degree n iff either

(a) α is an atomic type-function of degree n, and $m = 0$; or

(b) $\alpha \equiv \lambda x.\beta$, where β is a proper type-function of rank m and degree $n-1$ (and where, of course, $n > 0$); or

(c) $\alpha \equiv G\beta\gamma$, where β and γ are proper type-functions of degrees 0 and 1 respectively, $n = 0$ and $m = 1 + \text{rank}(\beta) + \text{rank}(\gamma)$.

The rank of a proper type-function α is the number of G's in α, not counting any G's in the arguments of θ's. The degree is the number of arguments $U_1,\ldots,U_n$ such that $\alpha U_1\ldots U_n$ is a type.

DEFINITION 16.26 (Basic generalized type-function, type). The term α is a basic generalized type-function (or, when the context is clear, a type-function) of rank m and degree n iff there is a proper type-function β of rank m and degree n such that $\alpha \rhd_* \beta$. A (basic generalized) type is a type-function of degree 0.

Type-functions were called G-obs in Seldin 1979.

Warning. 'Type' is used here in a different sense from Chapters 13-15; there, a type was a type-scheme without variables; here, a type may contain variables, but must have degree 0.

THEOREM 16.27. *The degree and rank of a type-function are unique.*

Proof. See Seldin 1979, Theorem 1.1. □

REMARK 16.28. It is easy to prove that type-functions have the following properties (where m, n, j, and k are ≥ 0); see Seldin 1979, Theorem 1.1.

T1. *If* α *is a type-function of rank* m *and degree* n *and if* β *is any term such that* $\alpha =_* \beta$, *then* β *is a type-function of rank* m *and degree* n.

T2. *If* α *is a type-function of rank* m *and degree* n, *then* $\lambda x.\alpha$ *is a type-function of rank* m *and degree* n + 1, *and conversely.*

T3. *If* α *is a type-function of rank* m *and degree* n + 1 *and* U *is any term, then* αU *is a type-function of rank* m *and degree* n.

T4. $G\alpha\beta$ *is a type-function of rank* m *and degree* 0 *iff* α *is a type-function of rank* j *and degree* 0, β *is a type-function of rank* k *and degree* 1, *and* $m = 1 + j + k$.

REMARK 16.29. A comparison of T1 - T4 with N1 - N3 and N6 in Remark 17.15 will show that G corresponds to a kind of restricted universal quantifier. In particular, $G\alpha\beta$ behaves similarly (but not identically) to $(\forall x \in \alpha)(\beta x)$, which Curry's writings and Chapter 17 call $\Xi\alpha\beta$, and the definition of F in terms of G is similar to the definition of implication (P in Chapter 17) in terms of Ξ. See §17.6. These similarities can be made precise, and form an extension of the formulas-as-types transformation in Chapters 14 and 15.

REMARK 16.30. It turns out that in order for basic generalized types to be truly more general than the types of ordinary type-assignment, we must have at least one type-constant whose degree is greater than 0; i.e. we cannot rely on Definition 16.25(b) to generate nontrivial type functions, we must postulate some too. To see this, define an F-type to be a type (not a type-scheme) as defined by Definition 14.1 except that $(\alpha \to \beta)$ is replaced by $F\alpha\beta$, where F is defined in 16.20. Then the following corollary can be easily proved by an induction on rank. (See Seldin 1979, Corollary 1.1.1.)

COROLLARY 16.27.1. *If every type-constant has degree* 0, *then every basic generalized type is equal to an* F-*type.*

16D2 BASIC GENERALIZED TYPE-ASSIGNMENT TO λ-TERMS

In this sub-section all terms are λ-terms, and statements involving $=_*$, $\rhd_*$ are valid for both β and $\beta\eta$ unless stated otherwise.

DEFINITION 16.31 (The type-assignment system TAG_λ). The system TAG_λ (generalized type-assignment to λ-terms) is a natural-deduction system whose formulas have form

$$M \,\varepsilon\, \alpha$$

for λ-terms M and basic generalized types α. TAG_λ has no axioms. Its rules are the following:

(G e) $$\frac{M \,\varepsilon\, G\alpha\beta \qquad N \,\varepsilon\, \alpha}{MN \,\varepsilon\, \beta N}$$

(G i) $$\frac{\begin{matrix}[x \,\varepsilon\, \alpha]\\ M \,\varepsilon\, \beta\end{matrix}}{\lambda x.M \,\varepsilon\, G\alpha(\lambda x.\beta)}$$ Condition: x is not free in α or any undischarged assumption.

Eq″ $$\frac{M \,\varepsilon\, \alpha \qquad \alpha =_* \beta}{M \,\varepsilon\, \beta}$$

($\equiv'_\alpha$) $$\frac{M \,\varepsilon\, \alpha \qquad M \equiv_\alpha N}{N \,\varepsilon\, \alpha.}$$

REMARK 16.32. (Cf. Note 16.9(b).) Note that rule ($\equiv'_\alpha$) is not restricted here the way that ($\equiv'_\alpha$) is in TA_λ (Definition 15.6), to the case in which $M \,\varepsilon\, \alpha$ is an assumption. This is because if this restriction were adopted here, then deductions would no longer be invariant under substitution, as the following example shows: let θ be a type-constant of degree 1, let x and z be distinct variables, and consider the deduction

$$\frac{\overset{1}{[z \,\varepsilon\, \theta x]}}{\lambda z.z \,\varepsilon\, G(\theta x)(\lambda z.\theta x).}\,(G\,i-1)$$

Suppose we substitute z for x in this deduction. Since x occurs free only in the predicate (type), and since

$$[z/x](G(\theta x)(\lambda z.\theta x)) \equiv G(\theta z)(\lambda u.\theta z),$$

where u is the first variable (in the given list of variables) distinct from x and z, we would expect a deduction of

$$\lambda z.z \,\varepsilon\, G(\theta z)(\lambda u.\theta z).$$

But without rule ($\equiv'_\alpha$) at the end of the deduction, this is impossible;

with $(\equiv'_\alpha)$ at the end, the required deduction is

$$\dfrac{\dfrac{\overset{1}{[u\ \varepsilon\ \theta z]}}{\lambda u.u\ \varepsilon\ G(\theta z)(\lambda u.\theta z)}(G\,i-1)}{\lambda z.z\ \varepsilon\ G(\theta z)(\lambda u.\theta z).}(\equiv'_\alpha)$$

It is not hard to prove that rule $(\equiv'_\alpha)$ can always be pushed down to the end of a deduction, though at the cost of introducing new Eq″-steps.

EXAMPLE 16.33. $\vdash_{TAG_\lambda}$ $K\ \varepsilon\ G_2\alpha\beta(K^2\alpha)$, where $K \equiv \lambda xy.x$.

Proof. Let x and y be distinct variables which do not occur free in α or β. Note that

$$K^2\alpha x \equiv K(K\alpha)x =_* K\alpha =_* (\lambda uy.u)\alpha =_* \lambda y.\alpha,$$

where u is distinct from y. Then we have the following deduction:

$$\dfrac{\dfrac{\dfrac{\overset{1}{[x\ \varepsilon\ \alpha]}}{\lambda y.x\ \varepsilon\ G(\beta x)(\lambda y.\alpha)}(G\,i-v)}{\lambda xy.x\ \varepsilon\ G\alpha(\lambda x.G(\beta x)(\lambda y.\alpha))}(G\,i-1)}{K\ \varepsilon\ G_2\alpha\beta(K^2\alpha).}\,Eq''$$

(The assumption discharged vacuously in the first (G i)-step is $y\ \varepsilon\ \beta x$.) □

EXAMPLE 16.34. $\vdash_{TAG_\lambda}$ $S\ \varepsilon\ G_3(G_2\alpha\beta\gamma)(K(G\alpha\beta))(K^2\alpha)(K(S\gamma))$, where $S \equiv \lambda xyz.xz(yz)$.

Proof. Let x, y, and z be distinct variables not in FV(αβγ), and note that

$$S\gamma y =_* (\lambda uvz.uz(vz))\gamma y =_* \lambda z.\gamma z(yz),$$

where u and v are distinct variables distinct from z. Then we have the following deduction:

$$\dfrac{\dfrac{\dfrac{\overset{3}{[x \,\varepsilon\, G_2\alpha\beta\gamma]} \qquad \overset{1}{[z \,\varepsilon\, \alpha]}}{xz \,\varepsilon\, (\lambda u.G(\beta u)(\gamma u))z}(G\,e)}{xz \,\varepsilon\, G(\beta z)(\gamma z)}Eq'' \qquad \dfrac{\overset{2}{[y \,\varepsilon\, G\alpha\beta]} \qquad \overset{1}{[z \,\varepsilon\, \alpha]}}{yz \,\varepsilon\, \beta z}(G\,e)}{xz(yz) \,\varepsilon\, \gamma z(yz)}(G\,e)$$

$$\dfrac{\dfrac{\dfrac{\dfrac{\dfrac{\dfrac{xz(yz) \,\varepsilon\, \gamma z(yz)}{\lambda z.xz(yz) \,\varepsilon\, G\alpha(\lambda z.\gamma z(yz))}(G\,i-1)}{\lambda z.xz(yz) \,\varepsilon\, G\alpha(S\gamma y)}Eq''}{\lambda yz.xz(yz) \,\varepsilon\, G(G\alpha\beta)(\lambda y.G\alpha(S\gamma y))}(G\,i-2)}{\lambda yz.xz(yz) \,\varepsilon\, G_2(G\alpha\beta)(K\alpha)(S\gamma)}Eq''}{S \,\varepsilon\, G(G_2\alpha\beta\gamma)(\lambda x.G_2(G\alpha\beta)(K\alpha)(S\gamma))}(G\,i-3)}{S \,\varepsilon\, G_3(G_2\alpha\beta\gamma)(K(G\alpha\beta))(K^2\alpha)(K(S\gamma)).}Eq''$$

□

EXERCISE 16.35. Let $=_*$ in rule Eq″ be $=_{\beta\eta}$. Prove that each of the following terms is assigned the indicated type-scheme in TAG_λ:

(a)* $B\ (\equiv \lambda xyz.x(yz))$ $\quad G_3(G\alpha\beta)(K(G\gamma(K\alpha)))(K^2\gamma)(K(B\beta))$,

(b)* $C\ (\equiv \lambda xyz.xzy)$ $\quad G_3(G_2\alpha(K\beta)(K\gamma))(K\beta)(K^2\alpha)(K(BK\gamma))$,

(c)* $W\ (\equiv \lambda xy.xyy)$ $\quad G_2(G_2\alpha(K\alpha)(K\beta))(K\alpha)(K\beta)$,

(d)* $\bar{0}\ (\equiv \lambda xy.y)$ $\quad G_2\alpha\beta(BK\beta)$,

(e)* $\bar{1}\ (\equiv \lambda xy.xy)$ $\quad G(G\alpha\beta)(K(G\alpha\beta))$,

(f)* $\bar{n}\ (\equiv \lambda xy.x^n y)\ \ (n \geq 2)$ $\quad G(G\alpha(K\alpha))(K(G\alpha(K\alpha)))$,

(g)* $\bar{\sigma}\ (\equiv \lambda xyz.y(xyz)))$ $\quad G_3(G_2(G\beta\gamma)\alpha(K^2\beta))(K(G\beta\gamma))(K\alpha)(B^2\gamma)$.

What is the relation between these type-schemes and the corresponding schemes in terms of F which are assigned to these same terms in TA_λ? Are any of these schemes equal to schemes in terms of F? (Cf. 15.7 - 8.)

REMARK 16.36. These examples, especially 16.33, show that when we discharge assumptions in a certain order, the subjects of the assumptions discharged later may appear in the types of the assumptions discharged earlier. Suppose we have been given a deduction of

$$(Xx_1 \ldots x_n) \; \varepsilon \; (\beta x_1 \ldots x_n)$$

from $x_1 \; \varepsilon \; \alpha_1$, $x_2 \; \varepsilon \; (\alpha_2 x_1)$, ..., $x_n \; \varepsilon \; (\alpha_n x_1 \ldots x_{n-1})$. From this, by (G i) (repeated), we can build a deduction giving

$$\vdash \; X \; \varepsilon \; \mathsf{G}_n \alpha_1 \ldots \alpha_n \beta.$$

Now, in TA_λ, if we have a deduction of $(Xx_1 \ldots x_n) \; \varepsilon \; \beta$ from some assumptions $x_1 \; \varepsilon \; \alpha_1$, ..., $x_n \; \varepsilon \; \alpha_n$, we can discharge these assumptions in any order we like, to give any of

$$\vdash \; X \; \varepsilon \; \alpha_1 \to \ldots \to \alpha_n \to \beta, \qquad \vdash \; X \; \varepsilon \; \alpha_n \to \ldots \to \alpha_1 \to \beta,$$

etc. In contrast, in TAG_λ, the order that assumptions can be discharged in is determined by which variables occur free in which types, and may even be unique (as it is in the above example if none of $\alpha_1, \ldots, \alpha_n$ contains K).

REMARK 16.37 (Subject-construction). Deductions in TAG_λ do not have as simple a subject-construction property as those in TA_λ, first because in rule $(\equiv'_\alpha)$ we cannot always restrict the left premise $M \; \varepsilon \; \alpha$ to be an assumption, and secondly because of rule Eq″, which cannot always be pushed down to the end of a deduction. However, if $\vdash X \; \varepsilon \; \alpha$, then the deduction of $X \; \varepsilon \; \alpha$ has the same tree-structure as the construction-tree of X, provided we regard steps by rules Eq″ and $(\equiv'_\alpha)$ as having zero length. (Further, rule $(\equiv'_\alpha)$ can always be pushed to the bottom.) This correspondence is close enough for useful conclusions to be drawn, for example the replacement lemma and subject-reduction theorem below. As a simpler example, the deductions in Examples 16.33 and 16.34 can be determined fairly easily by working backward from the conclusions.

DISCUSSION 16.38. The replacement lemma does not hold for TAG_λ in precisely the same form that it holds for TA_λ (Lemma 15.16). The reason for this is that the term of the minor (right) premise of an inference by (G e) occurs in the type of the conclusion. Thus, replacing a subterm of the term in the conclusion of an inference may also have an effect on its type. To take a simple example, if we take a single inference from the deduction in Example 16.34, namely

$$\frac{y \; \varepsilon \; G\alpha\beta \qquad\qquad z \; \varepsilon \; \alpha}{yz \; \varepsilon \; \beta z,} (G\,e)$$

and if we substitute a term N for z, then we get the inference

$$\frac{y \; \varepsilon \; G\alpha\beta \qquad\qquad N \; \varepsilon \; \alpha}{yN \; \varepsilon \; \beta N,} (G\,e)$$

and the type of the conclusion has been changed. However, if N β-converts to z (for example, if $N \equiv Iz$), then we will be able to use rule Eq″ to conclude $yN \; \varepsilon \; \beta z$. As it happens, the replacement lemma was only needed in Chapter 15 to prove the subject-reduction theorem, and in this special case the replaced term is replaced by a term convertible to it. And if we add this restriction to the hypotheses of the replacement lemma, as below, then it becomes provable for TAG_λ. (Though the only proof known to the authors is in unpublished notes by Seldin.)

LEMMA 16.39 (Replacement). *Let* $\mathcal{B}_1$ *be any basis, and let* $\mathcal{D}$ *be a deduction giving*

$$\mathcal{B}_1 \vdash_{TAG_\lambda} X \; \varepsilon \; \alpha.$$

Let V *be a term-occurrence in* X, *and let* $\lambda x_1, \ldots, \lambda x_n$ *be those* λ's *in* X *whose scope contains* V. *Let* $\mathcal{D}$ *contain a formula* $V \; \varepsilon \; \gamma$ *in the same position as* V *has in the construction tree of* X, *and let* $x_1 \; \varepsilon \; \delta_1, \ldots, x_n \; \varepsilon \; \delta_n$ *be the assumptions above* $V \; \varepsilon \; \gamma$ *that are discharged below it. Assume that* $V \; \varepsilon \; \gamma$ *is not in* $\mathcal{B}_1$. *Let* W *be a term such that* $W =_* V$ *and* $FV(W) \subseteq FV(V)$, *and let* $\mathcal{B}_2$ *be a basis not containing* $x_1, \ldots, x_n$ *free. Let* $\mathcal{D}_2$ *be a deduction of*

$$\mathcal{B}_2, \; x_1 \; \varepsilon \; \delta_1, \; \ldots, \; x_n \; \varepsilon \; \delta_n \vdash_{TAG_\lambda} W \; \varepsilon \; \gamma.$$

Let X^* *be the result of replacing* V *by* W *in* X. *Then*

$$\mathcal{B}_1 \cup \mathcal{B}_2 \vdash_{TAG_\lambda} X^* \; \varepsilon \; \alpha.$$

(In the above lemma, 'in the same position' means 'in the same position when Eq″ and $(\equiv_\alpha')$ steps are counted as having zero length'.)

DEFINITION 16.40. Monoschematic and normal-subjects bases are defined just as in Definitions 14.14 and 15.9. Recall that a β- [βη-] <u>normal-subjects basis</u> is one in which each subject is in β- [βη-] normal form and does not begin with a λ.

THEOREM 16.41 (<u>Subject-reduction theorem for</u> $\rhd_\beta$). *Let* $=_*$ *in rule* Eq″ *be* $=_\beta$. *Let* $\mathcal{B}$ *be a β-normal-subjects basis. If*

$$\mathcal{B} \vdash_{TAG_\lambda} M \varepsilon \alpha,$$

and $M \rhd_\beta M'$, *then*

$$\mathcal{B} \vdash_{TAG_\lambda} M' \varepsilon \alpha.$$

REMARK 16.42. Theorem 16.41 is Seldin 1979 Theorem 3.2. It cannot be extended to βη-reduction as can Theorem 15.17, its analogue for TA_λ; the following counter-example is due to C.-B. Ben-Yelles. Let θ be a type-constant of degree 1, and let

$$\eta \equiv G(\theta x)(\lambda y.\theta x).$$

Let ξ be any type in which x, y, z do not occur free, and let $K \equiv \lambda uv.u$ and $I \equiv \lambda w.w$. Then it is not hard to get

$$\vdash_{TAG_\lambda} \lambda x.KIx \ \varepsilon\ G\xi(\lambda x.\eta),$$

by first proving $KI \ \varepsilon\ G\xi(\lambda z.\eta)$ and then applying (G e) and (G i) once each. But

$$\nvdash_{TAG_\lambda} KI \ \varepsilon\ G\xi(\lambda x.\eta).$$

The closest approximation to the subject-reduction theorem for βη-reduction is the following result. (See Seldin 1979, Corollary 3.2.2.)

THEOREM 16.43 (<u>Restricted subject-reduction theorem for</u> $\rhd_{\beta\eta}$). *Let* $=_*$ *in rule* Eq″ *be* $=_\beta$ *or* $=_{\beta\eta}$. *Let* $\mathcal{B}$ *be a βη-normal-subjects basis. If*

$$\mathcal{B} \vdash_{TAG_\lambda} M \varepsilon \alpha,$$

where M *is in β-normal form, and if* $M \rhd_\eta M'$ *(Definition* 7.6), *then*

$$\mathcal{B} \vdash_{TAG_\lambda} M' \varepsilon \alpha.$$

As in TA_λ, the subject-reduction theorem is closely related to the theorem that every term with a type has a normal form. The latter can be proved using the following deduction-reductions.

DEFINITION 16.44 (Deduction-reductions for TAG_λ). A reduction of one deduction to another consists of a sequence of replacements by one of the following reduction-rules:

β-reductions. A deduction of the form

$$\dfrac{\dfrac{\dfrac{\begin{matrix}\overset{1}{[x \,\varepsilon\, \alpha]}\\ \mathcal{D}_1(x)\\ M \,\varepsilon\, \beta\end{matrix}}{\lambda x.M \,\varepsilon\, G\alpha(\lambda x.\beta)}(G i-1) \qquad G\alpha(\lambda x.\beta) =_* G\gamma\delta}{\lambda x.M \,\varepsilon\, G\gamma\delta}Eq'' \qquad \begin{matrix}\mathcal{D}_2\\ U \,\varepsilon\, \gamma\end{matrix}}{(\lambda x.M)U \,\varepsilon\, \delta U}(G e)$$
$$\mathcal{D}_3$$

reduces to

$$\dfrac{\begin{matrix}\dfrac{\begin{matrix}\mathcal{D}_2\\ U \,\varepsilon\, \gamma\end{matrix} \qquad \gamma =_* \alpha}{U \,\varepsilon\, \alpha}Eq''\\ \mathcal{D}_1(U)\\ [U/x]M \,\varepsilon\, [U/x]\beta\end{matrix} \qquad [U/x]\beta =_* \delta U}{[U/x]M \,\varepsilon\, \delta U}Eq''$$
$$\mathcal{D}_3',$$

where $\mathcal{D}_3'$ is obtained from $\mathcal{D}_3$ by replacing appropriate occurrences of $(\lambda x.M)U$ by $[U/x]M$. (Note that $[U/x]\beta =_* \delta U$ is true because

$$\delta U =_* (\lambda x.\beta)U =_* [U/x]\beta.)$$

η-reductions. A deduction of the form

$$\dfrac{\dfrac{\dfrac{\begin{matrix}\mathcal{D}_1 \\ M \varepsilon G\alpha\beta\end{matrix} \qquad \begin{matrix}1 \\ [x \varepsilon \alpha]\end{matrix}}{Mx \varepsilon \beta x}(G e) \qquad \beta x =_* \gamma}{Mx \varepsilon \gamma} Eq''}{\lambda x.Mx \varepsilon G\alpha(\lambda x.\gamma)}(G i - 1)$$
$$\mathcal{D}_2 ,$$

where $x \notin FV(M\alpha)$ and $\mathcal{D}_1$ is irreducible by β- steps, reduces to

$$\begin{matrix}\mathcal{D}_1' \\ M \varepsilon G\alpha(\lambda x.\gamma) \\ \mathcal{D}_2' ,\end{matrix}$$

where $\mathcal{D}_1'$ is the deduction whose existence is guarenteed by Theorem 16.43, and $\mathcal{D}_2'$ is obtained from $\mathcal{D}_2$ by replacing appropriate occurrences of $\lambda x.Mx$ by M.

(By the way, in η-reductions x may occur free in β, so we cannot assume that $\beta =_* \lambda x.\beta x$, and hence cannot deduce $M \varepsilon G\alpha(\lambda x.\gamma)$ directly from $M \varepsilon G\alpha\beta$ by Eq″.)

If the relation $=_*$ in rule Eq″ is $=_\beta$, then deductions are reduced using only β-reductions. If $=_*$ is $=_{\beta\eta}$, then η-reductions are used as well. If η-reductions are used, then clearly one can occur only if all possible β-reductions in the tree above it have been carried out first.

A β- [βη-] normal deduction is one which cannot be β- [βη-] reduced. If a deduction is β-normal, then it is easy to see that it can be η-reduced to a βη-normal deduction.

THEOREM 16.45 (Strong normalization theorem for deductions). *If* $=_*$ *is* β- [βη-] *equality, then every* β- [βη-] *reduction of a* TAG_λ*-deduction terminates in a* β- [βη-] *normal deduction.*

COROLLARY 16.45.1. *Let* $\mathcal{B}$ *be a* β- [βη-] *normal-subjects basis, and let* $=_*$ *be, respectively,* β- [βη-] *equality. If*

$$\mathcal{B} \vdash_{TAG_\lambda} M \varepsilon \alpha,$$

then M *is* SN.

(See Seldin 1979, Theorem 3.5 and Corollary 3.5.1.)

DEFINITION 16.46. The theory $TAG_{\lambda=}$ is obtained from the theory TAG_λ by adding rule Eq′, where the relation $=_*$ in rule Eq′ is the same as $=_*$ in rule Eq″.

Eq′
$$\frac{M \varepsilon \alpha \qquad M =_* N}{N \varepsilon \alpha.}$$

THEOREM 16.47 (Eq′-postponement theorem). *If*

$$\mathcal{B} \vdash_{TAG_{\lambda=}} M \varepsilon \alpha,$$

then there is a term N *such that* $N =_* M$ *and*

$$\mathcal{B} \vdash_{TAG_\lambda} N \varepsilon \alpha.$$

COROLLARY 16.47.1 (Normalization theorem). *If* $\mathcal{B}$ *is a* β- [βη-] *normal-subjects basis, and* $=_*$ *is* β- [βη-] *equality, and*

$$\mathcal{B} \vdash_{TAG_{\lambda=}} M \varepsilon \alpha,$$

then M *has a* β- [βη-] *normal form.*

(See Seldin 1979, Theorem 3.3 and Corollary 3.5.2.)

16D3 BASIC GENERALIZED TYPE-ASSIGNMENT TO CL-TERMS

In this subsection all terms are CL-terms, and $=_*$ is $=_{c\beta\eta}$. Because strong reduction will be mentioned, we shall assume throughout the section that I is an atom, as in Remarks 8.18 and 14.43, with the axiom-schemes

$$IX \rhd_w X, \qquad IX \succ_{\beta\eta} X.$$

(This assumption leaves the usual properties of both reductions still valid.) Abstraction is defined by the λ^* of Definition 2.14, but $\lambda^*x.x$ is now the atom I.

Sometimes an equation will be noted to hold for $=_{c\beta}$, which of course implies $=_{c\beta\eta}$. In fact the main results probably all hold for $=_\beta$ (and the λ^β of Definition 9.34), but no-one has ever checked this in detail. For weak equality, a G-system has been formulated in Seldin 1979 §2, and has very reasonable properties; but it is slightly more complicated than the system for $\beta\eta$, so we shall keep to $\beta\eta$ here.

DISCUSSION 16.48. To get a system like TAG_λ but based on CL-terms, we need axiom schemes for K, S and I which can be substituted for rule (G i). In view of Examples 16.33 and 16.34 and the deduction for $\lambda z.z$ in Remark 16.32, the natural axiom-schemes are as follows:

(G K) $\quad K \in G_2\alpha\beta(K^2\alpha)$,

(G S) $\quad S \in G_3(G_2\alpha\beta\gamma)(K(G\alpha\beta))(K^2\alpha)(K(S\gamma))$,

(G I) $\quad I \in G\alpha(K\alpha)$.

It is interesting to compare the first two schemes with (F K) and (F S) in Discussion 16.15. Scheme (G K) implies (using rules (G e) and Eq″ and the definition of G_n from 16.21):

(1) $\quad X \in \alpha,\ Y \in \beta X \ \vdash\ KXY \in \alpha$;

on the other hand, (F K) implies

(2) $\quad X \in \alpha,\ Y \in \beta \ \vdash\ KXY \in \alpha$.

It is easy to see that (F K) is β-equal to a substitution-instance of (G K), obtained by substituting $K\beta$ for β.

Now from (G S), using rules (G e) and Eq″, we can conclude

(3) $\quad X \in G_2\alpha\beta\gamma,\ Y \in G\alpha\beta,\ Z \in \gamma \ \vdash\ SXYZ \in S\gamma YZ$.

On the other hand, what we can conclude from (F S) is

(4) $\quad X \in G_2\alpha(K\beta)(K^2\gamma),\ Y \in G\alpha(K\beta),\ Z \in \gamma \ \vdash\ SXYZ \in \gamma$.

{To deduce (4), note that

$$F_2\alpha\beta\gamma =_{c\beta} G_2\alpha(K\beta)(K^2\gamma).\}$$

A routine calculation shows that (4) is β-equal to a substitution-instance of (3), obtained by substituting $K\beta$ for β and $K^2\gamma$ for γ. Furthermore, (F S) is β-equal to the corresponding substitution-instance of (G S).

{To prove this, note that by Exercise 9.39,

$$S(K^2\gamma) =_{c\beta} K^2\gamma.\}$$

The third scheme above, (G I), is identical, via the definition of F, to

$$I \varepsilon F\alpha\alpha,$$

which says $I \varepsilon \alpha \to \alpha$ in the $\to$-notation.

Unfortunately, the instances of (G K), (G S) and (G I) do not exhaust the axioms we need. The analogy between types and logical formulas suggests that extra axioms might be necessary, because types with G correspond approximately to formulas with universal quantification, and it is well known that in systems of first-order logic in which the only rule is modus ponens, no finite set of axiom-schemes is sufficient, but it is necessary to include in the definition of axiom the rule that if A is an axiom and x is free in A, then $(\forall x)A$ is an axiom. Here we need a corresponding axiom-generating rule, rule (G) in the following definition.

DEFINITION 16.49 (Axioms for TAG_C). An expression $X \varepsilon \xi$, where ξ is a basic generalized type, is called an axiom iff it is either an instance of one of the axiom-schemes

(G K) $K \varepsilon G_2\alpha\beta(K^2\alpha)$,

(G S) $S \varepsilon G_3(G_2\alpha\beta\gamma)(K(G\alpha\beta))(K^2\alpha)(K(S\gamma))$,

(G I) $I \varepsilon G\alpha(K\alpha)$,

or the conclusion of the following rule:

Rule (G) *If* $A \varepsilon \alpha$ *is any axiom, and* $x \varepsilon FV(\alpha)$, *and* β *is any type, then*

$$KA \varepsilon G\beta(\lambda^* x.\alpha)$$

is an axiom.

NOTES 16.50.

(a) The subject of any axiom is of the form K^nI, K^nK or K^nS $(n \geq 0)$.

(b) If $A \varepsilon \alpha$ is an axiom, then so is $A \varepsilon [U/x]\alpha$ for any term U. (The only variables occurring in axioms occur in the types.)

DEFINITION 16.51 (<u>The type-assignment system</u> TAG_C). Let I be a basic combinator along with K and S. The system TAG_C (<u>generalized type-assignment to CL-terms</u>) is a formal system whose formulas are expressions

$$X \; \varepsilon \; \alpha$$

where X is a CL-term and α is a basic generalized type. Its axioms are the axioms in Definition 16.49 above. Its rules are:

(G e)
$$\frac{X \; \varepsilon \; G\alpha\beta \qquad Y \; \varepsilon \; \alpha}{XY \; \varepsilon \; \beta Y,}$$

Eq″
$$\frac{X \; \varepsilon \; \alpha \qquad \alpha =_{c\beta\eta} \beta}{X \; \varepsilon \; \beta.}$$

REMARK 16.52. As in TAG_λ, deductions in TAG_C follow the construction of the term closely enough for useful conclusions to be drawn, for example the theorems below.

REMARK 16.53. Theorem 14.19 showed that rule (F i) is admissible in TA_C. Corresponding to this, we would naturally like to prove that rule (G i) is admissible in TAG_C. Unfortunately, it is not admissible. For a counterexample, let θ be a type-constant of degree 1, and let

$$\eta \equiv G(\theta x)(K(\theta x))$$

(as in Remark 16.42). Then by (G i) twice and (G e),

$$\vdash_{TAG_C} II \; \varepsilon \; \eta.$$

From this, (G i) would allow us to derive, for any type α,

$$K(II) \; \varepsilon \; G\alpha(\lambda^* x.\eta).$$

But a check of all possible cases shows that this formula is not provable in TAG_C.

The closest we can get to the admissibility of rule (G i) is the following theorem (Seldin 1979, Theorem 2.2).

THEOREM 16.54 (<u>Abstraction and types</u>). *Let* $\mathcal{B}$ *be any basis. If*

$$\mathcal{B},\ x \,\varepsilon\, \alpha \quad \vdash_{TAG_C} \quad X \,\varepsilon\, \beta,$$

where x *does not occur free in* $\mathcal{B}$ *or in* α, *then there is a term* Y *such that* $Yx \rhd_w X$, *and there is a deduction, not containing* x, *of*

$$\mathcal{B} \quad \vdash_{TAG_C} \quad Y \,\varepsilon\, G\alpha(\lambda^* x.\beta).$$

Recall (Definition 14.14) that a <u>weakly</u>- [<u>strongly</u>-] <u>normal-subjects basis</u> is one in which each subject is weakly [strongly] irreducible and does not begin with K, S, or I. (By 8.18, 'strongly irreducible' is here equivalent to 'in strong nf', 3.7.)

THEOREM 16.55 (<u>Subject-reduction theorem for weak reduction</u>). *Let* $\mathcal{B}$ *be a weakly-normal-subjects basis. If*

$$\mathcal{B} \quad \vdash_{TAG_C} \quad X \,\varepsilon\, \alpha$$

and $X \rhd_w X'$, *then*

$$\mathcal{B} \quad \vdash_{TAG_C} \quad X' \,\varepsilon\, \alpha.$$

(See Seldin 1979, Theorem 2.4.)

REMARK 16.56. The subject-reduction theorem cannot be extended to strong reduction. To get a counterexample, let θ and η be as in Remark 16.53, and construct, for any α, a deduction giving

$$\vdash_{TAG_C} \quad S(KI)(KI) \,\varepsilon\, G\alpha(\lambda^* x.\eta).$$

Thus S(KI)(KI) has a type that K(II) does not, by Remark 16.53. But

$$S(KI)(KI) \succ K(II).$$

THEOREM 16.57 (<u>SN theorem for weak reduction</u>). *Let* $\mathcal{B}$ *be a weakly-normal-subjects basis. If*

$$\mathcal{B} \quad \vdash_{TAG_C} \quad X \,\varepsilon\, \alpha,$$

then X *is* SN *with respect to weak reduction.*

This theorem is Seldin 1979, Corollary 2.8.1. The proof is via deduction-reductions, just like that of the corresponding result in Chapter 14 (Corollary 14.52.1). The reductions are set out in Seldin 1979 §2.4. As usual, the key step is to prove that all reductions of deductions terminate.

DEFINITION 16.58. The theory $TAG_{C=}$ is obtained from the theory TAG_C by adding the $\beta\eta$-version of Rule Eq′:

$$\text{Eq}\beta\eta' \qquad \frac{X \,\varepsilon\, \alpha \qquad X =_{c\beta\eta} Y}{Y \,\varepsilon\, \alpha.}$$

THEOREM 16.59 (Eq′-postponement theorem). *Let* $\mathcal{B}$ *be any basis. If*

$$\mathcal{B} \vdash_{TAG_{C=}} X \,\varepsilon\, \alpha,$$

then there is a term Y *such that* $Y =_{c\beta\eta} X$ *and*

$$\mathcal{B} \vdash_{TAG_C} Y \,\varepsilon\, \alpha.$$

(See Seldin 1979, Theorem 2.5.)

THEOREM 16.60 (Normalization theorem for strong reduction). *If* $\mathcal{B}$ *is a strongly-normal-subjects basis, and*

$$\mathcal{B} \vdash_{TAG_{C=}} X \,\varepsilon\, \alpha,$$

then X *has a strong normal form.*

(See Seldin 1979, §2.5. The proof does not use an SN theorem for reductions of deductions (which does not hold for strong reduction); instead, it is a less direct argument which depends on proving a cut-elimination theorem for a sequent-calculus version of $TAG_{C=}$.)

16D4 G-STRATIFICATION AND F-STRATIFICATION

We have seen that the theory of basic generalized type-assignment is rather similar to the theory of ordinary type-assignment. In fact, these two kinds of type-assignment are even more closely related than they first appear to be. Ben-Yelles 1981 proves the following result for both λ and CL, and all five equality relations.

THEOREM 16.61 (<u>Relation of</u> TAG <u>to</u> TA). *Let* X *be a pure term with* FV(X) = $x_1,\ldots,x_n$. *Then there are basic generalized types* $\alpha_1,\ldots,\alpha_n,\beta$ *such that*

$$x_1 \varepsilon \alpha_1, \ldots, x_n \varepsilon \alpha_n \quad \vdash_{TAG} \quad X \varepsilon \beta$$

iff there are types $\gamma_1,\ldots,\gamma_n,\delta$ *such that*

$$x_1 \varepsilon \gamma_1, \ldots, x_n \varepsilon \gamma_n \quad \vdash_{TA} \quad X \varepsilon \delta.$$

Recall that if every type-constant has degree 0, then every type is equal to an F-type (Corollary 16.27.1, after 16.30). By using this fact and modifying Ben-Yelles' proof, we can obtain the following result (see Seldin 1979, §2.2.3, p. 42, and §3.2.2, p. 56).

COROLLARY 16.61.1. *If every type-constant has degree* 0, *then under the hypotheses of the theorem,*

$$x_1 \varepsilon \alpha_1, \ldots, x_n \varepsilon \alpha_n \quad \vdash_{TAG} \quad X \varepsilon \beta$$

iff there are types $\gamma_1 =_* \alpha_1, \ldots, \gamma_n =_* \alpha_n, \delta =_* \beta$ *such that*

$$x_1 \varepsilon \gamma_1, \ldots, x_n \varepsilon \gamma_n \quad \vdash_{TA} \quad X \varepsilon \delta.$$

REMARK 16.62. The corollary shows that TAG is essentially a conservative extension of TA; the only real difference between the systems comes in deductions involving type-constants of positive degree. In fact, if we recall the properties of TA which fail to hold in TAG (restriction of rule ($\equiv'_\alpha$) to assumptions, the theorem on abstraction and types, and the subject-reduction theorems for $\beta\eta$-strong reduction and $\lambda\beta\eta$-reduction), an examination of the counterexamples to these properties for TAG (16.32, 16.42, 16.53, 16.56) shows that each example depends on the assumption of a type-constant of degree 1. If all type-constants have degree 0, these examples fail, and TAG reduces to TA (just as the predicate logic of 0-argument predicates reduces to propositional logic).

Ben-Yelles' theorem shows that even assuming type-constants of positive degree, the class of stratified terms (the terms to which types can be assigned) is no larger in TAG than in TA. This does not mean that TAG is trivial, however. If we view the types assigned to a term X as forming a description of X's behaviour in some sense, then the descriptive power of TAG is as far beyond that of TA, as predicate logic is beyond propositional.

16E EXTENSIONS OF TAG

In basic generalized type-assignment, if α is a type-function of degree k, then $\alpha U_1 \ldots U_k$ is a type, no matter what the terms $U_1, \ldots, U_k$ are. A natural way to increase the expressive power of this system is to introduce type-functions α such that $\alpha U_1 \ldots U_k$ is only a type when $U_1, \ldots, U_k$ have certain types.

In view of §B, there are two main approaches for extending the definition of type-function in this way:

(i) Include type-functions which have types as values when the arguments are numerical, with the idea of interpreting the infinite sequences of §B1.

(ii) Make the definition of type and type-function part of the deductive structure by postulating a 'large' type L of 'small' types. Then only the large types need be defined outside the structure. This is essentially the approach of Martin-Löf (§B4).

Let us take these approaches up in this order.

DISCUSSION 16.63. Suppose that for each natural number n, we can prove $X_n \; \varepsilon \; \alpha_n$ (in some extension of TAG_λ, say). Suppose further that there are a term X and a type-function α of degree 1 such that for each natural number m,

$$X\bar{m} \rhd_* X_m, \qquad \alpha\bar{m} \rhd_* \alpha_m.$$

Assume that N is a type and that $\bar{m} \; \varepsilon \; \mathsf{N}$ is provable for each natural number m. Then the type called $\Pi\alpha$ in §B1 can be taken to be $\mathsf{GN}\alpha$, since the rule (1) in §B1 can then be derived from $(\mathsf{G}\,\mathrm{e})$ as follows:

$$\frac{X \; \varepsilon \; \mathsf{GN}\alpha \qquad \bar{m} \; \varepsilon \; \mathsf{N}}{X\bar{m} \; \varepsilon \; \alpha\bar{m}.}\,(\mathsf{G}\,\mathrm{e})$$

However, to finish proving that X represents the infinite sequence $(X_0, X_1, \ldots, X_n, \ldots)$, we need to be able to use $X\bar{m} \; \varepsilon \; \alpha\bar{m}$ for each natural number m to deduce $X \; \varepsilon \; \mathsf{GN}\alpha$. If we postulate a rule which simply asserts this, i.e. that

$$\frac{X\bar{0} \; \varepsilon \; \alpha\bar{0}, \quad \ldots, \quad X\bar{n} \; \varepsilon \; \alpha\bar{n}, \quad \ldots}{X \; \varepsilon \; \mathsf{GN}\alpha,}$$

then we have a rule analogous to the ω-rule for first-order arithmetic (see Mendelson 1964, first edition, Appendix, p. 259, where the rule is called 'Infinite Induction'). This rule has infinitely many premises; if a more finitary rule is desired, the following form of the induction rule could be postulated instead:

$$\dfrac{X\bar{0}\ \varepsilon\ \alpha\bar{0} \qquad \begin{matrix}[x\ \varepsilon\ \mathsf{N},\ \ Xx\ \varepsilon\ \alpha x]\\ X(\bar{\sigma}x)\ \varepsilon\ \alpha(\bar{\sigma}x)\end{matrix}}{X\ \varepsilon\ \mathsf{GN}\alpha.}$$

In either case, such a system would be, under the formulas-as-types transformation, essentially equivalent to first-order arithmetic.

Unfortunately there appear to be technical difficulties in giving an adequate definition of type-function for such a system, and we, the authors, have not yet seen a definition.

DISCUSSION 16.64. The 'large' type L of small types is essentially what Martin-Löf calls his universe V_0 in his 1975a, and U_0 in his 1982. (The universes are not mentioned in Beeson 1985 Chapter XI.) Under the formulas-as-types transformation, systems of this type appear to correspond to systems of higher-order logic (especially systems such as HOPC of §17F). They can probably be expected to be the most interesting extensions of TAG because they promise to be the strongest.

As an example of the kind of strength such systems can be expected to have, it will now be proved that a sufficiently strong system of this kind is strong enough to interpret the system TAP of §B2. For convenience, we shall consider only a formulation for λβ, not for CL or λβη.

DEFINITION 16.65. The system TAGL is based on a system of λ-calculus with distinct non-redex constants $\mathsf{G}, \mathsf{L}, \theta_1, \theta_2, \ldots, \theta_n, \ldots$. The constants $\theta_1, \theta_2, \ldots, \theta_n, \ldots$ are called <u>type-function constants</u>: with each θ_i is associated a nonnegative integer $\underline{\mathrm{dg}}(\theta_i)$ called its <u>degree</u>. The <u>formulas</u> of TAGL have the form

$$M\ \varepsilon\ X,$$

where M and X are any terms. The <u>axioms</u> of TAGL are all the formulas of the form

$(\mathsf{L}\theta_i)$ $$\theta_i U_1 \ldots U_k\ \varepsilon\ \mathsf{L},$$

where $k = dg(\theta_i)$ and $U_1, \ldots, U_k$ are any terms. The rules of TAGL are:

Rules of type-formation:

(LG$_1$)
$$\dfrac{X \varepsilon L \qquad \begin{matrix}[x \varepsilon X] \\ Yx \varepsilon L\end{matrix}}{GXY \varepsilon L}$$
Condition: x is not free in XY or in any undischarged assumption;

(LG$_2$)
$$\dfrac{\begin{matrix}[x \varepsilon L] \\ Xx \varepsilon L\end{matrix}}{GLX \varepsilon L}$$
Condition: x is not free in X or in any undischarged assumption;

(Eq$'$L)
$$\dfrac{M \varepsilon L \qquad M =_{\lambda\beta} N}{N \varepsilon L.}$$

Rules of type-assignment:

(G e)
$$\dfrac{M \varepsilon GXY \qquad N \varepsilon X}{MN \varepsilon YN}$$

(G i*)
$$\dfrac{\begin{matrix}[x \varepsilon X] \\ M \varepsilon Y\end{matrix} \qquad X \varepsilon L}{\lambda x.M \varepsilon GX(\lambda x.Y)}$$
Condition: x is not free in X or in any undischarged assumption;

(GLi)
$$\dfrac{\begin{matrix}[x \varepsilon L] \\ M \varepsilon Y\end{matrix}}{\lambda x.M \varepsilon GL(\lambda x.Y)}$$
Condition: x is not free in any undischarged assumption;

Eq$''$
$$\dfrac{M \varepsilon X \qquad X =_{\lambda\beta} Y}{M \varepsilon Y}$$

($\equiv'_\alpha$)
$$\dfrac{M \varepsilon X \qquad M \equiv_\alpha N}{N \varepsilon X.}$$

This system appears similar in some respects to the system HOPC of §17F, but so far there is no exact comparison.

THEOREM 16.66. *Let a system* TAP *of second-order polymorphic type-assignment, without term-constants, be given (Definition* 16.8). *Let the system* TAGL *have enough type-function constants of degree* 0 *that the type-constants of* TAP *can be mapped into them in a one-to-one manner. Then there is a mapping* $*$, *from terms and types of* TAP *to terms of* TAGL, *such that if*

$$\mathcal{B} \vdash_{TAP} M \in \alpha,$$

and if $a_1,\ldots,a_p$ *are the type-variables which occur free in* M *or* α, *then*

$$\mathcal{B}^*,\ a_1^* \in \mathsf{L},\ \ldots,\ a_p^* \in \mathsf{L} \ \vdash_{TAGL} M^* \in \alpha^*$$

where $$\mathcal{B}^* = \{N^* \in \beta^* : N \in \beta \text{ is in } \mathcal{B}\}.$$

Proof. Divide the variables of TAGL into two disjoint countably infinite classes, and map the term-variables of TAP into one class and the type-variables of TAP into the other, both in a one-to-one manner. Then define the mapping $*$ as follows.

1. If x is a term-variable of TAP, then x^* is the corresponding variable in the first class of TAGL variables;
2. If a is a type-variable of TAP, then a^* is the corresponding variable in the second class of TAGL variables;
3. If θ is a type-constant of TAP, then θ^* is the corresponding type-function constant of degree 0 of TAGL;
4. $(MN)^*$ is M^*N^*;
5. $(\lambda x \in \alpha.\, M)^*$ is $\lambda x^*.M^*$;
6. $(M\alpha)^*$ is $M^*\alpha^*$;
7. $(\Lambda a.\, M)^*$ is $\lambda a^*.M^*$;
8. $(\alpha \to \beta)^*$ is $\mathsf{G}\alpha^*(\mathsf{K}\beta^*)$;
9. $(\Delta a.\, \alpha)^*$ is $\mathsf{GL}(\lambda a^*.\alpha^*)$.

Before continuing the proof, we need two lemmas.

LEMMA 16.67. *Let* α *be a type-scheme of* TAP *with free type-variables* $a_1,\ldots,a_p$. *Then there is a deduction* $\mathcal{D}_\alpha$ *of*

$$a_1^* \in \mathsf{L},\ \ldots,\ a_p^* \in \mathsf{L} \ \vdash_{TAGL} \alpha^* \in \mathsf{L}.$$

Proof. By induction on the structure of α according to Definition 16.2. There are the following cases.

(a) α is a type-constant. Then $p = 0$ and $\alpha^* \varepsilon \mathsf{L}$ is an axiom. $\mathscr{D}_\alpha$ consists of this axiom.

(b) α is a type-variable. Then $p = 1$, α is a_1, and $\mathscr{D}_\alpha$ is the one-step deduction

$$a_1^* \varepsilon \mathsf{L}.$$

(c) α is $\beta\to\gamma$. Then α^* is $\mathsf{G}\beta^*(\mathsf{K}\gamma^*)$, and $\mathscr{D}_\alpha$ is

$$\dfrac{\begin{array}{c}\mathscr{D}_\beta\\ \beta^* \varepsilon \mathsf{L}\end{array} \qquad \begin{array}{c}\mathscr{D}_\gamma\\ \dfrac{\gamma^* \varepsilon \mathsf{L}}{\mathsf{K}\gamma^* x \varepsilon \mathsf{L}}(\mathrm{Eq}'\mathsf{L})\end{array}}{\mathsf{G}\beta^*(\mathsf{K}\gamma^*) \varepsilon \mathsf{L},}(\mathsf{LG}_1 - \mathrm{v})$$

where $\mathscr{D}_\beta$ and $\mathscr{D}_\gamma$ exist by the induction hypothesis, and x is a variable which does not occur free in β^* or γ^*.

(d) α is $\Delta a.\beta$. Then α^* is $\mathsf{GL}(\lambda a^*.\beta^*)$, and $\mathscr{D}_\alpha$ is

$$\dfrac{\dfrac{\begin{array}{c}1\\ [a^* \varepsilon \mathsf{L}]\\ \mathscr{D}_\beta\\ \beta^* \varepsilon \mathsf{L}\end{array}}{(\lambda a^*.\beta^*)a^* \varepsilon \mathsf{L}}(\mathrm{Eq}'\mathsf{L})}{\mathsf{GL}(\lambda a^*.\beta^*) \varepsilon \mathsf{L}}(\mathsf{LG}_2 - 1)$$

where $\mathscr{D}_\beta$ exists by the induction hypothesis. □

LEMMA 16.68. *If* α *and* β *are type-schemes and* a *is a type-variable of* TAP, *then* $([\beta/a]\alpha)^* \equiv_\alpha [\beta^*/a^*]\alpha^*$.

Proof. By induction on the structure of α according to Definition 16.2. There are the following cases:

(a) α is a type-constant or a type-variable distinct from a. Then α^* is a type-function constant of degree 0, or else a variable of the second class of TAGL distinct from a^*. Hence,

$$([\beta/a]\alpha)^* \equiv \alpha^* \equiv [\beta^*/a^*]\alpha^*.$$

(b) $\alpha \equiv a$. Then $\alpha^* \equiv a^*$, and we have

$$([\beta/a]\alpha)^* \equiv \beta^* \equiv [\beta^*/a^*]\alpha^*.$$

(c) $\alpha \equiv \gamma \to \delta$. Then $\alpha^* \equiv G\gamma^*(K\delta^*)$, and

$$\begin{aligned}
([\beta/a]\alpha)^* &\equiv ([\beta/a]\gamma \to [\beta/a]\delta)^* \\
&\equiv G([\beta/a]\gamma)^*(K([\beta/a]\delta)^*) \\
&\equiv_\alpha G([\beta^*/a^*]\gamma^*)(K([\beta^*/a^*]\delta^*)) && \text{by induction hypothesis} \\
&\equiv [\beta^*/a^*](G\gamma^*(K\delta^*)) \\
&\equiv [\beta^*/a^*]\alpha^*.
\end{aligned}$$

(d) $\alpha \equiv \Delta b.\gamma$. Then $\alpha^* \equiv GL(\lambda b^*.\gamma^*)$, and

$$\begin{aligned}
([\beta/a]\alpha)^* &\equiv ([\beta/a](\Delta b.\gamma))^* \\
&\equiv (\Delta c.[\beta/a][c/b]\gamma)^* && \text{where } c \text{ is as below} \\
&\equiv GL(\lambda c^*.([\beta/a][c/b]\gamma)^*) \\
&\equiv_\alpha GL(\lambda c^*.[\beta^*/a^*][c^*/b^*]\gamma^*) && \text{by induction hypothesis} \\
&\equiv_\alpha GL(\lambda x.[\beta^*/a^*][x/b^*]\gamma^*) && \text{where } x \text{ is as below} \\
&\equiv [\beta^*/a^*](GL(\lambda b^*.\gamma^*)) \\
&\equiv [\beta^*/a^*]\alpha^*.
\end{aligned}$$

Here c is b if either b does not occur free in β or a does not occur free in γ, otherwise c is the first type-variable of TAP not free in the context. Also x is b^* if either b^* does not occur free in β^* or a^* does not occur free in γ^*, otherwise x is the first variable of TAGL which does not occur free in the context. (Identity fails here because we may not have $x \equiv c^*$.) □

Proof of Theorem 16.66, *continued.* Let $\mathfrak{D}$ be a deduction of

$$\mathcal{B} \vdash_{TAP} M \varepsilon \alpha,$$

and let $a_1,\ldots,a_p$ be the type-variables which occur free in M or α. We shall prove by induction on the structure of $\mathfrak{D}$ that there is a deduction $\mathfrak{D}^*$ of

$$\mathcal{B}^*,\ a_1^* \varepsilon L,\ \ldots,\ a_p^* \varepsilon L \vdash_{TAGL} M^* \varepsilon \alpha^*.$$

There are the following cases.

Case (a). $\mathcal{D}$ consists of the single formula $M \varepsilon \alpha$. Then $\mathcal{D}^*$ consists of the single formula $M^* \varepsilon \alpha^*$.

Case (b). The last inference in $\mathcal{D}$ is by $(\rightarrow e)$. Then $M \equiv M_1 M_2$, and $\mathcal{D}$ is

$$\dfrac{\begin{matrix}\mathcal{D}_1 & \qquad & \mathcal{D}_2 \\ M_1 \ \varepsilon\ \beta \rightarrow \alpha & \qquad & M_2 \ \varepsilon\ \beta\end{matrix}}{M_1 M_2 \ \varepsilon\ \alpha.} (\rightarrow e)$$

Let $b_1, \ldots, b_q$ be the type-variables free in β which do not occur free in M or α. By the induction hypothesis, there are deductions $\mathcal{D}_1^*$ and $\mathcal{D}_2^*$ of

$$\mathcal{B}^*,\ a_1^* \ \varepsilon\ L,\ \ldots,\ a_p^* \ \varepsilon\ L,\ b_1^* \ \varepsilon\ L,\ \ldots,\ b_q^* \ \varepsilon\ L \quad \vdash_{TAGL} \quad M_1^* \ \varepsilon\ G\beta^*(K\alpha^*)$$

and

$$\mathcal{B}^*,\ a_1^* \ \varepsilon\ L,\ \ldots,\ a_p^* \ \varepsilon\ L,\ b_1^* \ \varepsilon\ L,\ \ldots,\ b_q^* \ \varepsilon\ L \quad \vdash_{TAGL} \quad M_2^* \ \varepsilon\ \beta^*$$

respectively. Let θ be any type-function constant of degree 0. By substituting θ for each b_i^* in $\mathcal{D}_1^*$ and $\mathcal{D}_2^*$ and recognizing that $\theta \ \varepsilon\ L$ is an axiom, we get deductions $\mathcal{D}_1^{**}$ and $\mathcal{D}_2^{**}$ of

$$\mathcal{B}^*,\ a_1^* \ \varepsilon\ L,\ \ldots,\ a_p^* \ \varepsilon\ L \quad \vdash_{TAGL} \quad M_1^* \ \varepsilon\ G\beta^{**}(K\alpha^*)$$

and

$$\mathcal{B}^*,\ a_1^* \ \varepsilon\ L,\ \ldots,\ a_p^* \ \varepsilon\ L \quad \vdash_{TAGL} \quad M_2^* \ \varepsilon\ \beta^{**}$$

respectively. (Here β^{**} is $[\theta/b_1^*, \ldots, \theta/b_q^*]\beta^*$.) Then $\mathcal{D}^*$ is

$$\dfrac{\dfrac{\begin{matrix}\mathcal{D}_1^{**} & \qquad & \mathcal{D}_2^{**} \\ M_1^* \ \varepsilon\ G\beta^{**}(K\alpha^*) & \qquad & M_2^* \ \varepsilon\ \beta^{**}\end{matrix}}{M_1^* M_2^* \ \varepsilon\ K\alpha^* M_2^*} (G\,e)}{M_1^* M_2^* \ \varepsilon\ \alpha^*,} \mathrm{Eq}''$$

which is what we want, since $M^* \equiv M_1^* M_2^*$.

Case (c). The last inference in $\mathcal{D}$ is by ($\rightarrow$i). Then $\alpha \equiv \beta\rightarrow\gamma$, $M \equiv \lambda x \varepsilon \beta . N$, and $\mathcal{D}$ is

$$\dfrac{\begin{matrix} \overset{1}{[x \ \varepsilon \ \beta]} \\ \mathcal{D}_1 \\ N \ \varepsilon \ \gamma \end{matrix}}{\lambda x \varepsilon \beta . N \ \varepsilon \ \beta\rightarrow\gamma .}(\rightarrow i - 1)$$

Here $M^* \equiv \lambda x^* . N^*$ and $\alpha^* \equiv G\beta^*(K\gamma^*)$. Now x is a term-variable of TAP, not a type-variable, so $x \notin FV(\gamma)$. It follows that $x^* \notin FV(\gamma^*)$ (as can be proved by induction on the structure of γ). Hence, $\lambda x^* . \gamma^* =_\beta K\gamma^*$. By the induction hypothesis, there is a deduction $\mathcal{D}_1^*$ of

$$\mathcal{B}^*, \ a_1^* \ \varepsilon \ L, \ \ldots, \ a_p^* \ \varepsilon \ L, \ x^* \ \varepsilon \ \beta^* \ \vdash_{TAGL} \ N^* \ \varepsilon \ \gamma^* .$$

Also $a_1, \ldots, a_p$ include all the free type-variables of β, so by Lemma 16.67, there is a deduction $\mathcal{D}_\beta$ of

$$a_1^* \ \varepsilon \ L, \ \ldots, \ a_p^* \ \varepsilon \ L \ \vdash_{TAGL} \ \beta^* \ \varepsilon \ L.$$

Furthermore, since x does not occur free in $\mathcal{B}$, x^* does not occur free in $\mathcal{B}^*$. Also, since x^* is in the first class of TAGL variables, it is distinct from $a_1^*, \ldots, a_p^*$, which are in the second class. Hence, we can construct $\mathcal{D}^*$ thus:

$$\dfrac{\dfrac{\begin{matrix} \overset{1}{[x^* \ \varepsilon \ \beta^*]} \\ \mathcal{D}_1^* \\ N^* \ \varepsilon \ \gamma^* \end{matrix} \qquad \begin{matrix} \mathcal{D}_\beta \\ \beta^* \ \varepsilon \ L \end{matrix}}{\lambda x^* . N^* \ \varepsilon \ G\beta^*(\lambda x^* . \gamma^*)}(Gi^* - 1)}{\lambda x^* . N^* \ \varepsilon \ G\beta^*(K\gamma^*) .}\text{Eq}''$$

Case (d). The last inference of $\mathcal{D}$ is by (Δ e). Then $M \equiv N\beta$, $\alpha \equiv [\beta/a]\gamma$, and $\mathcal{D}$ is

$$\dfrac{\begin{matrix} \mathcal{D}_1 \\ N \ \varepsilon \ \Delta a . \gamma \end{matrix}}{N\beta \ \varepsilon \ [\beta/a]\gamma .}(\Delta \ e)$$

Now $M^* \equiv N^*\beta^*$, $\alpha^* \equiv_\alpha [\beta^*/a^*]\gamma^*$ (by Lemma 16.68), and $(\Delta a.\gamma)^* \equiv \mathsf{GL}(\lambda a^*.\gamma^*)$. By the induction hypothesis, if $b_1,\ldots,b_q$ are those of $a_1,\ldots,a_p$ which occur free in N or $\Delta a.\gamma$, there is a deduction $\mathscr{D}_1^*$ of

$$\mathscr{B}^*,\ b_1^* \varepsilon \mathsf{L},\ \ldots,\ b_q^* \varepsilon \mathsf{L} \quad \vdash_{\mathrm{TAGL}} \quad N^* \varepsilon \mathsf{GL}(\lambda a^*.\gamma^*).$$

Then $\mathscr{D}^*$ is

$$\dfrac{\dfrac{\begin{matrix}\mathscr{D}_1^* \\ N^* \varepsilon \mathsf{GL}(\lambda a^*.\gamma^*)\end{matrix} \qquad \begin{matrix}\mathscr{D}_\beta \\ \beta^* \varepsilon \mathsf{L}\end{matrix}}{N^*\beta^* \varepsilon (\lambda a^*.\gamma^*)\beta^*}\,(\mathrm{Le})}{N^*\beta^* \varepsilon \alpha^*,}\,\mathrm{Eq}''$$

since $\alpha^* \equiv_\alpha [\beta^*/a^*]\gamma^* =_\beta (\lambda a^*.\gamma^*)\beta^*$.

<u>Case</u> (e). The last inference of $\mathscr{D}$ is by $(\Delta\, \mathrm{i})$. Then $M \equiv \Lambda a.N$, $\alpha \equiv \Delta a.\beta$, and $\mathscr{D}$ is

$$\dfrac{\begin{matrix}\mathscr{D}_1 \\ N \varepsilon \beta\end{matrix}}{\Lambda a.N \varepsilon \Delta a.\beta.}\,(\Delta\, \mathrm{i})$$

Now $(\Lambda a.N)^* \equiv \lambda a^*.N^*$, $(\Delta a.\beta)^* \equiv \mathsf{GL}(\lambda a^*.\beta^*)$, and by the hypothesis of induction there is a deduction $\mathscr{D}_1^*$ of

$$\mathscr{B}^*,\ a_1^* \varepsilon \mathsf{L},\ \ldots,\ a_p^* \varepsilon \mathsf{L},\ a^* \varepsilon \mathsf{L} \quad \vdash_{\mathrm{TAGL}} \quad N^* \varepsilon \beta^*.$$

Clearly, a is not any of the a_i, and so we get the desired $\mathscr{D}^*$ by adding to $\mathscr{D}_1^*$ an inference by (GLi).

<u>Case</u> (f). The last inference of $\mathscr{D}$ is by $(\equiv_\alpha')$. Then $\mathscr{D}$ is

$$\dfrac{\begin{matrix}\mathscr{D}_1 \\ N \varepsilon \alpha\end{matrix}}{M \varepsilon \alpha,}\,(\equiv_\alpha')$$

where $N \equiv_\alpha M$. It is easy to prove by induction on the structure of N that $N^* \equiv_\alpha M^*$. Hence, $\mathscr{D}^*$ is

$$\dfrac{\begin{matrix}\mathscr{D}_1^* \\ N^* \varepsilon \alpha^*\end{matrix}}{M^* \varepsilon \alpha^*,}\,(\equiv_\alpha')$$

where $\mathcal{D}_1^*$ is the deduction whose existence is guaranteed by $\mathcal{D}_1$ and the hypothesis of induction.

Case (g). The last inference of $\mathcal{D}$ is by $(\equiv''_\alpha)$. Then $\mathcal{D}$ is

$$\dfrac{\begin{matrix}\mathcal{D}_1\\ M \;\varepsilon\; \beta\end{matrix}}{M \;\varepsilon\; \alpha,}(\equiv''_\alpha)$$

where $\beta \equiv_\alpha \alpha$. It is easy to prove by induction on the structure of β that $\beta^* \equiv_\alpha \alpha^*$. Hence, $\mathcal{D}^*$ is

$$\dfrac{\begin{matrix}\mathcal{D}_1^*\\ M^* \;\varepsilon\; \beta^*\end{matrix}}{M^* \;\varepsilon\; \alpha^*,}\mathrm{Eq}''$$

where $\mathcal{D}_1^*$ is the deduction whose existence is guaranteed by $\mathcal{D}_1$ and the hypothesis of induction. □

REMARK 16.69. This theorem can easily be extended to a system of TAP with term-constants by assuming the existence of appropriate constants in TAGL and postulating the right types for them.

REMARK 16.70. In their discussion of TAP (which they call the second-order λ-calculus), Fortune et al. 1983, p. 165 make the following point:

> 'There is a clear distinction in the second-order λ-calculus between expressions and types. In particular, a type is not itself a λ-expression; furthermore, it is not possible to construct an expression that represents a function with range the set of types. It would be interesting to consider the possibility of representing such functions in a λ-calculus. One problem is that it is not at all clear how to maintain the inductive type assignment and reduction closure properties of the second-order λ-calculus in such a system'.

(The inductive type assignment property is what this book has called the subject-construction property, and reduction closure is the subject-reduction theorem.)

Now TAGL can represent functions whose range is the set of types. Also, its type-assignment rules have the subject-construction property. Therefore, the interest expressed in the above quotation makes it worthwhile verifying that TAGL satisfies the subject-reduction theorem too.

THEOREM 16.71 (Subject-reduction theorem for λβ-reductions). *Let* $\mathcal{B}$ *be a* β-*normal-subjects basis. If*

$$\mathcal{B} \vdash_{TAGL} M \in X$$

and $M \rhd_{\lambda\beta} M'$, *then*

$$\mathcal{B} \vdash_{TAGL} M' \in X.$$

The proof is at present only in manuscript form. As with TAG_λ in 16.39 and 16.40, it is fairly straightforward, but depends on a technically messy replacement lemma.

HISTORICAL NOTE 16.72. A result like Theorem 16.66 was first proved by Martin-Löf in an early version of his 1975a. However, he used a postulate, corresponding to $L \in L$, which later turned out to be contradictory (as shown by Girard 1972). In his later work, Martin-Löf rejected, on philosophical grounds, not only $L \in L$ but also several other postulates of TAGL that are needed for Theorem 16.66. The system TAGL was constructed so that Theorem 16.66 could be proved, and is therefore only provisional. It may need some further postulates in order to be really useful.

NOTE 16.73 (added in proof). Recently Thierry Coquand has introduced a very interesting extension of TAGL and proved the normalization theorem for it; see his dissertation "Une Théorie des Constructions" (Thèse de III cycle, Université Paris VII, 1985). He shows the power of his system (including Theorem 16.66) in the dissertation, and in two joint papers with Gérard Huet ("Constructions: a higher-order proof-system for mechanizing mathematics", Springer LNCS 203, 1985; and "Concepts mathématiques et informatiques formalisés dans le calcul des constructions", presented to the Colloque de Logique d'Orsay, 1985).

In Coquand's unpublished preprint "An analysis of Girard's paradox" (1986), he analyzes the paradox of Girard 1972; he has used a computer to construct explicitly the term without a normal form given by that paradox, and he shows that his theory is the strongest for which normalization holds.

CHAPTER SEVENTEEN

LOGIC BASED ON COMBINATORS

17A INTRODUCTION

Both λ-calculus and combinatory logic were originally introduced as parts of systems of logic in the ordinary sense. Part of the interest in such systems can be seen by considering the inference in first-order predicate calculus from the general statement

$$(\forall x)A(x)$$

to a particular one,

$$A(t).$$

In carrying out this inference, it is necessary to substitute t for x, and to do so in a way which avoids conflicts of bound variables. But these complications of bound variables are really not part of the idea of generality, and it would therefore be nice to have a treatment which separates the properties of generality itself from the complications of substitution and bound variables. Such a treatment can be provided by the λ-calculus or CL as follows: regard the general formula $(\forall x)A(x)$ as representing a property not of the proposition $A(x)$ but of the propositional function $\lambda x.A(x)$. Then, since $A(t)$ can be obtained by reducing

$$(\lambda x.A(x))t,$$

the complications of substitution and bound variables are all taken care of by the properties of λ-calculus or CL. If we represent the property of generality by Π, and interpret $(\forall x)A(x)$ as $\Pi(\lambda x.A(x))$, then, writing X instead of $\lambda x.A(x)$, the inference at the beginning of the paragraph can be written

$$\Pi X \vdash Xt.$$

NOTATION 17.1. This chapter is written in the same neutral notation as Chapter 3, and its results will hold for both λ-calculus and combinatory logic. To the table in Notation 3.1 are added the following entries:

Notation	Meaning for λ	Meaning for CL
$X =_{\beta} Y$	$X =_{\beta} Y$	$X =_{c\beta} Y$
$X =_{\beta\eta} Y$	$X =_{\beta\eta} Y$	$X =_{c\beta\eta} Y$
$X \rhd_{\beta\eta} Y$	$X \rhd_{\beta\eta} Y$	$X \succ_{\beta\eta} Y$
$X =_{*} Y$	$X =_{\beta} Y$ or $X =_{\beta\eta} Y$	$X =_{w} Y$ or $X =_{c\beta} Y$ or $X =_{c\beta\eta} Y$.

The last line will be used in Sections A - C, in statements which will be true for all five kinds of equality.

In addition to talking about equality and reduction, we shall talk about provability of terms, for all of our 'formulas' will be terms. (There will be a term $\alpha \supset (\alpha \wedge \alpha)$, for example.) This is a departure from the notation of Curry and earlier works by Seldin, where formulas were given the form '$\vdash X$' to distinguish them from terms X. Here we shall give '$\vdash$' its standard meaning of provability; Notation 6.1.

In all systems below, we shall assume the following rule:

Eq $$\frac{X \qquad X =_{*} Y}{Y\,.}$$

A rule like this is necessary if we are to make use of the properties of λ or CL to deal with the complications of substitution. (Rule Eq′ of Chapter 14 and rule Eq″ of Chapter 16 are clearly special modifications of rule Eq.) This rule is really several rules, depending on which equality $=_{*}$ represents. They will be distinguished by writing

Eqw, Eqβ, Eqβη,

when it is necessary to be more explicit. In deductions, we shall often write the rule as

$$\begin{array}{l} X \\ \text{—Eq} \\ Y\,. \end{array}$$

A non-redex atom [non-redex constant] is, in CL, any atom [atomic constant] other than K and S, and, in λ, any atom [atomic constant].

Throughout this chapter, lower-case Greek letters will denote "canonical" terms, where the notion of a canonical term will be defined later.

LEMMA 17.2 (Invariance of form). *If* a, b *are non-redex atoms and* $=_*$ *is any of the five equalities in Notation* 17.1, *and*

$$aX_1 \ldots X_k =_* bY_1 \ldots Y_n,$$

then $a \equiv b$, $k = n$, *and* $X_i =_* Y_i$ *for* $i = 1, \ldots, n$.

Proof. For $\lambda\beta$, see Corollary 1.35.5 and its proof. For $\lambda\beta\eta$, use the Church-Rosser theorem (7.12) similarly. In CL: for weak, $\beta\eta$- and β-equalities, use 2.24.5, 8.15 - 17(d), and 9.38 respectively. □

17B CURRY'S PARADOX

It might seem that the natural way to base systems of logic on λ or CL is simply to introduce atomic constants for the connectives and quantifiers and to take as postulates those usual in, say, first-order logic. But this does not work. For some very simple properties of implication will lead to a contradiction in the presence of rule Eq. Suppose, for any terms X and Y, that $X \supset Y$ is the term representing the statement that X implies Y. Suppose, also, that in addition to rule Eq we have the following rule and axiom-scheme (both of which are essential to any standard logic including implication):

(1) $\quad X \supset Y,\ X \vdash Y,$ (modus ponens, or $\supset$e)

(2) $\quad (X \supset (X \supset Y)) \supset (X \supset Y).$

Then we have a contradiction in the sense that $\vdash Z$ holds for every term Z. For, given Z, define

$$X \equiv \mathsf{Y}(\lambda z.\, z \supset (z \supset Z)), \qquad z \notin FV(Z).$$

(Y can be any fixed-point combinator.) Then by Theorem 3.3,

(3) $\quad X =_* X \supset (X \supset Z)$

(4) $\quad\quad =_* (X \supset (X \supset Z)) \supset (X \supset Z).$

Hence, we have a deduction

$$(X \supset (X \supset Z)) \supset (X \supset Z) \quad \text{by (2);}$$

(5) $X \supset (X \supset Z)$ by (4), rule Eq;

(6) X by (3), rule Eq;

(7) $X \supset Z$ by (5), (6), rule (1);

Z by (7), (6), rule (1).

This argument is known as <u>Curry's Paradox</u>. (See Curry 1942a, Curry and Feys 1958 §8A p. 258 and Curry et al. 1972 §12B3 p. 180.)

This contradiction comes from such simple assumptions that many people have concluded that it is impossible to successfully use λ-calculus or combinatory logic as a basis for logic. For it is clear that we cannot set up logic in λ or CL without rule Eq, and (1) is the most essential rule known for implication. As for (2), it also is basic, since it follows from (1) and the generally accepted natural-deduction rule for introducing implication (Prawitz 1965 Chapter I, p. 20):

$$(\supset i) \qquad \dfrac{\begin{matrix}[X]\\ Y\end{matrix}}{X \supset Y}.$$

The proof of (2), using rule $(\supset i)$, is as follows:

$$\dfrac{\dfrac{\dfrac{\dfrac{\overset{1}{[X \supset (X \supset Y)]} \quad \overset{2}{[X]}}{X \supset Y}(\supset e) \quad \overset{2}{[X]}}{Y}(\supset e)}{X \supset Y}(\supset i - 2)}{(X \supset (X \supset Y)) \supset (X \supset Y)}(\supset i - 1).$$

It can thus be seen that there is a fundamental incompatibility between the unrestricted rules for implication and the property that $\lambda x.X$ exists for each term X, a property which Curry called <u>combinatory completeness</u>.

There is, however, a way to use λ-calculus or combinatory logic as a basis for logic despite Curry's paradox. That is to recognize that only certain terms can represent propositions (which, for our purposes, may be thought of as things which may meaningfully be called true or false). We shall call such terms <u>canonical</u>. The search for an appropriate system of logic involves a search for an appropriate definition of canonicalness; this is the approach adopted by Curry and his group, and it is the main subject of this chapter.

We shall start with first-order logic, although its theory is well known, in order to introduce the basic concepts and techniques, and end by sketching some higher-order systems. Since the work is still experimental, each system will have several variations.

17C LOGICAL CONSTANTS

Before we proceed further, it is worth saying a few words about representing the connectives and quantifiers in λ or CL. To represent the standard connectives and quantifiers, we shall adjoin new terms Λ, V, P Π, Σ, and $\bot$; to represent some other notions we shall adjoin the additional terms Ξ, F, E, G, and Q; some of these will be non-redex constants, and some will be defined in terms of the others. We shall adopt the following abbreviations:

$$
\begin{aligned}
X \wedge Y &\equiv \Lambda XY, \\
X \vee Y &\equiv \mathrm{V} XY, \\
X \supset Y &\equiv \mathrm{P} XY, \\
(\forall x)X &\equiv \Pi(\lambda x.X), \\
(\exists x)X &\equiv \Sigma(\lambda x.X), \\
\neg X &\equiv X \supset \bot \equiv \mathrm{P} X\bot, \\
(\forall x \in X)Y \equiv (\forall Xx)Y &\equiv \Xi X(\lambda x.Y), \\
X \to Y &\equiv \mathrm{F} XY, \\
X = Y &\equiv \mathrm{Q} XY.
\end{aligned}
$$

By Lemma 17.2, for each non-redex constant we have 'invariance of form under equality'; for example, if $\mathrm{P}XY =_* aZ_1 \ldots Z_n$, where a is any non-redex atom, then $a \equiv \mathrm{P}$, $n = 2$, $Z_1 =_* X$, and $Z_2 =_* Y$.

Although we are not yet in a position to define any of the systems we will consider, we have reached a point at which it is useful to see the rules from which these systems will be defined. If we use lower case Greek letters for canonical terms, we can list the rules (which are natural-deduction rules and use the conventions discussed earlier, in Discussion 15.1) as follows.

17.3. RULES FOR MINIMAL LOGIC. Rule Eq, plus the following:

($\wedge$e) $$\frac{\alpha_1 \wedge \alpha_2}{\alpha_i} \quad (i = 1,2)$$

($\wedge$i) $$\frac{\alpha_1 \qquad \alpha_2}{\alpha_1 \wedge \alpha_2}$$

($\vee$e) $$\frac{\alpha_1 \vee \alpha_2 \qquad \begin{matrix}[\alpha_1]\\ \beta\end{matrix} \qquad \begin{matrix}[\alpha_2]\\ \beta\end{matrix}}{\beta}$$

($\vee$i) $$\frac{\alpha_i}{\alpha_1 \vee \alpha_2} \quad (i = 1,2)$$

($\supset$e) $$\frac{\alpha \supset \beta \qquad \alpha}{\beta}$$

($\supset$i) $$\frac{\begin{matrix}[\alpha]\\ \beta\end{matrix}}{\alpha \supset \beta}$$

($\forall$e) $$\frac{\Pi\alpha}{\alpha X}$$

($\forall$i) $$\frac{\alpha x}{\Pi\alpha}$$

($\exists$e) $$\frac{\Sigma\alpha \qquad \begin{matrix}[\alpha x]\\ \beta\end{matrix}}{\beta}$$

($\exists$i) $$\frac{\alpha X}{\Sigma\alpha}\ .$$

To read the quantifier rules, recall that if $\alpha \equiv \lambda x.\gamma$, then $\Pi\alpha \equiv \Pi(\lambda x.\gamma) \equiv (\forall x)\gamma$ and $\alpha X \equiv (\lambda x.\gamma)X =_* [X/x]\gamma$. In these rules, X is any term; and x is a variable, called the characteristic variable of the rule, which does not occur free in α, β, or in any undischarged assumption.

17.4. RULES FOR INTUITIONISTIC LOGIC. All of the above, plus

($\perp$j) $$\frac{\perp}{\alpha}\ .$$

17.5. RULES FOR CLASSICAL LOGIC. The above rules for $\wedge$, $\supset$, $\forall$ ($\vee$ and $\exists$ being defined in the usual way), plus Eq, plus

($\perp$c) $$\frac{\begin{matrix}[\neg\alpha]\\ \perp\end{matrix}}{\alpha}\ .$$

See Prawitz 1965 Chapter I for a discussion of these rules.

Now let us look at some less-usual constants from Curry and Feys 1958.

17.6. THE CONSTANT Ξ. Look at the rules for $\forall$. They imply that we can prove $\Pi\alpha$ only if we can prove αX for every term X. But most uses of the universal quantifier in mathematics are over restricted ranges, such as over all natural numbers, or all real numbers, or all real-valued functions of real numbers, etc. In ordinary first-order logic, this does not matter, since we restrict (by the choice of individual constants and function symbols) the formation of terms in such a way that in any application of logic, we may assume that the universe is limited to the kinds of objects over which we are quantifying. But here, we have no such restriction on the formation of terms, and, as was shown in Chapter 4, we are automatically able to represent all natural numbers and all recursive functions, so an unrestricted $\forall$ here commits us to having these things in our domain of quantification. For this reason, we are less interested in formulas of the form $(\forall x)\beta$ and more interested in formulas of the form $(\forall x \in \alpha)\beta$, or, to use a notation more consistent with the rest of this chapter, $(\forall \alpha x)\beta$. For example, if N represents the predicate "being a natural number", then $(\forall Nx)(0 \leq x)$ would represent the true proposition that 0 is the least natural number (assuming a representation of 0 and $\leq$ as terms). To represent such restricted quantification, Curry introduced a new non-redex constant Ξ, giving $\Xi\alpha\beta$ the interpretation $(\forall x)(\alpha x \supset \beta x)$, or, in a common notation, $\alpha x \supset_x \beta x$. He then read

$$(\forall \alpha x)\beta$$

as an abbreviation for

$$\Xi\alpha(\lambda x.\beta);$$

or, more generally,

$$(\forall Xx)Y$$

as an abbreviation for

$$\Xi X(\lambda x.Y).$$

The natural deduction rules that he assumed for Ξ are as follows:

$$(\Xi e)\quad \frac{\Xi\alpha\beta \qquad \alpha X}{\beta X} \qquad\qquad (\Xi i)\quad \frac{\begin{matrix}[\alpha x]\\ \beta x\end{matrix}}{\Xi\alpha\beta}\ ,$$

where in (Ξi), x does not occur free in α, β, or in any undischarged assumption.

Note that Ξ can be defined in terms of Π and P:

$$\Xi \equiv \lambda xy.(\forall u)(xu \supset yu)$$
$$\equiv \lambda xy.\Pi(\lambda u.\mathsf{P}(xu)(yu)),$$

where the last form has been written without abbreviations. It is easy to show that if Ξ is defined this way, then its rules follow from those for $\forall$ and $\supset$.

Conversely, given Ξ, we can define P (and hence $\supset$) as follows:

$$\mathsf{P} \equiv \lambda xy.\Xi(\mathsf{K}x)(\mathsf{K}y),$$

The rules for $\supset$ then follow from those for Ξ.

We can also define Π (and hence $\forall$) in terms of Ξ and a universal predicate, E, such that $\mathsf{E}X$ is an axiom for each term X. The definition is then

$$\Pi \equiv \Xi\mathsf{E},$$

and the rules for $\forall$ then follow from those for Ξ and the axioms for E. (E was called 'ω' in Remark 14.54.)

17.7. THE CONSTANT F. As mentioned at the beginning of §16C, instead of writing types as $\alpha \to \beta$, Curry wrote them as $\mathsf{F}\alpha\beta$, where F is a non-redex constant. Thus, for Curry, types were terms. He also viewed types as propositional functions rather than as sets, and wrote

(8) $$\alpha X$$

instead of $X \in \alpha$. This notation made it easier for him to relate the types to systems of logic than does the $\to$-notation.

In terms of Curry's notation, the rules for F take the form

(F e) $$\frac{\mathsf{F}\alpha\beta X \qquad \alpha Y}{\beta(XY)}$$

(F i) $$\frac{\begin{array}{c}[\alpha x]\\ \beta(Xx)\end{array}}{\mathsf{F}\alpha\beta X},$$

where, in (F i), x is not free in α, β, X, or in any undischarged assumption.

Given either $\forall$ and $\supset$ and their rules, or else Ξ and its rules, F can be defined as follows:

$$\mathsf{F} \equiv \lambda xyz.(\forall xu)(y(zu))$$
$$\equiv \lambda xyz.(\forall u)(xu \supset y(zu)).$$

It is easy to see that the above forms of (F e) and (F i) follow from either the rules for $\forall$ and $\supset$ or else from the rules for Ξ.

17.8. THE CONSTANT G. In terms of the notation (8), the rules for G (§16C) take the form

(G e) $$\frac{G\alpha\beta X \qquad \alpha Y}{\beta Y(XY)}$$

(G i) $$\frac{\begin{matrix}[\alpha x] \\ \beta x(Xx)\end{matrix}}{G\alpha\beta X},$$

where in (G i), x does not occur free in α, β, X, or in any undischarged assumption. G can be defined in terms of $\forall$ and $\supset$ as follows:

$$G \equiv \lambda xyz.(\forall u)(xu \supset yu(zu)).$$

It is easy to see that the above forms of (G e) and (G i) can be proved from the rules for $\forall$ and $\supset$.

17.9. THE EQUALITY-CONSTANT Q. In a strong enough higher-order logic, equality is a definable concept. But in weaker systems, it is sometimes necessary to postulate it. We shall do this here, when necessary, by adjoining a new non-redex constant Q (reading QXY informally as X = Y), with an axiom scheme

(ρ) $$QXX$$

(for all terms X), and the rule

(Q e) $$\frac{QXY \qquad \alpha X}{\alpha Y}.$$

By scheme (ρ) and rule Eq, we can prove

(9) $$X =_* Y \vdash QXY,$$

so interconvertible terms will necessarily represent the same object in any "normal" model of Q.

REMARK 17.10. In some systems in the literature, it is possible to prove as a metatheorem the converse to (9), namely

(10) $$\vdash QXY \Rightarrow X =_* Y.$$

This gives us a consistency result, since it implies that $\nvdash$ QSK. This kind of consistency, which is fairly strong, is called Q-consistency. In fact it says that Q formalizes exactly $=_*$, and so it is not appropriate for systems formalizing, for example, set theory.

REMARK 17.11. Rule (Q e) gives us the property that equals can be substituted for equals, and it is well known that this property, together with the scheme (ρ), is sufficient to deduce all the usual properties of equality. In particular, if QXY is accepted as canonical for all terms X and Y, then the following derived rules can be proved from scheme (ρ) and rule (Q e):

(σ)	QXY $\vdash$ QYX,
(τ)	QXY, QYZ $\vdash$ QXZ,
(μ)	QXY $\vdash$ Q(ZX)(ZY),
(ν)	QXY $\vdash$ Q(XZ)(YZ) .

REMARK 17.12. Using (ρ), it is possible to define the universal predicate E as follows:

$$E \equiv \lambda x.Qxx.$$

17D FIRST-ORDER SYSTEMS

In this section we shall study first-order logic as a preparation for the higher-order theories we shall take up later.

From now on, equality and reduction will be βη, in λ or CL. (The results could probably be modified to hold for β without too much difficulty, however.) In CL, I is assumed to be an atom (cf. 8.18, 14.43).

Let us begin with the task of defining the class of canonical terms; i.e. those which are to be interpreted as propositions or propositional functions. We assume that the terms include some non-redex constants $\theta_1, \theta_2, \ldots,$ called canonical atoms (to serve as the atomic predicate symbols), each with an associated number $dg(\theta_i) \geq 0$, called its degree, which gives its intended number of arguments. One canonical atom is ⊥ with

$$dg(\bot) = 0.$$

Depending on the system, other canonical atoms may be E and Q with

$$dg(E) = 1, \qquad dg(Q) = 2.$$

From these canonical atoms, we shall define the canonical terms, called here canterms for short.

DEFINITION 17.13 (Canterms). The term α is said to be a canonical simplex of degree n iff

$$\alpha \equiv \theta U_1 \dots U_k$$

where θ is a canonical atom of degree $n+k$ and $U_1, \dots, U_k$ are any terms. The term α is a proper canterm of rank m and degree n iff

(a) α is a canonical simplex of degree n, and $m = 0$;

(b) $\alpha \equiv \lambda x.\beta$, where β is a proper canterm of rank m and degree $n-1$;

(c) $\alpha \equiv \beta \wedge \gamma$ $(\Lambda\beta\gamma)$, $\beta \vee \gamma$ $(V\beta\gamma)$, or $\beta \supset \gamma$ $(P\beta\gamma)$, where β and γ are proper canterms of degree 0, $n = 0$, and $m = 1 + \text{rank}(\beta) + \text{rank}(\gamma)$;

or

(d) $\alpha \equiv \Pi\beta$ or $\Sigma\beta$, where β is a proper canterm of degree 1, $n = 0$, and $m = 1 + \text{rank}(\beta)$.

A term α is a canterm of rank m and degree n iff there is a proper canterm β of rank m and degree n such that $\alpha \rhd_{\beta\eta} \beta$. Finally, a formula is a canterm of degree 0 (and any rank).

The degree of a canterm represents the number of intended arguments it takes, so that canterms of degree 0 represent propositions, and will be the only terms allowed to be steps of deductions. The rank of a canterm corresponds to the number of occurrences of connectives and quantifiers in a well-formed formula in ordinary logic; it is important because the standard proof-theoretic techniques for proving consistency, such as the normalization results of Prawitz 1965, depend on inductions on this number. For such proofs, the following result is crucial.

THEOREM 17.14. *The degree and rank of a canterm are unique.*

Proof. See Curry et al. 1972 §15B3 Theorem 4 and the subsequent discussion, also Seldin 1977a §1. □

REMARK 17.15. It is easy to prove that canterms have the following properties ($m, n, i, k \geq 0$). (For the proofs see Curry et al. 1972 §15B3 and Seldin 1977a §1.)

N1. *If* α *is a canterm of rank* m *and degree* n *and if* β *is any term such that* $\alpha =_{\beta\eta} \beta$, *then* β *is a canterm of rank* m *and degree* n.

N2. *If* α *is a canterm of rank* m *and degree* n, *then* $\lambda x.\alpha$ *is a canterm of rank* m *and degree* $n+1$ *and conversely.*

N3. *If* α *is a canterm of rank* m *and degree* $n+1$ *and* U *is any term, then* αU *is a canterm of rank* m *and degree* n.

N4. $\alpha \wedge \beta$, $\alpha \vee \beta$, *and* $\alpha \supset \beta$ *are canterms of rank* m *and degree* 0 *iff* α *and* β *are canterms of ranks* j *and* k *respectively and degree* 0 *and* $m = 1+j+k$.

N5. $\Pi\alpha$ *and* $\Sigma\alpha$ *are canterms of rank* $m+1$ *and degree* 0 *iff* α *is a canterm of rank* m *and degree* 1.

N6. $\Xi\alpha\beta$ *is a canterm of rank* m *and degree* 0 *iff* α *and* β *are canterms of ranks* j *and* k *respectively and degree* 1 *and* $m = 1+j+k$. ($\Xi \equiv \lambda xy.(\forall u)(xu \supset yu)$).

N7. $F\alpha\beta$ *is a canterm of rank* m *and degree* 1 *iff* α *and* β *are canterms of ranks* j *and* k *respectively and degree* 1 *and* $m = 1+j+k$. ($F \equiv \lambda xyz.(\forall u)(xu \supset y(zu))$).

N8. $G\alpha\beta$ *is a canterm of rank* m *and degree* 1 *iff* α *is a canterm of rank* j *and degree* 1, β *is a canterm of rank* k *and degree* 2, *and* $m = 1+j+k$. ($G \equiv \lambda xyz.(\forall u)(xu \supset yu(zu))$).

REMARK 17.16. Note that if one canterm is convertible to another, then they both reduce to convertible proper canterms. Note also that although two distinct proper canterms may be convertible, they will differ only in the arguments of the canonical atoms, and the corresponding arguments will be convertible. From this it follows that no canterm can convert to another with a different principal connective or quantifier. Furthermore, if $\alpha \wedge \beta =_{\beta\eta} \gamma \wedge \delta$, then $\alpha =_{\beta\eta} \gamma$ and $\beta =_{\beta\eta} \delta$, by Lemma 17.2.

DISCUSSION 17.17 (Proof-reductions and normalization). Theorem 17.14 makes it possible to prove proof-normalization results, just as Prawitz 1965 Chapter II - IV does for ordinary first-order logic. But because of Rule Eq, it is necessary to modify the reduction steps. These modified steps will be set out in detail here, as follows.

In each reduction-step, the deduction on the left is replaced by the one on the right.

We shall say that a deduction $\mathcal{D}_1$ reduces to a deduction $\mathcal{D}_2$ iff $\mathcal{D}_1$ can be transformed into $\mathcal{D}_2$ by a finite (perhaps empty) series of replacements by these steps.

In some reduction-steps, one formula will be called the cut formula; it will be indicated by special brackets ⟦ ⟧.

∧-reduction.

$$\begin{array}{c}\dfrac{\dfrac{\dfrac{\begin{array}{c}\mathcal{D}_1\\ \alpha_1\end{array}\qquad\begin{array}{c}\mathcal{D}_2\\ \alpha_2\end{array}}{[\![\alpha_1 \wedge \alpha_2]\!]}(\wedge\,i)}{\beta_1 \wedge \beta_2}\,\mathrm{Eq}}{\beta_i}(\wedge\,e)\\ \mathcal{D}_3\end{array}
\qquad\qquad
\begin{array}{c}\mathcal{D}_i\\ \dfrac{\alpha_i}{\beta_i}\,\mathrm{Eq}\\ \mathcal{D}_3\,.\end{array}\quad (i = 1 \text{ or } i = 2)$$

∨-reduction.

$$\begin{array}{c}\dfrac{\dfrac{\dfrac{\begin{array}{c}\mathcal{D}_0\\ \alpha_i\end{array}}{[\![\alpha_1 \vee \alpha_2]\!]}(\vee\,i)}{\beta_1 \vee \beta_2}\,\mathrm{Eq}\qquad\begin{array}{c}\overset{1}{[\beta_1]}\\ \mathcal{D}_1\\ \gamma\end{array}\qquad\begin{array}{c}\overset{2}{[\beta_2]}\\ \mathcal{D}_2\\ \gamma\end{array}}{\gamma}(\vee\,e - 1 - 2)\\ \mathcal{D}_3\end{array}
\qquad\qquad
\begin{array}{c}\mathcal{D}_0\\ \dfrac{\alpha_i}{\beta_i}\,\mathrm{Eq}\\ \mathcal{D}_i\\ \gamma\\ \mathcal{D}_3\,.\end{array}\quad (i = 1 \text{ or } i = 2)$$

⊃-reduction.

$$
\begin{array}{c}
\dfrac{\dfrac{\dfrac{\begin{array}{c}1\\ [\alpha]\\ \mathcal{D}_1\\ \beta\end{array}}{[\![\alpha \supset \beta]\!]}(\supset i - 1)}{\gamma \supset \delta}\text{Eq} \qquad \begin{array}{c}\mathcal{D}_2\\ \gamma\end{array}}{\delta}(\supset e)\\
\mathcal{D}_3
\end{array}
\qquad\qquad
\begin{array}{c}
\mathcal{D}_2\\
\dfrac{\gamma}{\alpha}\text{Eq}\\
\mathcal{D}_1\\
\dfrac{\beta}{\delta}\text{Eq}\\
\mathcal{D}_3\,.
\end{array}
$$

∀-reduction.

$$
\begin{array}{c}
\mathcal{D}_1(x)\\
\dfrac{\dfrac{\dfrac{\alpha x}{[\![\Pi\alpha]\!]}(\forall i)}{\Pi\beta}\text{Eq}}{\beta Z}(\forall e)\\
\mathcal{D}_2
\end{array}
\qquad\qquad
\begin{array}{c}
\mathcal{D}_1(Z)\\
\dfrac{\alpha Z}{\beta Z}\text{Eq}\\
\mathcal{D}_2\,.
\end{array}
$$

∃-reduction.

$$
\begin{array}{c}
\dfrac{\dfrac{\dfrac{\begin{array}{c}\mathcal{D}_1\\ \alpha Z\end{array}}{[\![\Sigma\alpha]\!]}(\exists i)}{\Sigma\beta}\text{Eq} \qquad \begin{array}{c}1\\ [\beta x]\\ \mathcal{D}_2(x)\\ \gamma\end{array}}{\gamma}(\exists e - 1)\\
\mathcal{D}_3
\end{array}
\qquad\qquad
\begin{array}{c}
\mathcal{D}_1\\
\dfrac{\alpha Z}{\beta Z}\text{Eq}\\
\mathcal{D}_2(Z)\\
\gamma\\
\mathcal{D}_3\,.
\end{array}
$$

<u>vE-reduction</u>. If R is an elimination rule or Eq, with γ as its major (left) premise, and $(\mathcal{D}_4)$ as the subproof(s) of its minor (right) premise(s),

$$\begin{array}{c}\dfrac{\dfrac{\begin{matrix}\mathcal{D}_1\\ \alpha\vee\beta\end{matrix}\quad\begin{matrix}1\\ [\alpha]\\ \mathcal{D}_2\\ \gamma\end{matrix}\quad\begin{matrix}2\\ [\beta]\\ \mathcal{D}_3\\ \gamma\end{matrix}}{\gamma}(\vee e-1-2)\quad(\mathcal{D}_4)}{\delta}R\\ \mathcal{D}_5\end{array}\qquad\begin{array}{c}\dfrac{\begin{matrix}\mathcal{D}_1\\ \alpha\vee\beta\end{matrix}\quad\dfrac{\begin{matrix}1\\ [\alpha]\\ \mathcal{D}_2\\ \gamma\end{matrix}\quad(\mathcal{D}_4)}{\delta}R\quad\dfrac{\begin{matrix}2\\ [\beta]\\ \mathcal{D}_3\\ \gamma\end{matrix}\quad(\mathcal{D}_4)}{\delta}R}{\delta}(\vee e-1-2)\\ \mathcal{D}_5\,.\end{array}$$

<u>∃E-reduction</u>. If R is an elimination-rule or Eq, with γ as its major (left) premise and $(\mathcal{D}_3)$ as the subproof(s) of its minor (right) premise(s),

$$\begin{array}{c}\dfrac{\dfrac{\begin{matrix}\mathcal{D}_1\\ \Sigma\alpha\end{matrix}\quad\begin{matrix}1\\ [\alpha x]\\ \mathcal{D}_2\\ \gamma\end{matrix}}{\gamma}(\exists e-1)\quad(\mathcal{D}_3)}{\delta}R\\ \mathcal{D}_4\end{array}\qquad\begin{array}{c}\dfrac{\begin{matrix}\mathcal{D}_1\\ \Sigma\alpha\end{matrix}\quad\dfrac{\begin{matrix}1\\ [\alpha x]\\ \mathcal{D}_2\\ \gamma\end{matrix}\quad(\mathcal{D}_3)}{\delta}R}{\delta}(\exists e-1)\\ \mathcal{D}_4\,.\end{array}$$

Before stating the normalization and consistency results and their consequences, we need to define our systems. We shall have three systems, for minimal logic, intuitionistic logic, and classical logic.

DEFINITION 17.18 (<u>Systems of first-order logic</u>).

(a) The system $\mathcal{F}_{31}^{M}$ (<u>minimal logic</u>), based on λ or CL, is the system in which the non-redex constants are Λ, V, P, Π, Σ, and the canonical atoms (including ⊥ and E but not Q). The axioms are those associated with canonical atoms; for example, they include

$$\mathsf{E}X$$

(for every term X). The rules are Eqβη, and the introduction and elimination rules for ∧, ∨, ⊃, ∀ and ∃, see 17.3. It is easy to see that all provable terms are canterms.

(b) The system $\mathcal{F}_{31}^{I}$ (<u>intuitionistic logic</u>) is obtained from $\mathcal{F}_{31}^{M}$ by adding the rule $(\bot j)$, see 17.4.

(c) The system $\mathcal{F}_{31}^{C}$ (<u>classical logic</u>) is obtained from $\mathcal{F}_{31}^{M}$ by dropping the constants V and Σ, dropping the introduction and elimination rules for $\vee$ and $\exists$, and adding the rule $(\bot c)$, see 17.5. In this system V and Σ are defined as follows:

$$\mathsf{V} \equiv \lambda xy.\neg(\neg x \wedge \neg y),$$

$$\Sigma \equiv \lambda x.\neg(\forall u)\neg xu.$$

(Of course, $\neg$ is defined in terms of P by

$$\neg \equiv \lambda x.x \supset \bot \equiv \lambda x.\mathsf{P}x\bot.)$$

REMARK 17.19. The system of names used in Definition 17.18 is essentially due to Curry. In his 1942b, he proposed a general name of $\mathcal{F}_2$ for systems based on Ξ, and $\mathcal{F}_3$ for systems based on P and Π. The series was begun by using $\mathcal{F}_1$ for systems based on F; the idea was to define Ξ as either $\lambda xy.\mathsf{F}xy\mathsf{I}$ or $\lambda xy.\mathsf{F}x\mathsf{I}y$, and to use Rule Eq in such a way that βY could be inferred from αX without having $\alpha =_* \beta$ and $X =_* Y$. (In particular, to derive Rule (Ξe), it is necessary to use Rule Eq to derive XU from either $Y(\mathsf{I}U)$ or $\mathsf{I}(YU)$, depending on which definition of Ξ is used.) The systems defined here are all related to systems called $\mathcal{F}_{31}$ in Curry et al. 1972, Chapter 16.

REMARK 17.20. In the classical system, it would be possible to drop Λ from the list of non-redex constants and define it instead by

$$\Lambda \equiv \lambda xy.\neg(x \supset \neg y).$$

This is not done here, because Prawitz 1965 Chapter III takes $\wedge$ as primitive, and we shall use some of Prawitz' proofs below.

LEMMA 17.21.

(a) *In* $\mathcal{F}_{31}^{C}$, *if Rule* $\bot c$ *is postulated only for canonical simplexes* α, *then it is admissible for all canterms.*

(b) *In* $\mathcal{F}_{31}^{I}$, *if Rule* $\bot j$ *is postulated only for canonical simplexes* α, *then it is admissible for all canterms.*

Proof.

(a) By induction on the rank of α, essentially following the proof of Prawitz 1965 Chapter III, Theorem 1.

(b) By induction on the rank of α. □

It is not hard to show that part (a) of this lemma would fail if $\bigvee$ and Σ were taken as primitive in $\mathcal{F}_{31}^{C}$.

Now let us turn to the normalization results. It turns out that we need to consider separately the cases of the classical and non-classical systems.

DISCUSSION 17.22 (Classical logic). Let us begin with the classical system. Because $\bigvee$ and Σ are defined in this system, we do not use the reduction-steps involving them. The only reduction-steps in $\mathcal{F}_{31}^{C}$ are $\wedge$-reductions, $\supset$-reductions, and $\forall$-reductions. A deduction is said to be normal iff it has no cut formulas.

THEOREM 17.23. *Every deduction in* $\mathcal{F}_{31}^{C}$ *can be reduced to a normal deduction.*

Proof. Follow the proof of Prawitz 1965 Chapter III, Theorem 2. Note that the proof here is an induction on the rank of the canterm that occurs as a cut formula with maximal rank. □

DISCUSSION 17.24 (Intuitionistic and minimal logic). Now let us turn our attention to the non-classical systems, for minimal and intuitionistic logic. These may be considered together, since the only difference is Rule ($\perp$j) in intuitionistic logic, and as we have seen (Lemma 17.21(b)), we need only postulate this rule when its conclusion is a canonical simplex.

The main complication of minimal and intuitionistic logic is that because of the elimination rules for $\vee$ and $\exists$, it is possible to have sequences of formula-occurrences in deductions which are identical, or, if rule Eq occurs in such a sequence, equal. Such sequences are called segments. More precisely, a segment is a sequence $\alpha_1,\ldots,\alpha_n$ of occurrences of formulas in a deduction, such that α_1 is not the conclusion of an inference by Eq, ($\vee$e), or ($\exists$e); α_n is not the major (left) premise for an inference by Eq or the minor (right) premise for an inference by (∨ e) or (∃ e); and for each i, $1 \leq i \leq n-1$, α_i is the major (left) premise for an inference by Eq or the minor (right) premise for an inference by (∨ e) or (∃ e), and α_{i+1} is the conclusion of that inference.

Note: (i) By the transitive law of equality, we may assume without loss of generality that we do not have two inferences by Eq in succession; (ii) a segment may consist of a single formula-occurrence; and (iii) this definition is like that of Prawitz 1965, p. 49, except that rule Eq may occur here.

It follows from this definition that all formulas of a segment are equal. A <u>maximum segment</u> is a segment in which the first formula is the conclusion of an inference by an i-rule and the last formula is the major (left) premise of an e-rule. (A maximum segment of length one is just a cut formula.)

A <u>normal deduction</u> in the minimal and intuitionistic systems is one in which there are no maximum segments. We obtain a normal deduction in these systems by using ∨E-reductions and ∃E-reductions to reduce the length of maximum segments to length one, when they can be eliminated by the other reductions.

THEOREM 17.25. *Every deduction in* $\mathcal{F}^{M}_{31}$ *and* $\mathcal{F}^{I}_{31}$ *can be reduced to a normal deduction.*

Proof. Like Prawitz 1965, Chapter IV, Theorem 1. □

COROLLARY 17.25.1. *In all three systems, in a normal deduction, no inference by an* i-*rule precedes an inference by* e-*rule in a given branch.*

Proof. Like Prawitz 1965, Chapter III Theorem 3 and Chapter IV Theorem 2. Note that this uses Prawitz' definition of a branch, so that a branch ends in the conclusion of the deduction or in a minor (right) premise for an inference by (⊃ e) or (Ξ e). (See Prawitz 1965, p. 41.) □

From this we can get all of the usual consequences of normalization, such as the subformula property (Prawitz 1965, Chapter III Corollary 1 and Chapter IV Corollary 1). The most important of these properties is consistency, which is proved as follows: if the conclusion of a deduction is a canonical simplex, then the main branch cannot have any inferences by an i-rule, and hence the term at the top of the branch cannot be discharged. Since $\perp$ is a canonical simplex in all three systems, this gives us the following result.

COROLLARY 17.25.2 (Consistency). *In all three systems,*

$$\nvdash \perp.$$

REMARK 17.26. Note that the axioms EX do not affect the normalization result. This is because these axioms are all canonical simplexes. In fact, checking the normalization proof shows that additional axioms consisting of canonical simplexes can be added without changing the normalization theorem or its consequences. Rules whose premises and conclusions are canonical simplexes can also be added in the same way. This fact can be used to add equality to the systems, as follows.

DEFINITION 17.27 (Systems with equality). If $\mathcal{F}_{31}$ is any of the three systems in Definition 17.18, then $\mathcal{F}_{31}$+Q is the system obtained from $\mathcal{F}_{31}$ by adding Q as a canonical atom of degree 2, adding the axiom scheme

(ρ) $$\mathrm{QXX}$$

(for each term X), and adding the rule

(Q e′) $$\frac{\mathrm{QXY} \qquad \alpha\mathrm{X}}{\alpha\mathrm{Y}}$$ Condition: α is a canonical simplex.

LEMMA 17.28. *If* $\mathcal{F}_{31}$ *is any of the three systems in Definition* 17.18, *then the following rule is admissible in* $\mathcal{F}_{31}$+Q *for each canterm* α:

(Q e) $$\frac{\mathrm{QXY} \qquad \alpha\mathrm{X}}{\alpha\mathrm{Y}\ .}$$

Proof. Similar to Seldin 1973 Corollary 3.1. □

By Remark 17.26, the normalization theorem is also true for the three systems $\mathcal{F}_{31}$+Q of Definition 17.27. As an immediate consequence, we have the following result:

THEOREM 17.29. *The three systems* $\mathcal{F}_{31}$+Q *are* Q-*consistent; i.e.* $\vdash$ QXY *only if* $X =_* Y$.

COROLLARY 17.29.1. *In the three systems* $\mathcal{F}_{31}$+Q, $\nvdash$ QSK.

It follows that in these systems, $\perp$ can be defined to be QSK.

These systems are clearly strong enough to interpret any standard system of first-order predicate calculus with equality (either minimal, intuitionistic, or classical), such as that of Mendelson 1964. It is only necessary to assume that to each individual constant c, each function symbol f, and each predicate symbol p in predicate calculus, there corresponds a non-redex constant c^*, f^*, or p^*, and that each such p^* is a canonical atom of the appropriate degree. (For example, if = is a predicate symbol of degree 2 representing equality, we have $=^* \equiv Q$.) The mapping $*$ is then extended to all formulas as follows:

$$f(t_1,\ldots,t_n)^* \equiv f^*t_1^*\ldots t_n^*,$$

$$p(t_1,\ldots,t_n)^* \equiv p^*t_1^*\ldots t_n^*,$$

$$(A \wedge B)^* \equiv A^* \wedge B^*,$$

$$(A \vee B)^* \equiv A^* \vee B^*,$$

$$(A \supset B)^* \equiv A^* \supset B^*,$$

$$((\forall x)A)^* \equiv \Pi(\lambda x.A^*),$$

$$((\exists x)A)^* \equiv \Sigma(\lambda x.A^*).$$

If Γ is any set of formulas, define

$$\Gamma^* = \{A^* : A \varepsilon \Gamma\}.$$

Then the following result is easy to prove:

THEOREM 17.30. *If* $\Gamma \vdash A$ *in minimal [intuitionistic, classical] predicate calculus, then* $\Gamma^* \vdash A^*$ *in* $\mathcal{F}^M_{31}[\mathcal{F}^I_{31}, \mathcal{F}^C_{31}]$.

REMARK 17.31. As was pointed out in Remark 17.19, Curry used $\mathcal{F}_3$ as a general name for systems in which Π and P are non-redex constants and Ξ is defined by $\Xi \equiv \lambda xy.(\forall u)(xu \supset yu)$, whereas he used $\mathcal{F}_2$ for systems in which Ξ is a non-redex constant and Π and P are defined by $\Pi \equiv \Xi E$ and $P \equiv \lambda xy.\Xi(Kx)(Ky)$. Curry originally assumed that there would be significant differences between $\mathcal{F}_2$ systems and $\mathcal{F}_3$ systems, but for first-order systems with E, such as the three systems $\mathcal{F}_{31}$, there is essentially no difference. It is, in fact, easy to define systems $\mathcal{F}^M_{21}$, $\mathcal{F}^I_{21}$, and $\mathcal{F}^C_{21}$ corresponding to, respectively, $\mathcal{F}^M_{31}$, $\mathcal{F}^I_{31}$, and $\mathcal{F}^C_{31}$, and equivalent to them. The first step is to modify Definition 17.13 of canterms by deleting the case for P in clause (c) and the case for Π in clause (d) and by adding the following new clause:

(e) $\alpha \equiv \Xi\beta\gamma$, where β and γ are proper canterms of degree 1, $n = 0$, and $m = 1 + \text{rank}(\beta) + \text{rank}(\gamma)$.

Theorem 17.14 (the uniqueness of the degree and rank of a canterm) and its consequences go through much as before; in particular, properties N4 - N6 are true as stated despite the change in definitions.

Then, for each of the three systems $\mathcal{F}_{31}$ of Definition 17.18, the corresponding system $\mathcal{F}_{21}$ is obtained by dropping Π and P from the list of non-redex constants, and the introduction and elimination rules for $\supset$ and $\forall$, and by adding in their place the non-redex constant Ξ and its introduction and elimination rules, 17.6. The rules for $\supset$ and $\forall$ now follow as derived rules (using the definitions of P and Π in terms of Ξ and E in 17.6). Lemma 17.21 (restriction of rules $\bot c$ and $\bot j$) holds for $\mathcal{F}^I_{21}$ and $\mathcal{F}^C_{21}$. Furthermore, if we modify the definition of proof-reduction steps (Discussion 17.17), we can obtain the normalization theorem and its consequences. The modification consists of dropping $\supset$- and $\forall$-reductions and replacing them by the following reduction step:

Ξ-reduction.

$$
\begin{array}{c}
\dfrac{\dfrac{\dfrac{\begin{array}{c}1\\ [\alpha x]\\ \mathcal{D}_1(x)\\ \beta x\end{array}}{[\![\,\Xi\alpha\beta\,]\!]}\;(\Xi i-1)}{\Xi\gamma\delta}\;\text{Eq} \qquad \begin{array}{c}\mathcal{D}_2\\ \gamma Z\end{array}}{\quad}\;(\Xi e)\\
\Downarrow\\
\mathcal{D}_3
\end{array}
\qquad\qquad
\begin{array}{c}
\mathcal{D}_2\\
\dfrac{\gamma Z}{\alpha Z}\;\text{Eq}\\
\mathcal{D}_1(Z)\\
\dfrac{\beta Z}{\Downarrow}\;\text{Eq}\\
\mathcal{D}_3\,.
\end{array}
$$

Then Theorem 17.23 (normalization) holds for $\mathcal{F}^{C}_{21}$, Theorem 17.25 (normalization) holds for $\mathcal{F}^{M}_{21}$ and $\mathcal{F}^{I}_{21}$, and Corollaries 17.25.1 - 2 (e-inferences before i-inferences in a branch, and consistency) hold for all three systems $\mathcal{F}_{21}$. Definition 17.27 (systems with equality) can be applied equally well to the three systems $\mathcal{F}_{21}$, and then Lemma 17.28 (admissibility of rule (Q e)), Theorem 17.29 (Q-consistency) and Corollary 17.29.1 (unprovability of QSK) can all be proved for the systems $\mathcal{F}_{21}$+Q. Finally, Theorem 17.30 (interpretation of first-order logic) holds as well for the systems $\mathcal{F}_{21}$.

17E MENTIONING PROPOSITIONS

We now turn our attention to systems stronger than those for first-order logic. While the systems of this section are not really higher-order systems, they represent a significant step in that direction. They generalize the systems of first-order logic by defining the set of canterms formally within the system. To do this we add a new non-redex constant H to represent propositions, and rules to ensure that HX is provable iff X is a canterm of degree 0. Then if L ≡ FEH (corresponding to the type E → H), L represents the set of all propositional functions of one variable, and LX is provable iff X is a canterm of degree 1. More generally, if we define

$$H_0 \equiv H, \qquad H_{n+1} \equiv FEH_n,$$

(H_{n+1} corresponds to the type $E \to H_n$), then H_nX is provable iff X is a canterm of degree n. Of course, the definition of canterm needs to be modified to take care of H.

Curry 1973 proposes such a system. That system, as originally proposed, had postulates only for Ξ, H, and E, but it is not difficult to extend the results to the other connectives and quantifiers.

We shall follow Curry in basing the system on Ξ rather than Π and P. To define it, we begin by modifying the definition of canterms.

DEFINITION 17.32 (Extended canterms). The extended canterms are defined as in 17.13, but deleting P from clause (c) and Π from (d) (cf. Remark 17.31), and adding these clauses:

(e) $\alpha \equiv \Xi\beta\gamma$, where β and γ are proper canterms of degree 1, $n = 0$, and $m = 1 + \text{rank}(\beta) + \text{rank}(\gamma)$;

(f) $\alpha \equiv H\beta$, where β is a proper canterm of degree 0, $n = 0$, and $m = 1 + \text{rank}(\beta)$.

This changed definition does not interfere with the proof of Theorem 17.14 (for extended canterms). Furthermore, extended canterms satisfy properties N1 - N8 and, in addition, the following:

N9. $H\alpha$ *is an extended canterm of rank* $m+1$ *and degree* 0 *iff* α *is an extended canterm of rank* m *and degree* 0.

We are now in a position to define the system itself.

DEFINITION 17.33 (Minimal logic mentioning propositions). The system $\mathcal{F}_{22}^{M}$ is the system whose non-redex constants are H, Ξ, Λ, V, Σ, and the canonical atoms (including ⊥, E, and Q). The terms F and L are defined as follows:

$$F \equiv \lambda xyz.(\forall u)(xu \supset y(zu)),$$
$$L \equiv FEH,$$

where ∀ and ⊃ are defined as usual in terms of Ξ and E:

$$\Pi \equiv \Xi E, \qquad P \equiv \lambda xy.\Xi(Kx)(Ky).$$

The system has the following axioms:

(EX)	EX	for all terms X,
(ρ)	QXX	for all terms X,
(H⊥)	H⊥,	
(FE)	LE.	

The rules are: Eqβη, (∧e), (∧i), (∨e), (∃e), (Ξe) from 17.3 and 17.6, modified by letting the lower-case Greek letters now denote arbitrary terms; and, in addition, the following:

(∨i*) $$\frac{X \qquad HY}{X \vee Y} \qquad \frac{Y \qquad HX}{X \vee Y}$$

(∃i*) $$\frac{XZ \qquad LX}{\Sigma X}$$

(Ξi*) $$\frac{\begin{matrix}[Xx] \\ Yx\end{matrix} \qquad LX}{\Xi XY}$$

Condition: x is not free in X, Y, or in any undischarged assumption;

(Qe*) $$\frac{QXY \qquad ZX \qquad LZ}{ZY}$$

(H) $$\frac{X}{HX}$$

(H∧) $$\frac{HX \qquad HY}{H(X \wedge Y)}$$

(H∨) $$\frac{HX \qquad HY}{H(X \vee Y)}$$

(HΞ) $$\frac{LX \qquad LY}{H(\Xi XY)}$$

(H∃) $$\frac{LX}{H(\Sigma X)}\ .$$

Finally, for each canonical atom θ of degree n, we include the rule

(H θ)
$$\frac{EX_1 \quad EX_2 \quad \dots \quad EX_n}{H(\theta X_1 X_2 \dots X_n)}\ .$$

As instances of this last rule, we have

(H E) $$\frac{EX}{H(EX)}$$ (HQ) $$\frac{EX \quad EY}{H(QXY)}$$

(although (H E) is redundant, since it is an instance of rule (H)).

DEFINITION 17.34 (Intuitionistic logic mentioning propositions). The system $\mathcal{F}^{I}_{22}$ is obtained from $\mathcal{F}^{M}_{22}$ by adding the rule

($\bot$ j*)
$$\frac{\bot \qquad HX}{X}\ .$$

DEFINITION 17.35 (Classical logic mentioning propositions). The system $\mathcal{F}^{C}_{22}$ is obtained from $\mathcal{F}^{M}_{22}$ by adding the rule

($\bot$ c*)
$$\frac{\begin{array}{c}[\neg X]\\ \bot\end{array} \qquad HX}{X}\ .$$

REMARK 17.36. Systems $\mathcal{F}_3$, based on Π and P as non-redex constants and equivalent to the three systems $\mathcal{F}_{22}$, are easy to define; it is only necessary to drop rules (Ξ e), (Ξ i*) and (H Ξ), and to add in their place rules ($\supset$ e), ($\forall$ e), ($\forall$ i), and the following:

($\supset$ i*)
$$\frac{\begin{array}{c}[X]\\ Y\end{array} \qquad HX}{X \supset Y}\ ,$$

(H $\supset$)
$$\frac{HX \qquad HY}{H(X \supset Y)}\ ,$$

(H∀)
$$\frac{LX}{H(\Pi X)}\,.$$

Since each of these systems is completely equivalent to the corresponding system $\mathcal{F}_{22}$, we shall not consider them further.

THEOREM 17.37. *If every undischarged assumption in a deduction in any of the systems* $\mathcal{F}_{22}$ *is an extended canterm of degree* 0, *then so is every term appearing as a step in the deduction.*

Proof. See Curry 1973, pp. 490-491. □

It follows from this result that the normalization theorem and its consequences (as given in §17D) apply to these systems; see Seldin 1975 and 1977a.

These systems also have interesting models. The models were first constructed in Scott 1975b, and then were studied more systematically in Aczel 1980, where they were given the name Frege structures. In fact, these models are models of even stronger systems (see Seldin 1976, pp. 84-85), which are defined as follows.

DEFINITION 17.38. If $\mathcal{F}_{22}$ is any of the systems defined in Definitions 17.33 - 17.35, then $\mathcal{F}_{22}^{+}$ is obtained from $\mathcal{F}_{22}$ by making the following replacements of rules:

(∨i*)	by	(∨i) from 17.3;
(∃i*)	by	(∃i) from 17.3;
(Qe*)	by	(Qe) from 17.9;
(H∧), (HΞ)	by	(H∧*), (HΞ*) below:

(H∧*)
$$\frac{HX \qquad \overset{[X]}{HY}}{H(X \wedge Y)}$$

(HΞ*)
$$\frac{LX \qquad \overset{[Xx]}{H(Yx)}}{H(\Xi XY)}$$
Condition: x is not free in X, Y or any undischarged assumption.

and, in the system $\mathcal{F}_{22}^{I}$,

(⊥j*)	by	(⊥j) from 17.4.

17F QUANTIFYING OVER PROPOSITIONS

While the systems of the previous section have some interesting models, they are still not substantially stronger than first-order logic. For although propositions may be mentioned in these systems, it is, for example, impossible to proceed from

(11) $$\mathcal{B},\ \mathsf{H}x \vdash X$$

to

(12) $$\mathcal{B} \vdash (\forall x)(\mathsf{H}x \supset X)$$

by the rules of the system (where, of course, x does not occur free in $\mathcal{B}$). One might suppose that we could make this inference possible by making H a canonical atom, adding

$$\mathsf{LH}$$

as an axiom, and applying rule (Ξ i*), but as is shown in Bunder and Meyer 1978, the resulting system is inconsistent. Furthermore, Aczel 1980 shows that the class of all propositions cannot be internally defined in a Frege structure, and it follows from this that there is no way to extend the system to allow inferences from (11) to (12).

Martin Bunder, in his 1983a, has proposed a system called HOPC (Higher Order Predicate Calculus), in which second- and higher-order logic can be interpreted. It is a type-theory, but Bunder has adopted the philosophy of Chapter 14 and made the rules of type-assignment part of the deductive structure of the theory, rather than incorporating types into the rules of term-formation as in Chapter 13. He has shown in his 1983b and 1983c that if some simple additional axioms relating to types and equality are assumed, then most of the axioms of Zermelo-Fraenkel set theory can be interpreted by provable formulas. (The main exceptions are the axiom of grounding, which is false, and extensionality and choice, which may be consistently added.) He has also given a weak 'consistency' proof for the basic system in his 1983d, by showing that there are some terms which are not provable; however, the proof does not show that any of these terms has the type of propositions.

One version of HOPC is formulated as follows.

DEFINITION 17.39 (Basic HOPC). The system basic HOPC has non-redex constants Ξ, H, and A. The constants P and F are defined as follows:

$$P \equiv \lambda xy.\Xi(Kx)(Ky),$$
$$F \equiv \lambda xyz.(\forall u)(xu \supset y(zu))$$
$$\equiv \lambda xyz.\Xi x(Byz).$$

Types are defined by taking as atomic types A (individuals) and H (propositions), and forming compound types with F as usual. There are no axioms (although, of course, axioms may be added to extend the system). The rules are Eq$\beta\eta$, (Ξ e) from 17.6 (from which, as we have seen, ($\supset$ e) follows), ($\supset$ i*) from 17.36, (H) from 17.33, and the following three:

(H$\supset$*)
$$\frac{\mathsf{H}X \qquad \begin{matrix}[X]\\ \mathsf{H}Y\end{matrix}}{\mathsf{H}(X \supset Y)}$$

($\Xi\alpha$ i)
$$\frac{\begin{matrix}[\alpha x]\\ Xx\end{matrix}}{\Xi\alpha X}$$
Condition: α is a type, x is not free in X or any undischarged assumption;

(H$\Xi\alpha$)
$$\frac{\begin{matrix}[\alpha x]\\ \mathsf{H}(Xx)\end{matrix}}{\mathsf{H}(\Xi\alpha X)}$$
Condition: as above.

REMARK 17.40. If we use $(\forall\alpha x)X$ as an abbreviation for $\Xi\alpha(\lambda x.X)$, and if we ignore uses of Rule Eq, we can rewrite the last two rules as follows:

$$\frac{\begin{matrix}[\alpha x]\\ X\end{matrix}}{(\forall\alpha x)X} \qquad \frac{\begin{matrix}[\alpha x]\\ \mathsf{H}X\end{matrix}}{\mathsf{H}((\forall\alpha x)X)},$$

where now x may occur free in X.

DISCUSSION 17.41 (Other connectives and quantifiers). As is well known (see, for example, Prawitz 1965, p. 67), the remaining connectives and quantifiers can be defined in terms of Ξ in second- and higher-order logic. These definitions work in HOPC as well: let us list them. For $\wedge$ and $\vee$ we define

$$\Lambda \equiv \lambda xy.(\forall Hz)((x \supset (y \supset z)) \supset z),$$
$$V \equiv \lambda xy.(\forall Hz)((x \supset z) \supset ((y \supset z) \supset z)),$$

and then abbreviate

$$X \wedge Y \equiv \Lambda XY, \qquad X \vee Y \equiv VXY.$$

For an existential quantifier of each type α, we define

$$(\exists\alpha x)X \equiv X\alpha(\lambda x.X),$$

where

$$X \equiv \lambda ux.(\forall Hz)((\forall uw)(xw \supset z) \supset z).$$

The usual introduction and elimination rules and rules like $(H\wedge)$ and $(H\vee)$ in 17.33 will follow from these definitions, except that some of them will require extra premises assigning types (such as H or L) to some of the terms involved. These are straightforward to work out.

Equality can be defined for each type α as follows:

$$Q\alpha \equiv \lambda xy.(\forall(F\alpha H)z)(zx \supset zy).$$

It is easy to prove that, for each type α, $Q\alpha$ satisfies (ρ) and a rule like $(Q\,e)$ in 17.9. In addition, it is straightforward to prove

$$\alpha X, \alpha Y \vdash H(Q\alpha XY).$$

In order to define negation, it is sufficient to define $\bot$ and to prove $H\bot$. We follow the usual practice in second-order logic and define

$$\bot \equiv (\forall Hx)x \equiv \Xi HI.$$

(Informally, this $\bot$ says that every proposition is true.) We have the following proof of $H\bot$, using rule $(H\Xi\alpha)$ from 17.39 in the special case $\alpha \equiv H$:

$$\dfrac{\dfrac{\overset{1}{[Hx]}}{H(Ix)}\,Eq}{H(\Xi HI).}\,(H\Xi H-1)$$

But this definition automatically gives us an intuitionistic system, since rule $(\bot\,j^*)$ in 17.34 follows, by $(\Xi\,e)$ (17.6). For a minimal logic, we can define

$$\bot_M \equiv Q(FH(FHH))\Lambda V.$$

For classical logic, we can either add the rule $\bot c$ of 17.5, or else add a single axiom such as

$$(\forall Hx)(\neg\neg x \supset x)$$

or

$$(\forall Hx)(\neg x \vee x).$$

DISCUSSION 17.42 (Generalized HOPC). A more general version of this system can be given by making the definition of a type part of the deductive structure. This can be done by adding a new atomic constant T, and taking for it the axioms

$$\mathsf{TA}, \qquad \mathsf{TH},$$

and the rule

$$\frac{\mathsf{TX} \qquad \mathsf{TY}}{\mathsf{T(FXY)}.}$$

Rules ($\Xi\alpha$i) and ($H\Xi\alpha$) are replaced by the rules

(Ξi**)

$$\frac{\begin{array}{cc}[\mathsf{Xx}] & \\ \mathsf{Yx} & \mathsf{TX}\end{array}}{\Xi\mathsf{XY}}$$

Condition: x is not free in X, Y or any undischarged assumption

and

($H\Xi$**)

$$\frac{\begin{array}{cc}[\mathsf{Xx}] & \\ \mathsf{H(Yx)} & \mathsf{TX}\end{array}}{\mathsf{H}(\Xi\mathsf{XY}).}$$

Condition: as above

Finally, we need the additional rules (ΞTi) and ($H\Xi$T), which are special cases of rules ($\Xi\alpha$i) and ($H\Xi\alpha$) with T in place of α.

17G PARTIALLY DEFINED OPERATIONS

A different approach has been developed by Solomon Feferman, who has studied systems which can be proved to be conservative extensions of standard systems whose consistency is known. This approach guarantees systems whose adequacy for standard theories is easy to establish, and some of these systems have features in common with those we have been considering.

In his 1975a, 1975b, 1978, and 1979, Feferman has studied systems in which combinators exist but are not necessarily defined for all arguments. In these systems, application is not a postulated binary operation on terms, but is instead a predicate App, with three arguments, App(X,Y,Z) being read as

$$\mathsf{XY} = \mathsf{Z}.$$

An application-term T_1T_2 is then regarded as an abbreviation, and is said to be defined just when $\vdash (\exists x)\mathrm{App}(T_1,T_2,x)$. The postulates for the basic combinators are that KXY and SXY are always defined when X and Y are defined, but SXYZ need not be always defined. (Similar ideas occur in Chauvin 1979, in Goodman 1970 and 1972, and in the theory of uniformly reflexive structures mentioned in Remark 12.72(IIa).)

This idea can be incorporated in Bunder's HOPC by assuming that the type A of individuals is the type of defined terms. Among the basic postulates for A we would include

$$\frac{AX \qquad AY}{A(KXY)} \qquad\qquad \frac{AX \qquad AY}{A(SXY)}\ ,$$

and perhaps rules of the form

$$\frac{AX_1 \quad AX_2 \quad \ldots \quad AX_n}{A(cX_1X_2\ldots X_n)}\ ,$$

where c is a non-redex constant. This would be for a system based on CL. For a system based on λ we might appear to need the axiom scheme

$$A(\lambda x.M),$$

but since the equations

$$\begin{aligned} I &= SKK, \\ \lambda x.U &= KU \qquad\qquad \text{if } x \notin FV(U), \\ \lambda x.MN &= S(\lambda x.M)(\lambda x.N) \end{aligned}$$

are all provable in λβ, the above rules for S and K would suffice here also.

It is important to note, however, that the resulting systems are still not equivalent to those of Feferman, since the latter have class variables which cannot be part of application terms. However, Feferman does have class abstraction terms, and he has introduced an intensional equivalence operator and shown that an unrestricted abstraction principle expressed with this operator is consistent (see Aczel and Feferman 1980). Bunder 1982 makes some observations on the set theory that can be developed within the system of Aczel and Feferman 1980 and on its relation to his HOPC, but he does not consider Feferman's partial combinatory operators, and the remarks are incomplete. The exact relationship between the systems of Feferman and Bunder's HOPC has not yet been completely determined.

CHAPTER EIGHTEEN

GÖDEL'S CONSISTENCY-PROOF FOR ARITHMETIC

18A PRIMITIVE RECURSIVE FUNCTIONALS OF FINITE TYPE

In this chapter we shall look at an example of an application of typed combinatory logic in proof theory, namely a consistency-proof for first-order intuitionist arithmetic taken from Gödel 1958.

Gödel's method has also been applied to stronger theories such as analysis, e.g. in Kreisel 1959, Spector 1962, Girard 1971 and Luckhardt 1973. However, the present account will stick to arithmetic for simplicity's sake, and will only treat enough detail to make the role of combinators clear. A fuller description of the method and its applications is in Troelstra 1973, Chapter III, §§5 - 7 and the references cited therein.

The key to the method is what is called the Dialectica interpretation; each formula of arithmetic is interpreted as an $\exists\forall$-statement about a certain set of functions, called the primitive recursive functionals of finite type. Roughly speaking, this set is the smallest set of functions which contains the successor-function σ, the number 0 (as a 'function' of no arguments), and is closed under the operations of λ-abstraction, application of a function to an argument, and primitive recursion.

We shall not need to make this more precise, as the consistency-proof does not formally involve the functions themselves, but only a system of notation for them. Such a system can be based on either λ-calculus, or combinatory logic. However, the full strength of $\lambda\beta$-equality is not needed so we shall use combinatory logic here, following Grzegorczyk 1964 and Sanchis 1967. (Some other notation-systems are in the references cited in Troelstra 1973, §1.6.18.)

DEFINITION 18.1. Types are defined as in 13.1, with only one atomic type, N (for the set of all natural numbers).

DEFINITION 18.2. Typed CL-terms are defined as in 13.16. The only atomic constants besides the combinators are the following:

(a) $\bar{0}^{N}$, $\bar{\sigma}^{(N\to N)}$ (zero and successor),

(b) for each type α, a distinct constant R_α with type

$$\alpha \to (N \to \alpha \to \alpha) \to N \to \alpha \; .$$

NOTATION 18.3. 'Term' here always means 'Typed CL-term' as defined above. For each α, R_α is called the recursion combinator corresponding to α; cf. the recursion combinators in Chapter 4.

Terms of form $\bar{\sigma}^{n}\bar{0}$ are called numerals, and we define

$$\bar{n} \equiv \bar{\sigma}^{n}\bar{0}.$$

(Each numeral has type N.)

Type-superscripts will be omitted except when needed for emphasis. Other conventions are as in Chapter 13. Recall that [N/x]M is only defined when N and x have the same type.

The definition of $\lambda^{*}x.M$ is assumed to be 13.20, but any of the usual alternatives would do instead.

DEFINITION 18.4. The theory of typed R-equality, called $(CL+R)_t$, is defined by adding to Definition 13.17 the following two axiom-schemes:

(a) $R_\alpha X^\alpha Y^{(N\to\alpha\to\alpha)}\bar{0} \quad = X,$

(b) $R_\alpha X^\alpha Y^{(N\to\alpha\to\alpha)}(\bar{\sigma}\bar{n}) = Y\bar{n}(R_\alpha XY\bar{n}).$

(It is easy to check that both sides of (a) and (b) are typed terms with the same type, namely α.)

DEFINITION 18.5. An R-redex is any term with form

$$R_\alpha X^\alpha Y^{(N\to\alpha\to\alpha)}\bar{n}.$$

Contracting an occurrence of an R-redex means replacing the left side of 18.4(a) or (b) by the right side. Reduction, $\rhd_{wR}$, is defined in the usual way by contracting weak and R-redexes (cf. Definition 2.7 for details). The equality relation, $=_{wR}$, is defined by contractions and reversed contractions as usual. A wR-normal form is a term containing no weak or R-redexes.

REMARK 18.6. The substitution lemmas (2.12 and 2.23) hold for $\triangleright_{wR}$ and $=_{wR}$. It is easy to show that two wR-equal terms always have the same type. Also

$$X =_{wR} Y \iff (CL+R)_t \vdash X = Y.$$

The following properties of reduction are less trivial.

THEOREM 18.7 (Church-Rosser theorem). *If* $X =_{wR} Y$, *then there exists* Z *such that* $X \triangleright_{wR} Z$ *and* $Y \triangleright_{wR} Z$.

Proof. Appendix 1, Theorem A1.11 and Remark A1.13. □

COROLLARY 18.7.1. *A term cannot have more than one* wR-*normal form.*

THEOREM 18.8 (Normalization theorem). *Every term has a* wR-*normal form.*

Proof. Appendix 2, Theorem A2.6. □

COROLLARY 18.8.1. *The relation* $=_{wR}$ *is decidable.*

REMARK 18.9. All the consistency-proof except the normalization theorem could be carried out in a formal system of first-order arithmetic, if we took the trouble to translate terms, etc. into natural numbers by some Gödel-numbering. Hence, by Gödel's well-known result on the unprovability of consistency, no proof of the normalization theorem can be formalized as a proof in arithmetic (unless arithmetic is actually inconsistent). See Appendix 2, Remark A2.10.

THEOREM 18.10. *Every closed term with type* N *reduces to a unique numeral.*

Proof. By Theorem 18.8, all we need prove is that every closed normal form X with type N must be a numeral. This is done by induction on X. First, every X has form

$$X \equiv aX_1 \dots X_n \qquad (n \geq 0),$$

where a is an atom. It is an easy exercise to show that if a was an $S_{\alpha,\beta,\gamma}$ or $K_{\alpha,\beta}$ or R_α, then X would either have compound type or contain a redex. Hence a can only be $\bar{0}$ or $\bar{\sigma}$. The result then follows by the induction hypothesis. □

DEFINITION 18.11. A typed term X <u>combinatorially defines</u> an n-argument total function ϕ of natural numbers, iff

(a) X has type $N \to N \to \dots N \to N$ (with n arrows),

(b) $X\bar{m}_1 \dots \bar{m}_n =_{wR} \overline{\sigma(m_1, \dots, m_n)}$ for all $m_1, \dots, m_n$.

REMARK 18.12. The above definition does not need to mention partial functions, because by Theorem 18.8, $X\bar{m}_1 \dots \bar{m}_n$ always has a normal form.

Not every recursive total function is definable; otherwise typed R-equality would be undecidable, by an argument like that in Chapter 5. Schwichtenberg 1975 shows that the definable functions are exactly those definable by transfinite recursion up to ordinals $< \varepsilon_0$. (Cf. Remark A2.10.)

On the other hand, $(CL+R)_t$ is not just a theory of functions of numbers; it has terms to represent higher-type functions too, and this is important.

THEOREM 18.13. *Every primitive recursive function* ϕ *can be combinatorially defined by a typed term* $\bar{\phi}$.

Proof. Like Lemma 4.5, but with the following changes.
For (I) and (II) use the atoms $\bar{\sigma}^{(N \to N)}$ and $\bar{0}^N$. For (V) use the atom R_N instead of the constructed R (cf. Remark 4.19). Give all variables the type N.

In all cases the resulting $\bar{\phi}$ is a typed term with type $N \to N \to \dots N \to N$ (n arrows). Also the proof of 4.5 gives

$$\bar{\phi}\bar{m}_1 \dots \bar{m}_n =_{wR} \overline{\phi(m_1, \dots, m_n)}. \qquad \square$$

COROLLARY 18.13.1. *There exist terms* $\bar{+}$, $\bar{\times}$, $\bar{\dot{-}}$, $\bar{E}$ *with type* $N \to N \to N$, *such that*

$$\bar{+}\,\bar{m}\,\bar{n} =_{wR} \overline{(m+n)};$$

$$\bar{\times}\,\bar{m}\,\bar{n} =_{wR} \overline{(m \times n)};$$

$$\bar{\dot{-}}\,\bar{m}\,\bar{n} =_{wR} \begin{cases} \overline{(m-n)} & \text{if } m \geq n, \\ \bar{0} & \text{if } m < n; \end{cases}$$

$$\bar{E}\,\bar{m}\,\bar{n} =_{wR} \begin{cases} \bar{1} & \text{if } m \neq n, \\ \bar{0} & \text{if } m = n. \end{cases}$$

Proof. Compare the second part of Exercise 4.24(a). □

DEFINITION 18.14. Corresponding to the propositional connectives $\wedge$, $\vee$, $\supset$, $\neg$, define terms $\bar{\wedge}$, $\bar{\vee}$, $\bar{\supset}$, $\bar{\neg}$ as follows:

$$\bar{\wedge} \equiv \bar{+},$$

$$\bar{\neg} \equiv \lambda^* x^N . (\bar{\dot{-}}\,\bar{1}\,x),$$

$$\bar{\supset} \equiv \lambda^* x^N y^N . \bar{\neg}(\bar{\wedge} x(\bar{\neg} y)),$$

$$\bar{\vee} \equiv \lambda^* x^N y^N . \bar{\neg}(\bar{\wedge}(\bar{\neg} x)(\bar{\neg} y)).$$

LEMMA 18.15. *If* X *and* Y *are closed terms with type* N, *then* $\bar{\wedge}XY$, $\bar{\neg}X$, $\bar{\supset}XY$, $\bar{\vee}XY$ *are also closed terms with type* N; *hence they reduce to numerals. And we have (where* $\rhd$ *is* $\rhd_{wR}$*)*:

$$\bar{\wedge}XY \rhd \bar{0} \iff X \rhd \bar{0} \text{ and } Y \rhd \bar{0};$$

$$\bar{\neg}X \rhd \bar{0} \iff X \rhd \overline{k+1} \text{ for some } k;$$

$$\bar{\vee}XY \rhd \bar{0} \iff X \rhd \bar{0} \text{ or } Y \rhd \bar{0};$$

$$\bar{\supset}XY \rhd \bar{0} \iff Y \rhd \bar{0} \text{ or } X \rhd \overline{k+1} \text{ for some } k.$$

Notation. $\bar{\wedge}XY$ will be called '$X\bar{\wedge}Y$' from now on, and similarly for $\bar{\vee}$, $\bar{\supset}$, $\bar{+}$, $\bar{\times}$, $\bar{\dot{-}}$.

LEMMA 18.16. *For each* α *there is a term* D_α *with type* $\alpha \to \alpha \to N \to \alpha$, *such that*

$$D_\alpha X^\alpha Y^\alpha \bar{0} =_{wR} X^\alpha,$$

$$D_\alpha X^\alpha Y^\alpha \overline{(k+1)} =_{wR} Y^\alpha.$$

Proof. Compare the construction of D^*, (4.21). □

REMARK 18.17.

(a) The $\bar{E}$ in Corollary 18.13.1 is an equality-test for closed terms of type N, because by 18.10, each such term is equal to a unique numeral. There is no equality-test in this system for terms with any other type.

(b) Instead of R_α we could have postulated atoms Z_α (one for each α) called iterators, with the axiom-schemes

$$Z_\alpha \bar{n} = \lambda^* x^{\alpha\to\alpha} y^\alpha . x^n y$$

(cf. 4.22)). From such Z_α, typed terms R_α satisfying 18.4(a) - (b) can be defined by a construction like (4.23).

This completes the basic combinatory apparatus for the consistency-proof. We shall now set up a formal theory of intuitionist arithmetic. The axioms and rules below are chosen especially to make the consistency-proof easy, and their equivalence to other formulations, for example Kleene 1952, pp. 82 and 101, is left as a tedious exercise (cf. Spector 1962, pp. 3 - 4).

DEFINITION 18.18 (Heyting arithmetic, HA). HA is a formal theory defined as follows.

Arithmetical terms: an infinity of variables and one constant, 0, are terms; if s and t are terms, then so are (s+t), (s×t), (s′).

Arithmetical formulas: for any terms s, t, the equation (s = t) is a formula; if $\mathcal{A}$ and $\mathcal{B}$ are formulas, then so are $(\mathcal{A}\wedge\mathcal{B})$, $(\mathcal{A}\vee\mathcal{B})$, $(\mathcal{A}\supset\mathcal{B})$, $(\neg\mathcal{A})$, $(\forall x\mathcal{A})$, $(\exists x\mathcal{A})$, where x is any variable.

Axiom-schemes:

$\mathcal{A}\supset(\mathcal{A}\wedge\mathcal{A})$,	$\neg(x'=0)$,
$(\mathcal{A}\vee\mathcal{A})\supset\mathcal{A}$,	$x=y\supset x'=y'$,
$(\mathcal{A}\wedge\mathcal{B})\supset(\mathcal{B}\wedge\mathcal{A})$,	$x'=y'\supset x=y$,
$(\mathcal{A}\vee\mathcal{B})\supset(\mathcal{B}\vee\mathcal{A})$,	$x=y\supset(x=z\supset y=z)$,
$(\mathcal{A}\wedge\mathcal{B})\supset\mathcal{B}$,	$x+0=x$,
$\mathcal{B}\supset(\mathcal{A}\vee\mathcal{B})$,	$x+y'=(x+y)'$,
$(0=0')\supset\mathcal{A}$,	$x\times 0=0$,
$\neg\mathcal{A}\supset(\mathcal{A}\supset 0=0')$,	$x\times y'=(x\times y)+x$,
$(\mathcal{A}\supset 0=0')\supset\neg\mathcal{A}$.	

Rules of inference:

$$\frac{\mathcal{A} \quad \mathcal{A} \supset \mathcal{B}}{\mathcal{B}} \qquad \frac{\mathcal{A} \supset \mathcal{B} \quad \mathcal{B} \supset \mathcal{C}}{\mathcal{A} \supset \mathcal{C}}$$

$$\frac{\mathcal{A} \supset (\mathcal{B} \supset \mathcal{C})}{(\mathcal{A} \wedge \mathcal{B}) \supset \mathcal{C}} \qquad \frac{(\mathcal{A} \wedge \mathcal{B}) \supset \mathcal{C}}{\mathcal{A} \supset (\mathcal{B} \supset \mathcal{C})}$$

$$\frac{\mathcal{B} \supset \mathcal{C}}{(\mathcal{A} \vee \mathcal{B}) \supset (\mathcal{A} \vee \mathcal{C})} \qquad \frac{\forall x \mathcal{A}(x)}{\mathcal{A}(t)}$$

$$\frac{\mathcal{C} \supset \mathcal{A}(x)}{\mathcal{C} \supset \forall x \mathcal{A}(x)} \qquad \frac{\mathcal{C} \supset \forall x \mathcal{A}(x)}{\mathcal{C} \supset \mathcal{A}(x)}$$

$$\frac{\mathcal{A}(x) \supset \mathcal{C}}{(\exists x \mathcal{A}(x)) \supset \mathcal{C}} \qquad \frac{(\exists x \mathcal{A}(x)) \supset \mathcal{C}}{\mathcal{A}(x) \supset \mathcal{C}}$$

$$\frac{\mathcal{A}(0) \qquad \forall x(\mathcal{A}(x) \supset \mathcal{A}(x'))}{\mathcal{A}(x)} .$$

(In the first quantifier-rule, t is any term free for x in $\mathcal{A}(x)$; and in the other four quantifier-rules, $x \notin FV(\mathcal{C})$.

Each arithmetical formula will be interpreted as a formula of the following kind.

DEFINITION 18.19. An $\exists\forall$-CL-formula is any expression

$$\exists y_1 \ldots y_m \; \forall z_1 \ldots z_n \; [X \rhd \bar{0}],$$

where $m, n \geq 0$, the variables $y_1,\ldots,y_m,z_1,\ldots,z_n$ are distinct (by the usual convention) and may have any types, and X is a CL-term with type N.

NOTATION 18.20. Sequences $x_1,\ldots,x_k$, $y_1,\ldots,y_m$, $z_1,\ldots,z_n$, $Y_1,\ldots,Y_m$ (k, m, $n \geq 0$) may be denoted by

$$\underline{x},\ \underline{y},\ \underline{z},\ \underline{Y}.$$

Variables may be displayed when writing CL-terms, for example

$$A\{\underline{x},\ \underline{y},\ \underline{z}\} \equiv A.$$

Substitution may be abbreviated in such a way that, for example,

$$A\{\underline{X},\ \underline{YX},\ \underline{z}\}$$

denotes

$$[X_1/x_1]\dots[X_k/x_k][(Y_1x_1\dots x_k)/y_1]\dots[(Y_mx_1\dots x_k)/y_m]A.$$

Whenever such a substitution is mentioned, it will be assumed that X_i has the same type as x_i, and the type of Y_j is chosen such that $Y_jx_1\dots x_k$ has the same type as y_j ($1 \le i \le k$, $1 \le j \le m$).

DEFINITION 18.21.

(a) A quantifier-free ∃∀-CL-formula

$$[A\{\underline{x}\} \rhd \bar{0}],$$

where $\underline{x} = FV(A)$, is called <u>true</u> iff, for all (sequences of) closed terms $\underline{X}$,

$$A\{\underline{X}\} \rhd_{wR} \bar{0}.$$

(b) An ∃∀-CL-formula of form

$$\exists\underline{y}\ \forall\underline{z}\ [A\{\underline{x},\ \underline{y},\ \underline{z}\} \rhd \bar{0}],$$

where $FV(A) = \{\underline{x},\ \underline{y},\ \underline{z}\}$, is called <u>true</u> iff there exist closed terms $\underline{Y}$ such that for all closed $\underline{X}$, $\underline{Z}$,

$$A\{\underline{X},\ \underline{YX},\ \underline{Z}\} \rhd_{wR} \bar{0}.$$

REMARK 18.22. In (a) above, the truth of the formula $[A \rhd \bar{0}]$ does not always imply that the term A wR-reduces to $\bar{0}$. For example, let

$$A \equiv R_N\bar{0}(\lambda^* uv.v)x$$

where u, v, x have type N; then for all closed X with type N,

$$[X/x]A \rhd_{wR} \bar{0},$$

but A itself does not reduce to $\bar{0}$.

So the truth of $[A \rhd \bar{0}]$ does not coincide with provability of the equation $A = \bar{0}$ in $(CL+R)_t$. However, it would so coincide if we added to $(CL+R)_t$ an induction rule and a little extra apparatus. (Cf. Troelstra 1973, §1.6.3 ff, the quantifier-free system qf-HA^ω.)

18B THE DIALECTICA INTERPRETATION

DEFINITION 18.23. We interpret each term t of HA as a CL-term t^*, and each formula $\mathcal{F}$ of HA as an $\exists\forall$-CL-formula $\mathcal{F}^*$, thus:

Terms:

(i) to each variable v, assign a distinct type-N variable v^*;

(ii) $0^* \equiv \bar{0}$;

(iii) $(t_1+t_2)^* \equiv t_1^*\bar{+}t_2^* \equiv \bar{+}t_1^*t_2^*$;

(iv) $(t_1 \times t_2)^* \equiv t_1^*\bar{\times}t_2^* \equiv \bar{\times}t_1^*t_2^*$;

(v) $(t')^* \equiv \bar{\sigma}t^*$.

Formulas:

(i) $(t_1=t_2)^* \equiv [(\bar{E}t_1^*t_2^*) \rhd \bar{0}]$ (see 18.13.1 for $\bar{E}$).

(ii) Suppose $\mathcal{A}^*, \mathcal{B}^*$ have already been defined. If they have any bound variables in common, change those in $\mathcal{B}^*$ to new ones. Now suppose

$$\mathcal{A}^* \equiv \exists\underline{y}\ \forall\underline{z}\ [A\{\underline{y}, \underline{z}\} \rhd \bar{0}],$$

$$\mathcal{B}^* \equiv \exists\underline{v}\ \forall\underline{w}\ [B\{\underline{v}, \underline{w}\} \rhd \bar{0}].$$

(Free variables in $\mathcal{A}^*$ and $\mathcal{B}^*$ are not displayed here.) Define

(a) $(\mathcal{A}\wedge\mathcal{B})^* \equiv \exists\underline{y}\,\underline{v}\ \forall\underline{z}\,\underline{w}\ [(A\{\underline{y}, \underline{z}\}\bar{\wedge}B\{\underline{v}, \underline{w}\}) \rhd \bar{0}]$;

(b) $(\mathcal{A}\vee\mathcal{B})^* \equiv \exists\underline{y}\,\underline{v}\,d\ \forall\underline{z}\,\underline{w}\ [(((\bar{E}d\bar{0})\bar{\wedge}A\{\underline{y}, \underline{z}\})\bar{\vee}((\bar{E}d\bar{1})\bar{\wedge}B\{\underline{v}, \underline{w}\})) \rhd \bar{0}]$,
where d is a new type-N variable;

(c) $(\neg\mathcal{A})^* \equiv \exists\underline{z}'\ \forall\underline{y}[\bar{\neg}A\{\underline{y}, \underline{z}'\underline{y}\} \rhd \bar{0}]$,
where $\underline{z}'$ are new variables whose types are such that $\underline{z}'\underline{y}$ have the same types as $\underline{z}$.

(d) $(\mathcal{A}\supset\mathcal{B})^* \equiv \exists\underline{z}'\underline{v}'\ \forall\underline{y}\,\underline{w}[(A\{\underline{y}, \underline{z}'\underline{y}\underline{w}\} \bar{\supset} B\{\underline{v}'\underline{y}, \underline{w}\}) \rhd \bar{0}]$,
where $\underline{z}', \underline{v}'$ are new variables such that $\underline{z}'\underline{y}\underline{w}, \underline{v}'\underline{y}$ have the same types as $\underline{z}, \underline{v}$ respectively;

(e) $(\exists x\mathcal{A})^* \equiv \exists x^*\underline{y}\ \forall\underline{z}\ [A\{\underline{y}, \underline{z}\} \rhd \bar{0}]$;

(f) $(\forall x\mathcal{A})^* \equiv \exists\underline{y}'\ \forall x^*\underline{z}[A\{\underline{y}'x^*, \underline{z}\} \rhd \bar{0}]$,
where $\underline{y}'$ are such that $\underline{y}'x^*$ have the same types as $\underline{y}$.

DISCUSSION 18.24 (Informal motivation). Suppose we use the logical symbols and variables informally for the moment, and assume that in general a formula $\forall x \exists y \mathcal{A}(x,y)$ is only true if there is a function y' that assigns to each x a value of y satisfying $\mathcal{A}$; that is, assume the 'axiom-scheme of choice':

$$\text{(AC)} \qquad \forall x\, \exists y\ \mathcal{A}(x,y) \iff \exists y'\, \forall x\ \mathcal{A}(x, y'(x)).$$

Then we can prove informally that $\mathcal{F}^*$ is equivalent to $\mathcal{F}$, as follows. (The cases in the 'proof' will correspond to the formula-cases in Definition 18.23.)

Case (i): $\mathcal{F}$ is $t_1 = t_2$. Let $FV(\mathcal{F}) = \underline{x}$. Then $\mathcal{F}^*$ is true iff $[\underline{X}/\underline{x}^*](\bar{E}t_1^* t_2^*)$ reduces to $\bar{0}$ for all closed type-N terms $\underline{X}$, i.e. iff t_1 and t_2 represent the same natural number, whatever numbers the variables $\underline{x}$ represent.

Case (ii): $\mathcal{F}$ is one of $\mathcal{A}\wedge\mathcal{B}$, $\mathcal{A}\vee\mathcal{B}$, $\neg\mathcal{A}$, $\mathcal{A}\supset\mathcal{B}$, $\exists x\mathcal{A}$, $\forall x\mathcal{A}$. Suppose that, informally, we are given

$$\mathcal{A} \iff \exists \underline{y}\ \forall \underline{z}\ [A\{\underline{y}, \underline{z}\} \rhd \bar{0}],$$

$$\mathcal{B} \iff \exists \underline{v}\ \forall \underline{w}\ [B\{\underline{v}, \underline{w}\} \rhd \bar{0}].$$

For simplicity, suppose $\mathcal{A}$ and $\mathcal{B}$ have no free variables. Then

(a):
$$
\begin{aligned}
& [\exists \underline{y}\ \forall \underline{z}\ [A \rhd \bar{0}]] \quad \wedge \quad [\exists \underline{v}\ \forall \underline{w}\ [B \rhd \bar{0}]] \\
\iff & \exists \underline{y}\,\underline{v}\ \forall \underline{z}\,\underline{w}\ [A \rhd \bar{0} \quad \wedge \quad B \rhd \bar{0}] \\
\iff & \exists \underline{y}\,\underline{v}\ \forall \underline{z}\,\underline{w}\ [(A \mathbin{\bar{\wedge}} B) \rhd \bar{0}] \qquad \text{by 18.15.}
\end{aligned}
$$

(b):
$$
\begin{aligned}
& [\exists \underline{y}\ \forall \underline{z}\ [A \rhd \bar{0}]] \quad \vee \quad [\exists \underline{v}\ \forall \underline{w}\ [B \rhd \bar{0}]] \\
\iff & \exists d\ [[d = 0 \ \wedge\ \exists \underline{y}\ \forall \underline{z}\ [A \rhd \bar{0}]] \ \vee\ [d = 1 \ \wedge\ \exists \underline{v}\ \forall \underline{w}\ [B \rhd \bar{0}]]] \\
\iff & \exists \underline{y}\,\underline{v}d\ \forall \underline{z}\,\underline{w}\ [[d = 0 \ \wedge\ A \rhd \bar{0}] \ \vee\ [d = 1 \ \wedge\ B \rhd \bar{0}]] \\
\iff & \exists \underline{y}\,\underline{v}d\ \forall \underline{z}\,\underline{w}\ [(((\bar{E}d\bar{0}) \mathbin{\bar{\wedge}} A) \mathbin{\bar{\vee}} ((\bar{E}d\bar{1}) \mathbin{\bar{\wedge}} B)) \rhd \bar{0}].
\end{aligned}
$$

(The role of d can be seen in the proof of Theorem 18.25 below, or Troelstra 1973 pp. 234-5.)

(c):
$$\neg\ \exists\underline{y}\ \forall\underline{z}\ [A\{\underline{y},\ \underline{z}\} \rhd \bar{0}]$$
$$\Longleftrightarrow\ \forall\underline{y}\ \exists\underline{z}\ \neg\ [A\{\underline{y},\ \underline{z}\} \rhd \bar{0}]$$
$$\Longleftrightarrow\ \exists\underline{z}'\ \forall\underline{y}\ \neg\ [A\{\underline{y},\ \underline{z}'\,\underline{y}\} \rhd \bar{0}] \qquad \text{by (AC)}$$
$$\Longleftrightarrow\ \exists\underline{z}'\ \forall\underline{y}\ [(\bar{\neg} A\{\underline{y},\ \underline{z}'\,\underline{y}\}) \rhd \bar{0}] \qquad \text{by 18.15.}$$

(d):
$$[\exists\underline{y}\ \forall\underline{z}\ [A \rhd \bar{0}]] \quad \supset \quad [\exists\underline{v}\ \forall\underline{w}\ [B \rhd \bar{0}]]$$
$$\Longleftrightarrow\ \forall\underline{y}\ [\forall\underline{z}\ [A \rhd \bar{0}] \quad \supset \quad \exists\underline{v}\ \forall\underline{w}\ [B \rhd \bar{0}]]$$
$$\Longleftrightarrow\ \forall\underline{y}\ \exists\underline{v}\ [\forall z\ [A \rhd \bar{0}] \ \supset\ \forall w\ [B \rhd \bar{0}]]$$
$$\Longleftrightarrow\ \forall\underline{y}\ \exists\underline{v}\ \forall\underline{w}\ \exists\underline{z}\ [A\{\underline{y},\ \underline{z}\} \rhd \bar{0} \quad \supset \quad B\{\underline{v},\ \underline{w}\} \rhd \bar{0}]$$
$$\Longleftrightarrow\ \exists\underline{v}'\ \underline{z}'\ \forall\underline{y}\,\underline{w}\ [A\{\underline{y},\ \underline{z}'\,\underline{y}\,\underline{w}\} \rhd \bar{0} \quad \supset \quad B\{\underline{v}'\,\underline{y},\ \underline{w}\} \rhd \bar{0}],$$

using (AC) three times for the last equivalence.

(e): trivial.

(f): from (AC).

A fuller discussion of the definition of $\mathcal{F}^*$ is in Troelstra 1973; p. 232, § 3.5.3 gives detailed motivation, and p. 239, Theorem 3.5.10(i) gives a formal proof of $\mathcal{F} \Leftrightarrow \mathcal{F}^*$ from suitable axioms. But to prove the consistency of arithmetic, all we need is the following theorem.

THEOREM 18.25. *If* $\mathcal{F}$ *is provable in* HA, *then* $\mathcal{F}^*$ *is true in the sense of Definition* 18.21.

COROLLARY 18.25.1. HA *is consistent.*

Proof of corollary. Let $\mathcal{F}$ be the equation $0' = 0$; then $\mathcal{F}^*$ is the closed formula

$$[\bar{E}\,\bar{1}\,\bar{0} \rhd \bar{0}].$$

If this was true, then $\bar{E}\,\bar{1}\,\bar{0}$ would reduce to $\bar{0}$; but in fact it reduces to $\bar{1}$. Hence $0' = 0$ cannot be proved in HA. □

Proof of Theorem 18.25. By induction on the length of the proof of $\mathcal{F}$ in HA, with cases according as $\mathcal{F}$ is an axiom or the conclusion of a rule. As examples we shall consider here one axiom-scheme, the first in Definition 18.18, and the most difficult rule, that of induction. The other cases will be left as a tedious but not very difficult exercise (cf. Troelstra 1973, Theorem 3.5.4).

For the first axiom-scheme, suppose $\mathcal{F}$ has the form

$$\mathcal{A} \supset (\mathcal{A} \wedge \mathcal{A}),$$

and for simplicity suppose $\mathcal{A}^*$ contains only three variables, one free and one bound by each quantifier; say

$$\mathcal{A}^* \equiv \exists y\ \forall z[A\{x,\ y,\ z\} \rhd \bar{0}].$$

To construct $\mathcal{F}^*$, we first make three 'copies' of $\mathcal{A}^*$ with distinct bound variables, but the same free variable:

$$\exists y_i\ \forall z_i\ [A\{x,\ y_i,\ z_i\} \rhd \bar{0}] \qquad (i = 1,2,3).$$

Using the last two copies, we then construct

$$(\mathcal{A} \wedge \mathcal{A})^* \equiv \exists y_2 y_3\ \forall z_2 z_3\ [(A\{x,y_2,z_2\} \bar{\wedge} A\{x,y_3,z_3\}) \rhd \bar{0}].$$

Then by Definition 18.23(iid),

$$\mathcal{F}^* \equiv \exists z_1' y_2' y_3'\ \forall y_1 z_2 z_3\ [G \rhd \bar{0}],$$

where

$$G \equiv (A\{x,\ y_1,\ z_1' y_1 z_2 z_3\}\ \bar{\supset}\ (A\{x,\ y_2' y_1,\ z_2\}\ \bar{\wedge}\ A\{x,\ y_3' y_1,\ z_3\})).$$

To show that $\mathcal{F}^*$ is true, first define closed terms $Z_1',\ Y_2',\ Y_3'$ by

$$Y_2' \equiv Y_3' \equiv \lambda^* x y_1 . y_1,$$

$$Z_1' \equiv \lambda^* x y_1 z_2 z_3 . D z_2 z_3 A_3 \qquad (\text{where } A_3 \equiv A\{x,y_1,z_3\}),$$

where D is from Lemma 18.16. Then for any closed $X,\ Y_1,\ Z_2,\ Z_3$, let

$$G^+ \equiv [X/x,\ Z_1'X/z_1',\ Y_2'X/y_2',\ Y_3'X/y_3',\ Y_1/y_1,\ Z_2/z_2,\ Z_3/z_3]G,$$

$$A_3^+ \equiv A\{X,\ Y_1,\ Z_3\}.$$

We must show that G^+ reduces to $\bar{0}$. First, G^+ reduces to the term

$$A\{X,\ Y_1,\ DZ_2Z_3A_3^+\}\ \bar{\supset}\ (A\{X,\ Y_1,\ Z_2\}\ \bar{\wedge}\ A_3^+).$$

Now A_3^+ and $A\{X, Y_1, Z_2\}$ are closed type-N terms, so by Theorem 18.10 they reduce to numerals.

<u>Case</u> 1: $A_3^+ \rhd_{wR} \bar{0}$. Then $DZ_2Z_3A_3^+ \rhd_{wR} Z_2$ by Lemma 18.16, so

$$G^+ \rhd_{wR} (A\{X, Y_1, Z_2\} \mathrel{\bar{\supset}} (A\{X, Y_1, Z_2\} \mathrel{\bar{\wedge}} \bar{0}))$$

$$\rhd_{wR} \bar{0} \qquad\qquad \text{by definition of } \bar{\supset}, \bar{\wedge}.$$

Case 2: $A_3^+ \rhd_{wR} \overline{k+1}$ for some k. Then $DZ_2Z_3A_3^+ \rhd_{wR} Z_3$, so

$$G^+ \rhd_{wR} (A_3^+ \mathrel{\bar{\supset}} (A\{X, Y_1, Z_2\} \mathrel{\bar{\wedge}} \overline{k+1}))$$

$$\rhd_{wR} (\overline{k+1} \mathrel{\bar{\supset}} (A\{X, Y_1, Z_2\} \mathrel{\bar{\wedge}} \overline{k+1}))$$

$$\rhd_{wR} \bar{0} \qquad\qquad \text{by definition of } \bar{\supset}.$$

Thus $\mathcal{F}^*$ is true, if $\mathcal{F}$ is an axiom from the first axiom-scheme.

Now let us look at the deduction-rule for induction. It says

$$\frac{\mathcal{A}(0) \qquad \forall x(\mathcal{A}(x) \supset \mathcal{A}(x'))}{\mathcal{A}(x)\ .}$$

Suppose for simplicity that $\mathcal{A}(x)$ has only one free variable u besides x, and that

(1) $$(\mathcal{A}(x))^* \equiv \exists y\ \forall z\ [A\{u, x, y, z\} \rhd \bar{0}].$$

(Here u, x are written for u^*, x^*.) Then from Definition 18.23 it can easily be shown that

(2) $$(\mathcal{A}(0))^* \equiv \exists y\ \forall z\ [A\{u, \bar{0}, y, z\} \rhd \bar{0}],$$

$$(\mathcal{A}(x'))^* \equiv \exists y\ \forall z\ [A\{u, \bar{\sigma}x, y, z\} \rhd \bar{0}].$$

To construct $(\forall x(A(x) \supset A(x')))^*$, we first make two 'copies',

$$A\{u, x, y_1, z_1\}, \qquad A\{u, \bar{\sigma}x, y_2, z_2\},$$

and then apply Definition 18.23(iid) to get

$$\exists z_1' y_2'\ \forall y_1 z_2\ [(A\{u, x, y_1, z_1'y_1z_2\} \mathrel{\bar{\supset}} A\{u, \bar{\sigma}x, y_2'y_1, z_2\}) \rhd \bar{0}].$$

Then by Definition 18.23(iif),

(3) $$(\forall x(\mathcal{A}(x) \supset \mathcal{A}(x')))^* \equiv \exists z_1'' y_2'' \forall x y_1 z_2\ [H \rhd \bar{0}],$$

where

$$H \equiv (A\{u, x, y_1, z_1''xy_1z_2\} \mathrel{\bar{\supset}} A\{u, \bar{\sigma}x, y_2''xy_1, z_2\}).$$

To deal with the induction rule, we assume that the formulas in (2) and (3) are true, and we must deduce that that in (1) is true. The assumption means that there exist closed terms Y, Z_1'', Y_2'' such that for all closed U, Z, X, Y_1, Z_2,

(4) $$A\{U, \bar{0}, YU, Z\} \rhd_{wR} \bar{0},$$

(5) $$(A\{U, X, Y_1, Z_1''UXY_1Z_2\} \supset A\{U, \bar{\sigma}X, Y_2''UXY_1, Z_2\}) \rhd_{wR} \bar{0}.$$

To show that the formula in (1) is true, first define

$$Y' \equiv \lambda^* u.R(Yu)(Y_2''u).$$

This Y' is closed; and for all U and all $m \geq 0$,

(6) $$\begin{cases} Y'U\bar{0} & \rhd_{wR} YU, \\ Y'U(\overline{m+1}) & =_{wR} Y_2''U\bar{m}(Y'U\bar{m}). \end{cases}$$

It is enough to show that for all closed U, X, Z,

$$A\{U, X, Y'UX, Z\} \rhd_{wR} \bar{0}.$$

But X has type N, so X reduces to a numeral by Theorem 18.10; hence it is enough to show that for all $k \geq 0$ and all closed U, Z,

(7) $$A\{U, \bar{k}, Y'U\bar{k}, Z\} \rhd_{wR} \bar{0}.$$

This is done by induction on k, as follows.

<u>If</u> $k = 0$: then $Y'U\bar{k} \rhd_{wR} YU$ by (6), so (7) follows by (4).

<u>If</u> $k = m+1$: then by (6), the left term in (7) converts to

(8) $$A\{U, \overline{m+1}, Y_2''U\bar{m}(Y'U\bar{m}), Z\}.$$

This is a closed type-N term, so it reduces to a numeral $\bar{p}$; we must show $p = 0$. Now (5) holds for all closed U, X, Y_1, Z_2. Take $X \equiv \bar{m}$, $Y_1 \equiv Y'U\bar{m}$, $Z_2 \equiv Z$; by (5) the term

(9) $$(A\{U, \bar{m}, Y'U\bar{m}, Z_1''U\bar{m}(Y'U\bar{m})Z\} \supset A\{U, \overline{m+1}, Y_2''U\bar{m}(Y'U\bar{m}), Z\})$$

reduces to $\bar{0}$. The second part of this term is (8), which reduces to $\bar{p}$. The first part is like the left term in (7), but with $\bar{m}$ instead of $\bar{k}$ and $Z_1''U\bar{m}(Y'U\bar{m})Z$ instead of Z. But the induction-hypothesis says that (7) holds for $\bar{m}$ and for all Z, in particular for $Z_1''U\bar{m}(Y'U\bar{m})Z$, so

(9) reduces to

$$(\bar{0} \mathbin{\bar{\supset}} \bar{p}).$$

Since (9) also reduces to $\bar{0}$, the definition of $\bar{\supset}$ implies $p = 0$. This completes the induction-step, and shows that the formula in (1) is true.

Thus we seem to have finished the induction rule. But this is not quite so; we made the assumption that A^* has only one $\exists$-quantified variable y, and unfortunately the general case is not just a trivial extension of this special case.

To deal with it, suppose that instead of (1) we have

$$\text{(10)} \qquad (A(x))^* \equiv \exists y_1 \ldots y_n \, \forall z \, [A\{u, x, y_1, \ldots, y_n, z\} \rhd \bar{0}].$$

Then the previous argument will work if we can find closed terms $Y'_1, \ldots, Y'_n$ such that for each $i = 1, \ldots, n$,

$$\text{(11)} \qquad \begin{cases} Y'_i u \bar{0} =_{wR} A_i u, \\ Y'_i u (\overline{m+1}) =_{wR} B_i u \bar{m} (Y'_1 u \bar{m}) \ldots (Y'_n u \bar{m}), \end{cases}$$

where $A_1, \ldots, A_n$ are n analogues of the Y in (4) and (6), and $B_1, \ldots, B_n$ are n analogues of the Y''_2 in (5) and (6).

So, instead of satisfying a single recursion (6), we must now find $Y'_1, \ldots, Y'_n$ satisfying an n-fold simultaneous recursion.

If $y_1, \ldots, y_n$ all have the same type, this can easily be done using the D's in Lemma 18.16.

If $y_1, \ldots, y_n$ have different types, we can use the following construction, due to K. Schütte (in correspondence, 1969). It proceeds by induction on n.

<u>For</u> $n = 1$: define $Y'_1 \equiv \lambda^* u.R(A_1 u)(B_1 u)$.

<u>For</u> $n > 1$: first define

$$C \equiv \lambda^* u v_1 \ldots v_{n-1}.R(A_n u)(\lambda^* x.B_n u x (v_1 x) \ldots (v_{n-1} x)),$$

and for $i = 1, \ldots, n-1$, define

$$E_i \equiv \lambda^* u x v_1 \ldots v_{n-1} w.D(v_i w)(B_i u x (v_1 x) \ldots (v_{n-1} x)(C u v_1 \ldots v_{n-1} x))(w \mathbin{\dot{-}} x).$$

(D comes from Lemma 18.16.) By the induction hypothesis, we can construct terms $G_1, \ldots, G_{n-1}$ such that

$$(12) \quad \begin{cases} G_i u\bar{0} & =_{wR} (\lambda^* ux.A_i u)u, \\ G_i u(\overline{m+1}) & =_{wR} E_i u\bar{m}(G_1 u\bar{m})\ldots(G_{n-1} u\bar{m}). \end{cases}$$

Using these, define

$$Y_i' \equiv \lambda^* ux.G_i uxx \quad \text{for } i = 1,\ldots,n-1,$$

$$Y_n' \equiv \lambda^* ux.Cu(G_1 ux)\ldots(G_{n-1} ux)x.$$

The proof that these satisfy (11) is straightforward. (Note that $G_i u(\overline{m+1})\bar{m}$ converts to $G_i u\bar{m}\bar{m}$.) □

APPENDIX ONE

THE CHURCH-ROSSER THEOREM

The Church-Rosser theorem for a transitive relation $\rhd$ says that $\rhd$ has what is called the Church-Rosser property:

(CR) $\quad P \rhd M,\ P \rhd N \ \Rightarrow\ (\exists T)\ M \rhd T,\ N \rhd T$

(cf. Figure A1.1(a)). It is valid for all the reducibility relations in this book.

This appendix will present the shortest known proof of (CR) for λβ-reduction. It is due to Per Martin-Löf, having been adapted in 1981 from an unpublished proof for weak reduction by William W. Tait. Proofs for other reductions will be sketched after the β-proof.

Figure A1.1

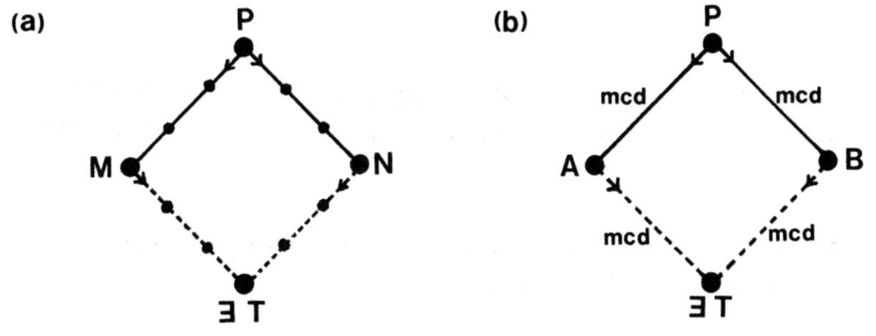

NOTATION A1.1. In this Appendix, redex always means a particular occurrence of a redex in a given term; e.g. 'let R, S be redexes in P' means 'let R, S be occurrences of redexes in P'. (For occurrence, see the note after Definition 1.6.)

Contractions, reductions, length of a reduction are defined in 3.14. An α-contraction or α-step is a change of a single bound variable (cf. Definition 1.16). Recall that a β-reduction may include α-steps. (Definition 1.22.)

THEOREM A1.2 (Theorem 1.29). (CR) *holds for λ-terms and* $\triangleright_\beta$.

This is the original 'Church-Rosser theorem', and its first proof is described in Church 1941 §7. There have been many others since. An informative modern one is very nicely presented in Barendregt 1981 §11.1, and the deeper underlying properties that make all the proofs work are exposed, investigated and explained very clearly in Klop 1980.

Before we begin the Martin-Löf-Tait proof, we shall need some definitions and lemmas.

The first definition describes what happens to a redex S in a term P when another redex R in P is contracted. Of its cases, only 1, 2, 3 will actually be needed in the proof (these are the simple cases); but 4 is often used elsewhere in reduction-theory, so it is included here.

DEFINITION A1.3 (Residuals). Let R, S be β-redexes in a λ-term P. When R is contracted, let P change to P′. The residuals of S with respect to R are redexes in P′, defined as follows (cf. Curry et al. 1958 §4B1):

Case 1: R, S are non-overlapping parts of P. Then contracting R leaves S unchanged. This unchanged S in P′ is called the residual of S.

Case 2: R ≡ S. Then contracting R is the same as contracting S. We say S has no residual in P′.

Case 3: R is part of S and R ≢ S. Then S has form (λx.M)N and R is in M or in N. Contracting R changes M to M′ or N to N′, and S to (λx.M′)N or (λx.M)N′ in P′; this is the residual of S.

Case 4: S is part of R and S ≢ R. Then R has form (λx.M)N and S is in M or in N. Contracting R changes (λx.M)N to [N/x]M.

<u>Subcase</u> 4a: S <u>is in</u> M. When $[N/x]M$ is formed from M, then S is changed to a redex S' with one of the forms

$$[N/x]S, \qquad [N/x][z_1/y_1]\ldots[z_n/y_n]S, \qquad S,$$

depending on how many times clause 1.11(f) is used in determining $[N/x]M$, and on whether S is in the scope of a λx in M. This S' is called the residual of S. (It is clearly a β-redex.)

<u>Subcase</u> 4b: S <u>is in</u> N. When $[N/x]M$ is formed, there is an occurrence of S in each substituted N. These are called the residuals of S.

Note that except in case 4b, S has at most one residual. (In 4b, S may have many or none.)

DEFINITION A1.4 (<u>MCD's</u>). Let $R_1,\ldots,R_n$ $(n \geq 0)$ be redexes in a term P. An R_i is called <u>minimal</u> iff it properly contains no other R_j. We say

$$P \rhd_{mcd} Q$$

iff Q is obtained from P by the following process, called a <u>minimal complete development</u> (<u>MCD</u>) of the set $\{R_1,\ldots,R_n\}$. First contract any minimal R_i (say $i = 1$ for convenience). By the note above, this leaves at most $n-1$ residuals $R'_2,\ldots,R'_n$, of $R_2,\ldots,R_n$. Contract any minimal R'_j. This leaves at most $n-2$ residuals. Repeat until no residuals are left. Then make as many α-contractions as you like. (This process is not unique.)

NOTES A1.5.

(a) In any non-empty set of redexes, there is always a minimal member.

(b) If $n = 0$, an MCD is just a perhaps-empty series of α-steps.

(c) A single β-contraction is an MCD of a one-member set.

(d) Non-MCD's exist; for example the reduction

$$(\lambda x.xy)(\lambda z.z) \rhd_{1\beta} (\lambda z.z)y \rhd_{1\beta} y.$$

(e) The relation $\rhd_{mcd}$ is not transitive; e.g. in (d) there is clearly no MCD from $(\lambda x.xy)(\lambda z.z)$ to y.

(f) If $M \rhd_{mcd} M'$ and $N \rhd_{mcd} N'$, then $MN \rhd_{mcd} M'N'$.

(g) It is fairly easy to show that, modulo congruence, Q is determined uniquely by the set $\{R_1,\ldots,R_n\}$. (This fact will not be needed here, however.)

LEMMA A1.6. *For* $\lambda\beta$: *if* $P \rhd_{mcd} Q$ *and* $P \equiv_\alpha P^*$, *then* $P^* \rhd_{mcd} Q$.

Proof. A boring induction on $lgh(P)$ or the number of β-steps from P to Q; cf. the proof of Lemma 3 in Appendix 1 of Hindley, Lercher and Seldin 1972. It is best to use the restricted α-conversion in Remark 1.21. □

LEMMA A1.7. *For* $\lambda\beta$: *if* $M \rhd_{mcd} M'$ *and* $N \rhd_{mcd} N'$, *then*

$$[N/x]M \rhd_{mcd} [N'/x]M'.$$

Proof. By Lemmas 1.19 and A1.6, we may assume that no variable bound in M is free in xN, and the given MCD's have no α-steps.

We proceed by induction on M. Let $R_1,\ldots,R_n$ be the redexes developed in the given MCD of M.

Case 1: $M \equiv x$. Then $n = 0$ and $M' \equiv x$, so

$$[N/x]M \equiv N \rhd_{mcd} N' \equiv [N'/x]M'.$$

Case 2: $x \notin FV(M)$. Then $x \notin FV(M')$ by Lemma 1.28(a), so

$$[N/x]M \equiv M \rhd_{mcd} M' \equiv [N'/x]M'.$$

Case 3: $M \equiv \lambda y.M_1$. Then each β-redex in M is in M_1, so M' has form $\lambda y.M_1'$ where $M_1 \rhd_{mcd} M_1'$. (We have assumed the MCD of M has no α-steps.) Hence

$$\begin{aligned}[N/x]M &\equiv [N/x](\lambda y.M_1) & \\ &\equiv \lambda y.[N/x]M_1 & \text{by 1.11(e) since } y \notin FV(xN)\\ &\rhd_{mcd} \lambda y.[N'/x]M_1' & \text{by induction hypothesis}\\ &\equiv [N'/x]M' & \text{by 1.11(e) since } y \notin FV(xN').\end{aligned}$$

<u>Case</u> 4: $M \equiv M_1 M_2$ and each R_i is in M_1 or in M_2. Then M' has form $M_1' M_2'$ where $M_j \rhd_{mcd} M_j'$ for $j = 1,2$. Hence

$$\begin{aligned} [N/x]M &\equiv ([N/x]M_1)([N/x]M_2) \\ &\rhd_{mcd} ([N'/x]M_1')([N'/x]M_2') \quad \text{by induction hypothesis and A1.5(f)} \\ &\equiv [N'/x]M'. \end{aligned}$$

<u>Case</u> 5: $M \equiv (\lambda y.L)Q$ and one R_i, say R_1, is M itself, and the others are in L or Q. In the given MCD of M, the residual of R_1 must be contracted last, because R_1 contains the other R's. Hence the MCD has form

$$\begin{aligned} M \equiv (\lambda y.L)Q &\rhd_{mcd} (\lambda y.L')Q' \qquad (L \rhd_{mcd} L',\ \ Q \rhd_{mcd} Q') \\ &\rhd_{1\beta} [Q'/y]L' \\ &\equiv M'. \end{aligned}$$

By the induction-hypothesis we have MCD's of $[N/x]L$ and $[N/x]Q$; each one may have some α-steps at the end, say

$$[N/x]L \rhd_{mcd} L^* \equiv_\alpha [N'/x]L',$$
$$[N/x]Q \rhd_{mcd} Q^* \equiv_\alpha [N'/x]Q',$$

where the MCD's to L^* and Q^* have no α-steps. Hence

$$\begin{aligned} [N/x]M &\equiv (\lambda y.[N/x]L)([N/x]Q) && \text{by 1.11(e) since } y \notin FV(xN) \\ &\rhd_{mcd} (\lambda y.L^*)Q^* && \text{without } \alpha\text{-steps} \\ &\rhd_{1\beta} [Q^*/y]L^* \\ &\equiv_\alpha [([N'/x]Q')/y][N'/x]L' && \text{by above and 1.19} \\ &\equiv_\alpha [N'/x][Q'/y]L' && \text{by 1.15(c) and 1.18} \\ &\equiv [N'/x]M'. \end{aligned}$$

This reduction is an MCD, as required. □

LEMMA A1.8. *For* $\lambda\beta$: *if* $P \rhd_{mcd} A$ *and* $P \rhd_{mcd} B$, *then there exists* T *such that* $A \rhd_{mcd} T$ *and* $B \rhd_{mcd} T$. (*Figure* A1.1(b).)

Proof. By Lemma A1.6, we may assume that the given MCD's have no α-steps. We shall use induction on P.

Case 1: $P \equiv x$. Then $A \equiv B \equiv P$. Choose $T \equiv P$.

Case 2: $P \equiv \lambda x.P_1$. Then all β-redexes in P are in P_1, and we have assumed the MCD's have no α-steps, so

$$A \equiv \lambda x.A_1, \qquad B \equiv \lambda x.B_1,$$

where $P_1 \rhd_{mcd} A_1$ and $P_1 \rhd_{mcd} B_1$. By the induction-hypothesis there is a T_1 such that

$$A_1 \rhd_{mcd} T_1, \qquad B_1 \rhd_{mcd} T_1.$$

Choose $T \equiv \lambda x.T_1$.

Case 3: $P \equiv P_1P_2$ and all the redexes developed in the MCD's are in P_1, P_2. Then the induction-hypothesis gives us T_1, T_2, and we choose $T \equiv T_1T_2$.

Case 4: $P \equiv (\lambda x.M)N$ and just one of the given MCD's involves contracting P's residual; say it is $P \rhd_{mcd} A$. Then that MCD has form

$$\begin{aligned} P \equiv\ & (\lambda x.M)N \\ \rhd_{mcd}\ & (\lambda x.M')N' && (M \rhd_{mcd} M',\ N \rhd_{mcd} N') \\ \rhd_{1\beta}\ & [N'/x]M' \\ \equiv\ & A. \end{aligned}$$

And the other MCD has form

$$\begin{aligned} P \equiv\ & (\lambda x.M)N \\ \rhd_{mcd}\ & (\lambda x.M'')N'' && (M \rhd_{mcd} M'',\ N \rhd_{mcd} N'') \\ \equiv\ & B. \end{aligned}$$

The induction-hypothesis applied to M, N gives us M^+, N^+ such that

$$M' \rhd_{mcd} M^+, \quad M'' \rhd_{mcd} M^+;$$

$$N' \rhd_{mcd} N^+, \quad N'' \rhd_{mcd} N^+.$$

Choose $T \equiv [N^+/x]M^+$. Then there is an MCD from A to T, thus:

$$\begin{aligned} A &\equiv [N'/x]M' \\ &\rhd_{mcd} [N^+/x]M^+ \qquad \text{by Lemma A1.7.} \end{aligned}$$

To construct an MCD of B, first split the MCD's of M'' and N'' into β-part and α-part, thus:

$$M'' \rhd_{mcd} M^* \equiv_\alpha M^+, \qquad N'' \rhd_{mcd} N^* \equiv_\alpha N^+,$$

where the reductions to M^* and N^* have no α-steps. Then

$$\begin{aligned} B &\equiv (\lambda x.M'')N'' \\ &\rhd_{mcd} (\lambda x.M^*)N^* \qquad \text{without } \alpha\text{-steps} \\ &\rhd_{1\beta} [N^*/x]M^* \\ &\equiv_\alpha [N^+/x]M^+ \qquad \text{by 1.19.} \end{aligned}$$

Case 5: $P \equiv (\lambda x.M)N$ and both the given MCD's contract P's residual. Then these MCD's have form

$$\begin{aligned} P &\equiv (\lambda x.M)N \\ &\rhd_{mcd} (\lambda x.M')N' \\ &\rhd_{1\beta} [N'/x]M' \\ &\equiv A, \end{aligned} \qquad \begin{aligned} P &\equiv (\lambda x.M)N \\ &\rhd_{mcd} (\lambda x.M'')N'' \\ &\rhd_{1\beta} [N''/x]M'' \\ &\equiv B. \end{aligned}$$

Apply the induction-hypothesis to M and N as in Case 4, and choose $T \equiv [N^+/x]M^+$. Then Lemma A1.7 gives the result, as above. □

Figure A1.2.

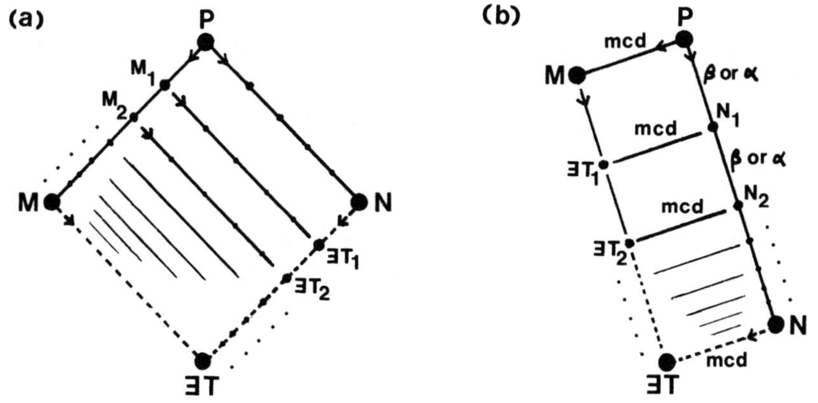

PROOF OF THEOREM A1.2. Let $P \rhd_\beta M$ and $P \rhd_\beta N$. We must find a T such that $M \rhd_\beta T$ and $N \rhd_\beta T$. By induction on the length of the reduction from P to M (cf. Figure A1.2(a)), it is enough to prove

(1) $$P \rhd_{1\beta} M,\ P \rhd_\beta N \Rightarrow (\exists T)\ M \rhd_\beta T,\ N \rhd_\beta T.$$

To prove (1), note that a single β-step is an MCD. Hence, (1) follows from

(2) $$P \rhd_{mcd} M,\ P \rhd_\beta N \Rightarrow (\exists T)\ M \rhd_\beta T,\ N \rhd_{mcd} T.$$

But (2) comes from Lemma A1.8 by induction on the number of β-steps from P to N, cf. Figure A1.2(b). □

THEOREM A1.9 (Theorem 7.12). (CR) *holds for λ-terms and* $\rhd_{\beta\eta}$.

Proof notes. For $\rhd_{\beta\eta}$, see Definitions 7.6 and 7.7. This theorem was first proved by Curry (Curry et al. 1958 §4D4), and there is another proof in Hindley 1974 §9 or Barendregt 1981 §3.3.

Exercise. Make a third proof by following the β-proof and adding extra cases for η-redexes. Hint: Definition A1.3 must be extended as follows.

Case 3: If $S \equiv (\lambda x.Lx)N$ and $R \equiv \lambda x.Lx$, then contracting R changes S to LN; we say S has no residual.

If $S \equiv \lambda x.(\lambda y.L)x$ and $R \equiv (\lambda y.L)x$, then contracting R changes S to $\lambda x.[x/y]L$; we say S has no residual.

If $S \equiv \lambda x.Mx$ and R is in M, then contracting R changes S to $\lambda x.M'x$. By Lemma 1.28(a), $x \notin FV(M')$. Take $\lambda x.M'x$ to be the residual of S.

Case 4: See Curry et al. 1958 §4B2. (Not needed in the proof.) □

THEOREM A1.10 (Theorem 2.13). (CR) *holds for* CL-*terms and* $\rhd_w$.

Proof notes. A proof was first discovered by Rosser before the λ-proof, and his account is still perhaps the best in print, though his basic combinators are not K and S. (Rosser 1935 Part I, p. 144 Theorem 12.) There is another account in Curry et al. 1972 §11B2, for any set of basic combinators.

<u>Exercise</u>. Make a proof by following the above λ-proof. <u>Hints</u>: The definition of residuals should be obvious; if not, see Curry et al. 1972 §11B2. Instead of 'MCD', use 'simultaneous contraction of a set of non-overlapping redexes'. (If we tried to simplify the λ-proof in this way, Lemma A1.8 would fail.) Omit all mention of α-contraction, and omit Lemmas A1.6, A1.7. Thus the absence of bound variables greatly simplifies the proof. □

THEOREM A1.11. (CR) *holds if we add to* λβ *or* λβη *or* CLw *the extra atomic constants* R, $\bar{0}$, $\bar{\sigma}$ *and the axiom-schemes*

(a) $$R X Y \bar{0} \rhd X, \qquad R X Y (\overline{k+1}) \rhd Y \bar{k} (R X Y \bar{k}),$$

or Z, $\bar{0}$, $\bar{\sigma}$ *and the axioms*

(b) $$Z \bar{n} \rhd Z_n.$$

Proof notes. The motivation for (a) and (b) is in Discussion 4.19; $\bar{n}$ is $\bar{\sigma}^n \bar{0}$ and Z_n is the Church numeral representing n.

For the proof, add extra cases to the proofs for $\rhd_\beta$, $\rhd_{\beta\eta}$ and $\rhd_w$. Alternatively, the β and weak-reduction results are special cases of a general Church-Rosser theorem, Hindley 1974 Theorem 2A. (Cf. Hindley 1978a §5 for a re-statement of that theorem's conditions in a λ or CL context.) The βη-result can be deduced from the β-result by Hindley 1974 §9 or Barendregt 1981 §3.3. □

WARNING A1.12. The following axiom-schemes for R seem at least as natural as (a) above:

(a′) $$R \bar{0} X Y \rhd X, \qquad R (\overline{k+1}) X Y \rhd Y \bar{k} (R \bar{k} X Y).$$

But if these were added to λβη, (CR) would fail. We would have

$$\lambda y . R \bar{0} x y \rhd_\eta R \bar{0} x, \qquad \lambda y . R \bar{0} x y \rhd \lambda y . x,$$

but $R\bar{0}x$ and $\lambda y.x$ could not be further reduced.

REMARK A1.13 (<u>Typed terms</u>). The proofs of the above Church-Rosser theorems are valid also for typed terms. To see this, all we need to check is that if P has type α and a redex R in P is contracted, then the resulting P' also has type α; and this property comes from the definitions of redexes in the typed systems (Definitions 13.9 - 11, 13.17 - 19, 18.4 - 5).

REMARK A1.14. Church-Rosser theorems have been proved and used in other contexts besides λ and CL. In general, their proofs have two parts; a 'global' topological part like the last part of the proof of Theorem A1.2, and a 'local' part like Lemma A1.8. In λ and CL the local lemma is so strong that the global part needed is almost trivial, but this is not always so.

Some examples of (CR) in non-λ-CL-contexts are in Downey and Sethi 1976, Knuth and Bendix 1967, Huet 1981, and the surveys Huet and Oppen 1980, Book 1983 and the references they cite.

Proofs for an infinitary λ-system are in Maass 1974 and Mitschke 1980.

Abstract discussions of (CR) are in Staples 1975, Rosen 1973, Sethi 1974, Huet 1980, Hindley 1969b, 1974, and Klop 1980, Chapters 2 and 3.

APPENDIX TWO

THE STRONG NORMALIZATION THEOREM FOR TYPES

As we have seen in Chapter 13, the main property of typed systems not possessed by untyped systems is that all reductions are finite, and hence each typed term has a normal form. In this appendix we shall prove this theorem for the various typed systems considered earlier.

The proofs will be variations on a method due to Tait 1967. (Cf. also Troelstra 1973 §§2.2.1 - 2.3.13.) Simpler methods are known for pure λ and CL, but Tait's has so far been the easiest to extend to new systems.

We begin with two definitions which have meaning for any reduction-concept defined by sequences of replacements (cf. Definition 3.14).

DEFINITION A2.1 (Strongly normalizable (SN) terms). A typed or untyped CL- or λ-term X is strongly normalizable with respect to a given reduction concept, iff all reductions starting at X are finite. (Cf. before Theorem 14.42, SN w.r. to weak reduction.) It is normalizable iff it reduces to a normal form. (Of course this does not always imply SN.)

DEFINITION A2.2 (Strongly computable (SC) terms). For typed CL- or λ-terms, strong computability is defined by induction on the number of occurrences of '→' in the term's type:

(a) a term of atomic type is SC iff it is SN;

(b) a term $X^{\alpha\to\beta}$ is SC iff, for every SC term Y^{α}, the term $(XY)^{\beta}$ is SC.

THEOREM A2.3 (Theorem 13.12). *In the system of typed λ-terms, there are no infinite β-reductions.*

Proof. Let 'term' mean 'typed λ-term'; then the theorem says that all terms are SN (with respect to $\triangleright_\beta$). We shall prove that

(i) each SC term is SN,

(ii) every term is SC.

The proof will consist of five simple notes and three lemmas, and the last lemma will be equivalent to the theorem.

Note 1. Each type α can be written in a unique way in the form $\alpha_1 \to \ldots \to \alpha_n \to \theta$, where θ is atomic.

Note 2. It follows immediately from A2.2(b) that if

$$\alpha \equiv \alpha_1 \to \ldots \to \alpha_n \to \theta$$

where θ is an atomic type, then X^α is SC iff, for all SC terms

$$X_1^{\alpha_1}, \ldots, X_n^{\alpha_n},$$

$(XX_1 \ldots X_n)^\theta$ is SC. And by A2.2(a), $(XX_1 \ldots X_n)^\theta$ is SC iff it is SN.

Note 3. If X^α is SC, then every term which differs from X^α only by changes of bound variables is SC. And the same holds for SN.

Note 4. By A2.2(b), if $X^{\alpha\to\beta}$ is SC and Y^α is SC, then $(XY)^\beta$ is SC.

Note 5. If X^α is SN, then every subterm of X^α is SN, because any infinite reduction from a subterm of X gives rise to an infinite reduction from X.

LEMMA 1. *Let* α *be any type*:

(a) *Every term* $(aX_1 \ldots X_n)^\alpha$, *where* a *is an atom and* $X_1, \ldots, X_n$ *are all* SN, *is* SC.

(b) *Every* SC *term of type* α *is* SN.

Proof. By induction on the number of occurrences of '$\to$' in α.

<u>Basic step</u>: α is an atomic type.

(a) Since $X_1,\ldots,X_n$ are SN, $(aX_1\ldots X_n)$ must be SN. Hence it is SC by Definition A2.2(a).

(b) By Definition A2.2(a).

<u>Induction step</u>: $\alpha \equiv \beta\to\gamma$.

(a) Let Y^β be SC. By the induction hypothesis (b), Y^β is SN. Using the induction hypothesis (a), we get that the term $(aX_1\ldots X_nY)^\gamma$ is SC. Hence, so is $(aX_1\ldots X_n)^\alpha$, by A2.2(b).

(b) Let X^α be SC, and let x^β not occur (free or bound) in X^α. By the induction hypothesis (a) with $n = 0$, x is SC. Hence, by Note 4 above, $(Xx)^\gamma$ is SC. By the induction hypothesis (b), $(Xx)^\gamma$ is SN. But then by Note 5, X^α is SN as well. □

LEMMA 2. *If* $[N^\alpha/x^\alpha]M^\beta$ *is* SC, *then so is* $(\lambda x^\alpha.M^\beta)N^\alpha$; *provided that* N^α *is* SC *if* x^α *is not free in* M^β.

<u>Remark</u>. This lemma says that if the contractum of a typed redex R is SC, and if all terms (if any) that are cancelled when R is contracted are SC, then R is SC.

Proof. Let $\beta \equiv \beta_1\to\ldots\to\beta_n\to\theta$, where θ is atomic, and let

$$M_1^{\beta_1},\ldots,M_n^{\beta_n}$$

be SC terms. Since $[N^\alpha/x^\alpha]M^\beta$ is SC, it follows by Note 2 that

(1) $$([N/x]M)M_1\ldots M_n)^\theta$$

is SN. The lemma will follow by Note 2, if we can prove that

(2) $$((\lambda x.M)NM_1\ldots M_n)^\theta$$

is SN.

Now since (1) is SN, so are all of its subterms; these include $[N/x]M$, $M_1,\ldots,M_n$. Also, by hypothesis and the preceding lemma, N is SN if it does not occur in $[N/x]M$. Therefore, an infinite reduction from (2) cannot consist entirely of contractions in $M, N, M_1,\ldots,M_n$. Hence, such a reduction must have the form

$$(\lambda x.M)NM_1\ldots M_n \rhd_\beta (\lambda x.M')N'M_1'\ldots M_n' \qquad (M \rhd M',\ \text{etc.})$$
$$\rhd_{1\beta} ([N'/x]M')M_1'\ldots M_n'$$
$$\rhd_\beta \ldots\ .$$

From the reductions $M \rhd M'$ and $N \rhd N'$ we get $[N/x]M \rhd [N'/x]M'$; hence, we can construct an infinite reduction from (1):

$$([N/x]M)M_1\ldots M_n \rhd_\beta ([N'/x]M')M_1'\ldots M_n'$$
$$\rhd_\beta \ldots\ .$$

This contradicts the fact that (1) is SN. Hence, (2) must be SN. □

LEMMA 3. *For every typed term* M^β:

(a) M^β *is* SC;

(b) *For all* $x_1^{\alpha_1},\ldots,x_n^{\alpha_n}$ *and all* SC *terms* $N_1^{\alpha_1},\ldots,N_n^{\alpha_n}$, *the term* $M^{*\beta} \equiv [N_1/x_1]\ldots[N_n/x_n]M$ *is* SC.

Proof. Part (a) is all that is needed to prove the theorem, but the extra strength of (b) is needed to make the proof of the lemma work. (Part (a) is a special case of (b), namely $N_i \equiv x_i$).

We prove (b) by induction on the construction of M.

<u>Case</u> 1. M is a variable x_i and $\beta \equiv \alpha_i$. Then M^* is N_i and the result is immediate.

<u>Case</u> 2. M is an atom distinct from $x_1,\ldots,x_n$. Then

$$M^* \equiv M,$$

which is SC by Lemma 1.

<u>Case</u> 3. $M \equiv M_1M_2$. Then

$$M^* \equiv M_1^*M_2^*.$$

By the induction hypothesis, M_1^* and M_2^* are SC, and so M^* is SC by Note 4.

Case 4. $M^\beta \equiv (\lambda x^\gamma . M_1^\delta)$, where $\beta \equiv \gamma \to \delta$. Then

$$M^* \equiv \lambda x . M_1^*$$

if we neglect changes in bound variables.

To show that M^* is SC, we must prove that for all SC terms N^γ, the term M^*N is SC. But

$$\begin{aligned} M^*N &\equiv (\lambda x . M_1^*)N \\ &\rhd_{1\beta} [N/x]M_1^*, \end{aligned}$$

which is SC by the induction hypothesis applied with $n+1$ instead of n. Then M^*N is SC by Lemma 2. □

This completes the proof of Theorem A2.3. With a minor change, we can extend the proof to $\lambda\beta\eta_t$, as follows.

THEOREM A2.4 (Remark 13.14). *In the system of typed λ-terms, there are no infinite* $\beta\eta$*-reductions.*

Proof. The same as Theorem A2.3, except that in Lemma 2, near the end of the proof, we need to allow for the possibility that an infinite reduction of $(\lambda x.M)NM_1 \ldots M_n$ has the form

$$\begin{aligned} (\lambda x.M)NM_1 \ldots M_n &\rhd_{\beta\eta} (\lambda x.M')N'M_1' \ldots M_n' && (M \rhd M', \text{ etc.}) \\ &\equiv (\lambda x.Px)N'M_1' \ldots M_n' && (x \notin FV(P)) \\ &\rhd_{1\eta} PN'M_1' \ldots M_n' \\ &\rhd_{\beta\eta} \ldots . \end{aligned}$$

But in this case we can construct an infinite reduction from $([N/x]M)M_1 \ldots M_n$ as follows:

$$\begin{aligned} ([N/x]M)M_1 \ldots M_n &\rhd_{\beta\eta} ([N'/x]M')M_1' \ldots M_n' \\ &\equiv PN'M_1' \ldots M_n' \\ &\rhd_{\beta\eta} \ldots . \end{aligned}$$

And this contradicts the fact that (1) is SN. □

Now let us take up CLw_t.

THEOREM A2.5 (<u>Theorem</u> 13.22). *In the system of typed* CL-*terms, there are no infinite weak reductions.*

Proof. Modify the proof of Theorem A2.3 as follows. In Lemma 1 (a), we need to specify that a is not an $S_{\alpha,\beta,\gamma}$ or $K_{\alpha,\beta}$. The proof of Lemma 1 is unchanged. Lemma 2 must be replaced by the following.

LEMMA 2'.

(a) *If* $X^{\alpha\to\beta\to\gamma}Z^{\alpha}(Y^{\alpha\to\beta}Z^{\alpha})$ *is* SC, *then so is* $S_{\alpha,\beta,\gamma}XYZ$.

(b) *If* X^{α} *and* Y^{β} *are* SC, *then so is* $K_{\alpha,\beta}XY$.

Proof of Lemma 2'.

(a) The type of $XZ(YZ)$ is γ. Let

$$\gamma \equiv \gamma_1 \to \ldots \to \gamma_n \to \theta,$$

where θ is atomic, and let

$$U_1^{\gamma_1}, \ldots, U_n^{\gamma_n}$$

be SC terms. Since $XZ(YZ)$ is SC, it follows by Note 2 that

$$XZ(YZ)U_1 \ldots U_n$$

is SN, and hence X, Y, Z, $U_1, \ldots, U_n$ are all SN.

We prove that $SXYZU_1 \ldots U_n$ is SN. If there were an infinite reduction from $SXYZU_1 \ldots U_n$, it would have to have the form

$$\begin{aligned} SXYZU_1 \ldots U_n &\rhd_w SX'Y'Z'U_1' \ldots U_n' \qquad (X \rhd X',\ \text{etc.}) \\ &\rhd_{1w} X'Z'(Y'Z')U_1' \ldots U_n' \\ &\rhd_w \ldots . \end{aligned}$$

But this gives rise to an infinite reduction from $XZ(YZ)U_1 \ldots U_n$, contrary to assumption.

(b) Let $\alpha \equiv \alpha_1 \to \ldots \to \alpha_n \to \theta$, where θ is atomic, and let

$$U_1^{\alpha_1}, \ldots, U_n^{\alpha_n}$$

be SC terms; since X is SC, it follows by Note 2 that

$$XU_1 \ldots U_n$$

is SN, and hence $X, U_1, \ldots, U_n$ are SN. Also, Y is SC, so Y is SN by Lemma 1(b).

The proof that $KXYU_1 \ldots U_n$ is SN is like (a). This completes the proof of Lemma 2′. □

To complete the proof of Theorem A2.5, we must prove Lemma 3(a). (Part (b) turns out not to be needed for CL-terms.)

The proof is as before, except that Case 4 is now not needed and Case 2 needs the following two extra subcases.

<u>Subcase</u> 2a. $M^\beta \equiv S_{\gamma,\delta,\varepsilon}$ and $\beta \equiv (\gamma \to \delta \to \varepsilon) \to (\gamma \to \delta) \to (\gamma \to \varepsilon)$. For any SC terms $X^{(\gamma \to \delta \to \varepsilon)}$, $Y^{(\gamma \to \delta)}$, Z^γ, the term $XZ(YZ)$ is SC by Note 4. Hence $S_{\gamma,\delta,\varepsilon}XYZ$ is SC by Lemma 2′. Since X, Y, and Z are arbitrary SC terms of the given types, $S_{\gamma,\delta,\varepsilon}$ is SC by A2.2(b).

<u>Subcase</u> 2b. $M^\beta \equiv K_{\gamma,\delta}$ and $\beta \equiv (\gamma \to \delta \to \gamma)$. For any SC terms X^γ and Y^δ, the term $K_{\gamma,\delta}XY$ is SC by Lemma 2′. Hence $K_{\gamma,\delta}$ is SC by A2.2(b). This completes the proof of Theorem A2.5. □

Let us now turn to $(CL + R)_t$ (Definition 18.4).

THEOREM A2.6 (<u>Implies</u> 18.8). *In the system of typed* CL-*terms of Definitions* 18.1 - 4, *there are no infinite* wR-*reductions.*

Proof. Modify the proof of Theorem A2.3 as follows.

In Lemma 1(a), specify that a is not an $S_{\alpha,\beta,\gamma}$ or $K_{\alpha,\beta}$ or R_α (although a is permitted to be $\bar{0}^N$ or $\bar{\sigma}^{N \to N}$). The proof of the lemma remains unchanged.

Lemma 2 must be replaced by two lemmas; Lemma 2′ above, and the following one.

LEMMA 2″. *Let* X^{α}, $Y^{N\to\alpha\to\alpha}$ *and* Z^{N} *be* SC:

(a) *If* Z *does not reduce to a numeral, then* $R_{\alpha}XYZ$ *is* SC.

(b) *If* $Z \rhd_{wR} \bar{0}$, *then* $R_{\alpha}XYZ$ *is* SC.

(c) *If* $Z \rhd_{wR} \overline{k+1}$ *and* $Y\bar{k}(R_{\alpha}XY\bar{k})$ *is* SC, *then so is* $R_{\alpha}XYZ$.

Proof of Lemma 2″. Let $\alpha \equiv \alpha_1 \to \ldots \to \alpha_n \to \theta$, where θ is atomic. (In fact, $\theta \equiv N$ because N is the only atomic type, but the proof would work for systems in which other atomic types are postulated.) Let

$$U_1^{\alpha_1},\ldots,U_n^{\alpha_n}$$

be any SC terms. Note that X, Y, Z, $U_1,\ldots,U_n$, being SC, are SN by Lemma 1.

(a) Just as in the proof of Lemma 2, it is enough to prove that

$$R_{\alpha}XYZU_1\ldots U_n$$

is SN. But Z does not reduce to a numeral, so the only possible reductions in this term are inside X, Y, Z, $U_1,\ldots,U_n$. The result follows.

(b) It is enough to prove that

$$R_{\alpha}XYZU_1\ldots U_n$$

is SN, and this follows by the argument of Lemma 2.

(c) We must prove that

$$R_{\alpha}XYZU_1\ldots U_n$$

is SN, given that $Z \rhd_{wR} \overline{k+1}$, and that

$$Y\bar{k}(R_{\alpha}XY\bar{k})U_1\ldots U_n$$

is SN. But an infinite reduction from $R_{\alpha}XYZU_1\ldots U_n$ would have to have the form

$$\begin{aligned} RXYZU_1\ldots U_n &\rhd_{wR} RX'Y'\overline{k+1}\,U_1'\ldots U_n' \qquad (X \rhd X',\ \text{etc.})\\ &\rhd_{1wR} Y'\bar{k}(RX'Y'\bar{k})U_1'\ldots U_n'\\ &\rhd_{wR} \ldots . \end{aligned}$$

This would imply that $Y\bar{k}(RXY\bar{k})U_1\ldots U_n$ is not SN, contrary to hypothesis. This completes the proof of Lemma 2″. □

To complete the proof of Theorem A2.6, we must prove Lemma 3(a). The proof is the same as before, except that Case 1 is omitted, subcases 2a and 2b (see earlier) are added to prove that $S_{\alpha,\beta,\gamma}$ and $K_{\alpha,\beta}$ are SC, and the following new subcase is added to prove that each R_α is SC.

Subcase 2c. $M^\beta \equiv R_\alpha$, where $\beta \equiv \alpha \to (N \to \alpha \to \alpha) \to N \to \alpha$. Let X^α, $Y^{N \to \alpha \to \alpha}$ and Z^N be SC. We must prove that $R_\alpha XYZ$ is SC.

If Z does not reduce to a numeral, the result follows from Lemma 2″(a). If Z reduces to a numeral $\bar{k}$, we proceed by induction on k.

Basis ($k = 0$): The result follows by Lemma 2″(b).

Induction step (k to $k+1$): By the induction hypothesis, $R_\alpha XYW$ is SC for all SC terms W^N that reduce to $\bar{k}$. In particular, $R_\alpha XY\bar{k}$ is SC. Hence the result follows by Lemma 2″(c).

This completes the proof of Lemma 3, and hence of Theorem A2.6. □

REMARK A2.7. This proof has been written out for the system in Chapter 18, in which N is the only atomic type and reduction is defined in CL. But it would also apply if there were other atomic types besides N, and it can easily be modified to apply to systems defined in λ-calculus (obtained by adding R and its reduction-axioms to $\lambda\beta_t$ or $\lambda\beta\eta_t$). See Troelstra 1973 §§2.2.27 - 31.

REMARK A2.8 (Combinatory strong reduction, cf. 8.16). This is not covered by the above theorems. No doubt a non-trivial definition of SN could be given, using the axioms in Hindley 1967, but no SN theorem has yet been attempted. In contrast, the normal-form or normalization theorem (Theorem 14.42) was probably the earliest to be published. (Curry et al. 1958 §9F, Corollary 9.2.)

REMARK A2.9 (Normalization theorems). A normalization theorem says that every typed or stratified term has a normal form. Historically, such theorems came before SN theorems: for pure λβ, Turing sketched a proof in 1942, see Gandy 1980a (cf. also Andrews 1971, Proposition 2.7.3); for pure combinatory strong reduction, see Curry et al. 1958 §9F, as above. Weak reduction came later; two different proofs are in Sanchis 1967 (with R included), and Seldin 1977b (without R).

The proofs for $\lambda\beta$, $\lambda\beta\eta$ and CL_w determine a specific order for contracting redexes in a term, which guarantees to reach a normal form. For a comparison of proofs, see Troelstra 1973 §2.2.35.

REMARK A2.10 (Arithmetizability). As noted in Remark 18.9, no proof of the normalization theorem for $(CL+R)_t$ can be translated into a valid proof in first-order arithmetic. Hence the proof of the SN theorem, A2.6 above, is also not arithmetizable. One non-arithmetical part of that proof is the definition of SC, A2.2; see Troelstra 1973 §2.3.11. Another is the statement of the SN theorem itself, because infinite reductions are not a first-order concept.

For systems without R, we can avoid these problems by proving the normalization theorem directly, using the non-R proofs in Remark A2.9, which can all be arithmetized. The SN theorem can then be deduced, if desired, by Gandy 1980b §§1 - 3. But that theorem now says that for each typed term X there is a number n(X) such that all finite reductions starting at X have less than n(X) steps. (This indirect method was not used for Theorems A2.3 - 5, because the Tait method is shorter and more adaptable to stronger systems.)

Systems with R have the property that a term $R_\alpha XYZ$ which is not a redex (because Z is not a numeral) may become a redex when Z is reduced, and this fouls up the definition of the order for contracting redexes given in the non-R normal-form proofs, when we try to apply them directly to R.

The unarithmetizability of Theorem A2.6's proof can be characterized in terms of ordinal numbers. Let ε_0 be the least ordinal ε for which $\omega^\varepsilon = \varepsilon$; ε_0 is countable, so there are ways of coding the ordinals $\leq \varepsilon_0$ as natural numbers. Also there are proofs of the normalization theorem that use induction up to ε_0 and no stronger principles. (Eg. Schütte 1977 Chapter VI §16 has a proof for a system equivalent to the present one.) Hence there is no way of coding the ordinals which makes the principle of induction up to ε_0 provable in first-order arithmetic. On the other hand, Gentzen has shown that induction up to α can be formalized in arithmetic for each ordinal $\alpha < \varepsilon_0$, so induction up to ε_0 is a 'minimal' unarithmetizable principle.

For more on arithmetizability, see Troelstra 1973, §§2.3.11 - 13.

APPENDIX THREE

CARE OF YOUR PET COMBINATOR

Contributed by C.J. Hindley

Combinators make ideal pets.

Housing They should be kept in a suitable axiom-scheme, preferably shaded by Böhm trees. They like plenty of scope for their contractions and a proved extensionality is ideal for this.

Diet To keep them in strong normal form a diet of mixed free variables should be given twice a day. Bound variables are best avoided as they can lead to contradictions. The exotic R combinator needs a few Church numerals added to its diet to keep it healthy and active.

House-training If they are kept well supplied with parentheses, changed daily (from the left), there should be no problems.

Exercise They can safely be let out to contract and reduce if kept on a long corollary attached to a fixed point theorem, but do watch that they don't get themselves into a logical paradox whilst playing around it.

Discipline Combinators are generally well behaved but a few rules of inference should be enforced to keep their formal theories equivalent.

Health For those feeling less than weakly equal a check up at a nearby lemma is usually all that is required. In more serious cases a theorem (Church-Rosser is a good general one) should be called in. Rarely a trivial proof followed by a short remark may be needed to get them back on their feet.

Travel If you need to travel any distance greater than the length of M with your combinators try to get a comfortable Cartesian Closed Category. They will feel secure in this and travel quite happily.

Choosing your combinator Your combinators should be obtained from a reputable combinatory logic monograph. Make sure that you are given the full syntactic identity of each combinator. A final word: *do* consider obtaining a recursive function; despite appearances, they can make charming pets!

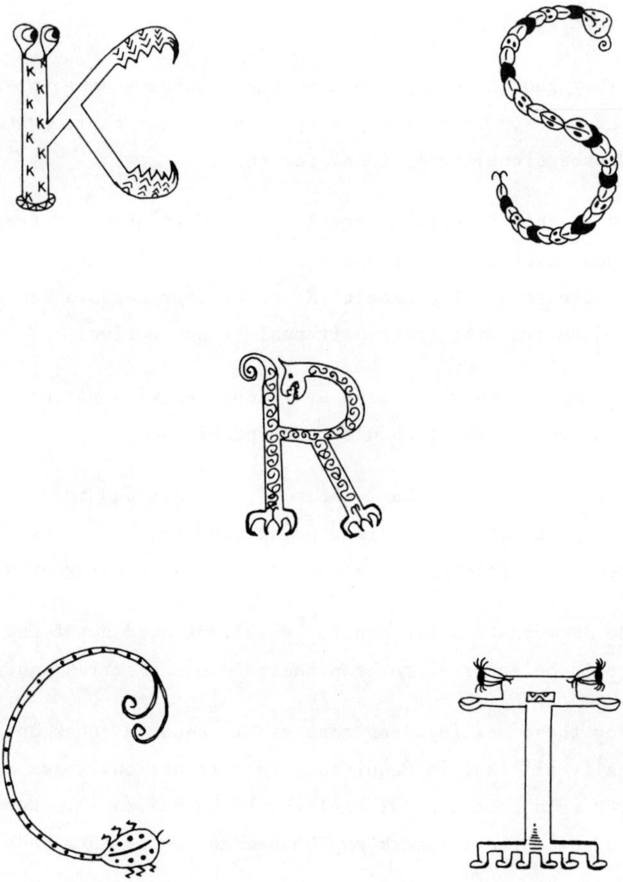

APPENDIX FOUR

ANSWERS TO STARRED EXERCISES

1.4. (a) (i) $D(\lambda x.x^2) = \lambda x.2x$. (ii) $(D(\lambda x.x^2))3 = 6$.

(iii) $(D(\lambda x.f(x^2)))x \neq (Df)(x^2)$.

(b) (i) $(((ux)(yz))(\lambda v.(vy)))$.

(ii) $((((\lambda x.(\lambda y.(\lambda z.((xz)(yz)))))u)v)w)$.

(iii) $(((w(\lambda x.(\lambda y.(\lambda z.((xz)(yz))))))u)v)$.

1.7. (a) No; $ux(yz)$ is really $((ux)(yz))$.

(b) In (ii) but not in (iii); see unabbreviated versions above.

1.13. (a) $\lambda y.(\lambda y.xy)(\lambda x.x)$. (b) $y(\lambda z.(\lambda y.vy)z)$.

1.20. Let $\lambda x.M \equiv \lambda x.(\lambda y.yx)$. Then $\lambda y.[y/x]M \equiv \lambda y.(\lambda z.zy)$ by 1.11(f). And one step is not enough to change this to $\lambda x.M$.

1.31. $(\lambda x.xx)(\lambda x.xx)$.

1.37. (a) Let $K \equiv \lambda xy.x$. Then $KPQ =_\beta P$ for all P, Q. Hence

(i) $P =_\beta KPQ =_\beta (\lambda xy.y)PQ$ by new axiom, $=_\beta Q$.

(ii) $P =_\beta KP(KQ) =_\beta (\lambda xy.yx)(KQ)(KP)$

$=_\beta (\lambda x.x)(KQ)(KP)$ by new axiom

$=_\beta KQ(KP) =_\beta Q$.

2.6. (a) Let $\mathcal{S}$ abbreviate $[U_1/x_1,\ldots,U_n/x_n]$; then define:

$\mathcal{S}x_i \equiv U_i$; $\mathcal{S}a \equiv a$ if a is an atom not in $\{x_1,\ldots,x_n\}$;

$\mathcal{S}(VW) \equiv (\mathcal{S}V\,\mathcal{S}W)$.

Example: $[y/x,x/y](xy) \equiv yx$, $[y/x][x/y](xy) \equiv [y/x](xx) \equiv yy$.
The identity is true if $(\forall i)(U_i$ contains none of $x_1,\ldots,x_{i-1})$.

(b) To define $[U_1/x_1,\ldots,U_n/x_n]Y$, first change all bound variables in Y that occur free in $x_1\ldots x_nU_1\ldots U_n$. The definition then becomes straightforward, as Stoughton 198- shows.

2.11. $I \equiv SKK$, $W \equiv SS(KI)$. There are other answers, of course.

2.26. (a) Here is the usual answer, due to Church. (There are others, e.g. in (4.7).)

$D \equiv \lambda^* xyz.zxy$, $D_i \equiv \lambda^* u.u(\lambda^* x_1 x_2 . x_i)$ $(i = 1,2)$.

(b) If A exists, then $S =_w AK =_w A(KKK) =_w K$, contrary to 2.24.3.

(c) $WWW \rhd_w WWW \rhd_w WWW \rhd_w \ldots$.

3.5. First, by analogy with Exercise 2.26(a), define an ordered k-tuple combinator $D^{(k)}$ and its projections $D_i^{(k)}$ $(1 \le i \le k)$ such that

$$D_i^{(k)}(D^{(k)} x_1 \ldots x_k) \rhd x_i .$$

Then solve, by Corollary 3.3.1, the equation

$$xy_1 \ldots y_n = D^{(k)} Z_1 \ldots Z_k .$$

Let X be the solution. Then define

$$X_i \equiv \lambda y_1 \ldots y_k . D_i^{(k)}(Xy_1 \ldots y_k).$$

3.11. (a) By induction on Definition 3.7, prove that if $Y \in$ strong nf, then $(\exists Z \in \text{strong nf})(Yx \rhd_w Z)$. (In the λ^*-step we must prove that if $X \in$ strong nf, then so is $[x/v]X$, by induction on lgh(X).) Note $Z \in$ weak nf. If $Yx =_w x(Yx)$, then $Z =_w xZ$, contrary to Corollary 2.24.3.

(b) (i) Choose $n = 3$, and $L_1 \equiv \lambda x.x$, $L_2 \equiv \lambda uxy.y$, $L_3 \equiv \lambda uxy.x$.

(ii) Choose $n = 4$, and $L_1 \equiv \lambda x.x$, $L_2 \equiv \lambda u.u(\lambda xy.y)$, $L_3 \equiv \lambda uxy.x$, $L_4 \equiv \lambda xy.y$.

5.9. (a) Hint: if the range is finite, then the set of all terms is the union of a finite number of recursively enumerable sets. Full answer: Barendregt 1981, Theorem 20.2.5.

(b) K occurs in D. For a counterexample, let $\mathcal{A}$ be the set of all closed terms and $\mathcal{B}$ be its complement.

9.18. (b) $S_\lambda K_\lambda =_\beta \lambda yz.z$, $K_\lambda I_\lambda =_\beta \lambda yz.z$. By Corollary 2.24.3, $SK \neq_w KI$.

(c) $S_\lambda(K_\lambda I_\lambda) =_\beta \lambda xy.xy =_\eta \lambda x.x \equiv I_\lambda$. If $S(KI) =_{c\beta} I$, then by above $\lambda xy.xy =_\beta \lambda x.x$, contrary to Corollary 1.35.3.

14.8. (a)

$$\dfrac{\overset{(\to K)}{K \;\varepsilon\; (\alpha\to\alpha)\to\beta\to(\alpha\to\alpha)} \qquad \overset{(\text{by } 14.6)}{I \;\varepsilon\; \alpha\to\alpha}}{KI \;\varepsilon\; \beta\to\alpha\to\alpha.}\,(\to e)$$

(b) Let $\xi \equiv \beta\to\gamma$, $\eta \equiv \alpha\to\beta$, $\zeta \equiv \alpha\to\gamma$; then we have

$$\dfrac{\overset{(\to S)}{S \;\varepsilon\; (\xi\to\eta\to\zeta)\to(\xi\to\eta)\to\xi\to\zeta} \qquad \overset{(\text{by } 14.7)}{B \;\varepsilon\; \xi\to\eta\to\zeta}}{SB \;\varepsilon\; (\xi\to\eta)\to\xi\to\zeta.}\,(\to e)$$

(c) Let $\xi \equiv \alpha\to\alpha\to\beta$, $\eta \equiv \alpha\to\alpha$, $\zeta \equiv \alpha\to\beta$; then we have

$$\dfrac{\dfrac{\overset{(\to S)}{S \;\varepsilon\; (\xi\to\eta\to\zeta)\to(\xi\to\eta)\to\xi\to\zeta} \qquad \overset{(\to S)}{S \;\varepsilon\; \xi\to\eta\to\zeta}}{SS \;\varepsilon\; (\xi\to\eta)\to\xi\to\zeta} \qquad \overset{(\text{by } 14.8(\text{a}))}{KI \;\varepsilon\; \xi\to\alpha\to\alpha}}{SS(KI) \;\varepsilon\; \xi\to\zeta.}$$

16.35. (a)

$$\dfrac{\dfrac{\dfrac{\dfrac{\overset{3}{[x \;\varepsilon\; G\alpha\beta]} \qquad \dfrac{\dfrac{\overset{2}{[y \;\varepsilon\; G\gamma(K\alpha)]} \qquad \overset{1}{[z \;\varepsilon\; \gamma]}}{yz \;\varepsilon\; K\alpha z}\,(G\,e)}{yz \;\varepsilon\; \alpha}\,\text{Eq}''}{x(yz) \;\varepsilon\; \beta(yz)}\,(G\,e)}{\lambda z.x(yz) \;\varepsilon\; G\gamma(\lambda z.\beta(yz))}\,(G\,i-1)}{\lambda yz.x(yz) \;\varepsilon\; G(G\gamma(K\alpha))(\lambda y.G\gamma(\lambda z.\beta(yz)))}\,(G\,i-2)$$

$$\dfrac{\dfrac{\lambda yz.x(yz) \;\varepsilon\; G(G\gamma(K\alpha))(\lambda y.G\gamma(\lambda z.\beta(yz)))}{B \;\varepsilon\; G(G\alpha\beta)(\lambda x.G(G\gamma(K\alpha))(\lambda y.G\gamma(\lambda z.\beta(yz))))}\,(G\,i-3)}{B \;\varepsilon\; G_3(G\alpha\beta)(K(G\gamma(K\alpha)))(K^2\gamma)(K(B\beta)).}\,\text{Eq}''$$

(b)

$$\dfrac{\overset{3}{[x \;\varepsilon\; G_2\alpha(K\beta)(K\gamma)]} \qquad \overset{1}{[z \;\varepsilon\; \alpha]}}{xz \;\varepsilon\; (\lambda u.G(K\beta u)(K\gamma u))z}\,(\mathrm{G\,e})$$

$$\dfrac{xz \;\varepsilon\; (\lambda u.G(K\beta u)(K\gamma u))z}{xz \;\varepsilon\; G\beta\gamma}\;\mathrm{Eq}''$$

$$\dfrac{xz \;\varepsilon\; G\beta\gamma \qquad \overset{2}{[y \;\varepsilon\; \beta]}}{xzy \;\varepsilon\; \gamma y}\,(\mathrm{G\,e})$$

$$\dfrac{xzy \;\varepsilon\; \gamma y}{\lambda z.xzy \;\varepsilon\; G\alpha(\lambda z.\gamma y)}\,(\mathrm{G\,i-1})$$

$$\dfrac{\lambda z.xzy \;\varepsilon\; G\alpha(\lambda z.\gamma y)}{\lambda yz.xzy \;\varepsilon\; G\beta(\lambda y.G\alpha(\lambda z.\gamma y))}\,(\mathrm{G\,i-2})$$

$$\dfrac{\lambda yz.xzy \;\varepsilon\; G\beta(\lambda y.G\alpha(\lambda z.\gamma y))}{C \;\varepsilon\; G(G_2\alpha(K\beta)(K\gamma))(\lambda x.G\beta(\lambda y.G\alpha(\lambda z.\gamma y)))}\,(\mathrm{G\,i-3})$$

$$\dfrac{C \;\varepsilon\; G(G_2\alpha(K\beta)(K\gamma))(\lambda x.G\beta(\lambda y.G\alpha(\lambda z.\gamma y)))}{C \;\varepsilon\; G_3(G_2\alpha(K\beta)(K\gamma))(K\beta)(K^2\alpha)(K(BK\gamma)).}\;\mathrm{Eq}''$$

(c)

$$\dfrac{\overset{2}{[x \;\varepsilon\; G_2\alpha(K\alpha)(K\beta)]} \qquad \overset{1}{[y \;\varepsilon\; \alpha]}}{xy \;\varepsilon\; (\lambda u.G(K\alpha u)(K\beta u))y}\,(\mathrm{G\,e})$$

$$\dfrac{xy \;\varepsilon\; (\lambda u.G(K\alpha u)(K\beta u))y}{xy \;\varepsilon\; G\alpha\beta}\;\mathrm{Eq}''$$

$$\dfrac{xy \;\varepsilon\; G\alpha\beta \qquad \overset{1}{[y \;\varepsilon\; \alpha]}}{xyy \;\varepsilon\; \beta y}\,(\mathrm{G\,e})$$

$$\dfrac{xyy \;\varepsilon\; \beta y}{\lambda y.xyy \;\varepsilon\; G\alpha(\lambda y.\beta y)}\,(\mathrm{G\,i-1})$$

$$\dfrac{\lambda y.xyy \;\varepsilon\; G\alpha(\lambda y.\beta y)}{\lambda y.xyy \;\varepsilon\; G\alpha\beta}\;\mathrm{Eq}''$$

$$\dfrac{\lambda y.xyy \;\varepsilon\; G\alpha\beta}{W \;\varepsilon\; G(G_2\alpha(K\alpha)(K\beta))(\lambda x.G\alpha\beta)}\,(\mathrm{G\,i-2})$$

$$\dfrac{W \;\varepsilon\; G(G_2\alpha(K\alpha)(K\beta))(\lambda x.G\alpha\beta)}{W \;\varepsilon\; G_2(G_2\alpha(K\alpha)(K\beta))(K\alpha)(K\beta).}\;\mathrm{Eq}''$$

(d)

$$\dfrac{\overset{1}{[y \;\varepsilon\; \beta x]}}{\lambda y.y \;\varepsilon\; G(\beta x)(\lambda y.\beta x)}\,(\mathrm{G\,i-1})$$

$$\dfrac{\lambda y.y \;\varepsilon\; G(\beta x)(\lambda y.\beta x)}{\lambda xy.y \;\varepsilon\; G\alpha(\lambda x.G(\beta x)(\lambda y.\beta x))}\,(\mathrm{G\,i-v})$$

$$\dfrac{\lambda xy.y \;\varepsilon\; G\alpha(\lambda x.G(\beta x)(\lambda y.\beta x))}{\bar{0} \;\varepsilon\; G_2\alpha\beta(BK\beta).}\;\mathrm{Eq}''$$

(e)

$$\dfrac{\dfrac{\dfrac{\dfrac{\overset{2}{[x \;\varepsilon\; G\alpha\beta]} \qquad \overset{1}{[y \;\varepsilon\; \alpha]}}{xy \;\varepsilon\; \beta y}\,\text{(G e)}}{\lambda y.xy \;\varepsilon\; G\alpha(\lambda y.\beta y)}\,\text{(G i - 1)}}{\lambda xy.xy \;\varepsilon\; G(G\alpha\beta)(\lambda x.G\alpha(\lambda y.\beta y))}\,\text{(G i - 2)}}{\bar{1} \;\varepsilon\; G(G\alpha\beta)(K(G\alpha\beta)).}\,\text{Eq}''$$

(f) For example, let

$$\dfrac{\dfrac{\dfrac{\dfrac{\dfrac{\overset{2}{[x \;\varepsilon\; G\alpha(K\alpha)]} \qquad \dfrac{\dfrac{\overset{1}{[x \;\varepsilon\; G\alpha(K\alpha)]} \qquad \overset{1}{[y \;\varepsilon\; \alpha]}}{xy \;\varepsilon\; K\alpha y}\,\text{(G e)}}{xy \;\varepsilon\; \alpha}\,\text{Eq}''}{x(xy) \;\varepsilon\; K\alpha(xy)}\,\text{(G e)}}{x^2y \;\varepsilon\; \alpha}\,\text{Eq}''}{\lambda y.x^2y \;\varepsilon\; G\alpha(\lambda y.\alpha)}\,\text{(G i - 1)}}{\lambda xy.x^2y \;\varepsilon\; G(G\alpha(K\alpha))(\lambda x.G\alpha(\lambda y.\alpha))}\,\text{(G i - 2)}}{\bar{2} \;\varepsilon\; G(G\alpha(K\alpha))(K(G\alpha(K\alpha))).}\,\text{Eq}''$$

(g)

$$\dfrac{\dfrac{\dfrac{\dfrac{\dfrac{\overset{2}{[y \;\varepsilon\; G\beta\gamma]} \qquad \dfrac{\dfrac{\dfrac{\dfrac{\overset{3}{[x \;\varepsilon\; G_2(G\beta\gamma)\alpha(K^2\beta)]} \quad \overset{2}{[y \;\varepsilon\; G\beta\gamma]}}{xy \;\varepsilon\; (\lambda u.G(\alpha u)(K^2\beta u))y}\,\text{(G e)}}{xy \;\varepsilon\; G(\alpha y)(K\beta)}\,\text{Eq}'' \qquad \overset{1}{[z \;\varepsilon\; \alpha y]}}{xyz \;\varepsilon\; K\beta z}\,\text{(G e)}}{xyz \;\varepsilon\; \beta}\,\text{Eq}''}{y(xyz) \;\varepsilon\; \gamma(xyz)}\,\text{(G e)}}{\lambda z.y(xyz) \;\varepsilon\; G(\alpha y)(\lambda z.\gamma(xyz))}\,\text{(G i - 1)}}{\lambda yz.y(xyz) \;\varepsilon\; G(G\beta\gamma)(\lambda y.G(\alpha y)(\lambda z.\gamma(xyz)))}\,\text{(G i - 2)}}{\bar{\sigma} \;\varepsilon\; G(G_2(G\beta\gamma)\alpha(K^2\beta))(\lambda x.G(G\beta\gamma)(\lambda y.G(\alpha y)(\lambda z.\gamma(xyz))))}\,\text{(G i - 3)}}{\bar{\sigma} \;\varepsilon\; G_3(G_2(G\beta\gamma)\alpha(K^2\beta))(K(G\beta\gamma))(K\alpha)(B^2\gamma).}\,\text{Eq}''$$

BIBLIOGRAPHY

As far as possible, references have not been made to unpublished manuscripts. The following abbreviations will be used:

JSL = Journal of Symbolic Logic;
NDJ = Notre Dame Journal of Formal Logic;
TCS = Theoretical Computer Science;
ZML = Zeitschrift für Mathematische Logik und Grundlagen der Mathematik;
SLNM = Lecture-notes in Mathematics, Springer-Verlag, Berlin, Germany;
SLNCS = Lecture-notes in Computer Science, Springer-Verlag;
N-H = North-Holland Publishing Co., Amsterdam, Netherlands;
ACM = Association for Computing Machinery; 11, W.42 St., New York, N.Y. 10036, U.S.A.

ACZEL, P. [1980]. Frege structures and the notions of proposition, truth and set. *In* The Kleene Symposium, ed. J. Barwise et al., N-H, pp. 31-59.

ACZEL, P., FEFERMAN, S. [1980]. Consistency of the unrestricted abstraction principle using an intensional equivalence operator. *In* Hindley and Seldin 1980, pp. 67-98.

ANDREWS, P.B. [1965]. A transfinite type theory with type variables, N-H.

ANDREWS, P.B. [1971]. Resolution in type theory. JSL 36, pp. 414-32.

ANDREWS, P.B. [1974a]. Resolution and the consistency of analysis. NDJ 15, pp. 73-84.

ANDREWS, P.B. [1974b]. Provability in elementary type-theory. ZML 20, pp. 411-18.

BACKUS, J. [1978]. Can programming be liberated from the von Neumann style? Comm. ACM 21(8), pp. 613-41.

BARENDREGT, H.P. [1973]. Combinatory logic and the axiom of choice. Indag. Math. 35, pp. 203-21.

BARENDREGT, H.P. [1974]. Pairing without conventional restraints. ZML 20, pp. 289-306.

BARENDREGT, H.P. [1981(I)],[1981(II)]. The Lambda Calculus, its Syntax and Semantics. I: 1st ed., N-H 1981. II: 2nd ed., N-H 1984.

BARENDREGT, H.P., COPPO, M., DEZANI, M. [1983]. A filter lambda model and the completeness of type assignment. JSL 48, pp. 931-40.

BARENDREGT, H.P., KOYMANS, K. [1980]. Comparing some classes of lamba-calculus models. *In* Hindley and Seldin 1980, pp. 287-301.

BARENDREGT, H.P., LONGO, G. [1980]. Equality of λ-terms in the model T^{ω}. *In* Hindley and Seldin 1980, pp. 303-37.

BARNES, J.G.P. [1982]. Programming in Ada. Addison-Wesley.

BEESON, M.J. [1985]. Foundations of constructive mathematics. Springer-Verlag.

BEN-YELLES, C-B. [1979]. Type-assignment in the Lambda-calculus. Ph.D. thesis, University College, Swansea SA2 8PP, Britain.

BEN-YELLES, C-B. [1981]. G-stratification is equivalent to F-stratification. ZML 27, pp. 141-50.

BERRY, G. [1978]. Stable models of typed λ-calculi. SLNCS 62, pp. 72-89.

BERRY, G. [1981]. On the definition of λ-calculus models. SLNCS 107, pp. 218-30.

BÖHM, C., GROSS, W. [1966]. Introduction to the CUCH. *In* Automata Theory, ed. E.R. Caianiello, Academic Press, pp. 35-65.

BÖHM, C. [1968]. Alcune proprietà delle forme βη-normali del λ-K-calcolo. Pubblicazione No. 696, Istituto per le Applicazioni del calcolo, Piazzale delle Scienze 7, Rome.

BÖHM, C. (ed.) [1975]. λ-calculus and computer science theory. SLNCS 37.

BÖHM, C., BERARDUCCI, A. [1985]. Automatic synthesis of typed λ-programs on term algebras. TCS, to appear.

BÖHM, C., DEZANI, M., PERETTI, P., RONCHI, S. [1979]. A discrimination algorithm inside λβ-calculus. TCS 8, pp. 271-91.

BOOK, R.V. [1983]. Thue systems and the Church-Rosser property. *In* Combinatorics on Words: Progress and Perspectives. Ed. L.J. Cummings, Academic Press, pp. 1-38.

BRUCE, K., LONGO, G. [1984]. A note on combinatory algebras and their expansions. TCS 31, pp. 31-40.

BRUCE, K., LONGO, G. [1985]. Provable isomorphisms and domain equations in models of typed languages. *In* 17th ACM Symp. on Theory of Computing, publ. ACM.

BRUCE, K.B., MEYER, A.R. [1984]. The semantics of second order polymorphic lambda calculus. SLNCS 173, pp. 131-44.

de BRUIJN, N.G. [1970]. The mathematical language AUTOMATH, its usage, and some of its extensions. SLNM 125, pp. 29-61.

de BRUIJN, N.G. [1980]. A survey of the project AUTOMATH. *In* Hindley and Seldin 1980, pp. 579-606.

BUNDER, M.W.V. [1969]. Set theory based on combinatory logic. Thesis, Univ. Amsterdam, publ. in 5 parts in NDJ: 11(1970), pp. 467-470; 14(1973), pp. 53-54, 341-346; 15(1974), pp. 25-34, 192-206.

BUNDER, M.W.V. [1982]. Some results in Aczel-Feferman logic and set theory. ZML 28, pp. 269-76.

BUNDER, M.W.V. [1983a]. Predicate calculus of arbitrarily high finite order. Archiv. Math. Logik 23, pp. 1-10.

BUNDER, M.W.V. [1983b]. A one axiom set theory based on higher order predicate calculus. Archiv. Math. Logik 23, pp. 99-107.

BUNDER, M.W.V. [1983c]. Set theory in predicate calculus with equality. Archiv. Math. Logik 23, pp. 109-13.

BUNDER, M.W.V. [1983d]. A weak absolute consistency proof for some systems of illative combinatory logic. JSL 48, pp. 771-76.

BUNDER, M.W.V., MEYER, R.K. [1978]. On the inconsistency of systems similar to $\mathcal{F}_{21}^{*}$. JSL 34, p. 1-2.

BURSTALL, R.M., MACQUEEN, D.B., SANNELLA, D.T. [1980]. Hope: an experimental applicative language. *In* Conference Record of the 1980 LISP CONFERENCE, ACM, pp. 136-43.

BYERLY, R. [1982a]. An invariance notion in recursion theory. JSL 47, pp. 48-66.

BYERLY, R. [1982b]. Recursion theory and the lambda calculus. JSL 47, pp. 67-83.

CARDELLI, L., MACQUEEN, D. (eds.). [1983+]. Polymorphism, the ML/LCF/Hope Newsletter; informally publ. by the above, A.T. & T., Bell Labs., Murray Hill, N.J. 07974, U.S.A.

CHAUVIN, A. [1979]. Theory of objects and set theory: introduction and semantics. NDJ 20, pp. 37-54. (Engl. trans. of part of thesis Théorie des Objets et Théorie des Ensembles, 1974, Univ. de Clermont-Ferrand, France.

CHURCH, A. [1936a]. A note on the entscheidungsproblem. JSL 1, pp. 40-1, 101-2.

CHURCH, A. [1936b]. An unsolvable problem of elementary number theory. American J. Math. 58, pp. 345-63.

CHURCH, A. [1940]. A formulation of the simple theory of types. JSL 5, pp. 56-68.

CHURCH, A. [1941]. The Calculi of Lambda Conversion. Princeton Univ. Press, reprinted 1963 by University Microfilms Inc., Ann Arbor, Michigan, U.S.A.

CONSTABLE, R.L. [1980]. Programs and types. 21st I.E.E.E. Symposium on Foundations of Comput. Sci., pp. 118-28.

COPPO, M. [1983]. On the semantics of polymorphism. Acta Informatica 20, pp. 159-70.

COPPO, M. [1984]. Completeness of type assignment in continuous lambda models. TCS 29, pp. 309-24.

COPPO, M., DEZANI, M. [1978]. A new type assignment for λ-terms. Archiv. Math. Logik 19, pp. 139-56.

COPPO, M., DEZANI, M. [1981]. Functional characters of solvable terms. ZML 27, pp. 45-58.

COPPO, M., DEZANI, M., HONSELL, F., LONGO, G. [1984]. Extended type structures and filter lambda models. *In* Logic. Colloq. 82, ed. G. Lolli et al. N-H, pp. 241-62.

COPPO, M., DEZANI, M., RONCHI, S. [1978]. (Semi)-separability of finite sets of terms in Scott's D_∞-models of the λ-calculus. SLNCS 62, pp. 142-64.

COPPO, M., DEZANI, M., SALLÉ, P. [1979]. Functional characterization of some semantic equalities inside λ-calculus. SLNCS 71, pp. 133-46.

COPPO, M., DEZANI, M., VENNERI, B. [1980]. Principal type-schemes and λ-calculus semantics. *In* Hindley and Seldin 1980, pp. 535-60.

COPPO, M., DEZANI, M., ZACCHI, M. [198-]. Type theories, normal forms and D_∞-lambda models. Information and Control, to appear.

CURIEN, P.-L. [1985]. Typed categorical combinatory logic. SLNCS 185, pp. 157-72.

CURRY, H.B. [1930]. Grundlagen der Kombinatorischen Logik. American J. Math. 52, pp. 509-36, 789-834.

CURRY, H.B. [1934]. Functionality in combinatory logic. Proc. Nat. Acad. Sci. U.S.A. 20, pp. 584-90.

CURRY, H.B. [1936]. First properties of functionality in combinatory logic. Tohoku Math. J. 41, pp. 371-401.

CURRY, H.B. [1942a]. The inconsistency of certain formal logics. JSL 7, pp. 115-17.

CURRY, H.B. [1942b]. Some advances in the combinatory theory of quantification. Proc. Nat. Acad. Sci. U.S.A. 28, pp. 564-69.

CURRY, H.B. [1963]. Foundations of Mathematical Logic. McGraw Hill Co., reprinted by Dover Publications Inc., New York, 1976.

CURRY, H.B. [1969a]. Modified basic functionality in combinatory logic. Dialectica 23, pp. 83-92.

CURRY, H.B. [1969b]. The undecidability of λK-conversion. *In* Foundations of Mathematics: Symposium Papers Commemorating the Sixtieth Birthday of Kurt Gödel, ed. J.J. Bulloff et al., Springer, pp. 10-14.

CURRY, H.B. [1973]. The consistency of a system of combinatory restricted generality. JSL 38, pp. 489-92.

CURRY, H.B., FEYS, R. [1958]. Combinatory Logic, Vol. I. N-H.

CURRY, H.B., HINDLEY, J.R., SELDIN, J.P. [1972]. Combinatory Logic, Vol. II. N-H.

van DAALEN, D.T. [1980]. The Language Theory of AUTOMATH. Thesis, Tech. Hogeschool Eindhoven, Netherlands.

DAMAS, L., MILNER, R. [1982]. Principal type-schemes for functional programs. 9th ACM Symposium on Princs. of Prog. Langs. ACM, pp. 207-12.

DEZANI, M., MARGARIA, I. [198-]. A characterization of F-complete type assignments. MS 1984, submitted to TCS.

DONAHUE, J. [1979]. On the semantics of data type. SIAM J. Comput. 8, pp. 546-60.

DOWNEY, P.J., SETHI, R. [1976]. Correct computation rules for recursive languages. SIAM J. Comput. 5, pp. 378-401.

DUNGAN, D.M. [1979]. Bibliography on data types. SIGPLAN Notices 14(11), November 1979, pp. 31-59.

ENGELER, E. [1981]. Algebras and combinators. Algebra Universalis 13, pp. 389-92.

FEFERMAN, S. [1975a]. A language and axioms for explicit mathematics. SLNM 450, pp. 87-139.

FEFERMAN, S. [1975b]. Non-extensional type-free theories of partial operations and classifications. SLNM 500, pp. 73-118.

FEFERMAN, S. [1977]. Categorical foundations and foundations of category theory. *In* Logic, Foundations of Mathematics and Computability Theory, ed. R.E. Butts et al., D. Reidel Co., pp. 149-69.

FEFERMAN, S. [1978]. Recursion theory and set theory: a marriage of convenience. *In* Generalized Recursion Theory II, ed. J.E. Fenstad et al. N-H, pp. 55-98.

FEFERMAN, S. [1979]. Constructive theories of functions and classes. *In* Logic Colloq.'78, ed. M. Boffa et al. N-H, pp. 159-224.

FEFERMAN, S. [1984]. Towards useful type-free theories, I. JSL 49, pp.75-111.

FITCH, F.B. [1936]. A system of formal logic without an analogue to the Curry W operator. JSL 1, pp. 92-100.

FITCH, F.B. [1963]. The system CΔ of combinatory logic. JSL 28, pp.87-97.

FITCH, F.B. [1967]. A complete and consistent modal set theory. JSL 32, pp. 93-103.

FITCH, F.B. [1974]. Elements of Combinatory Logic. Yale Univ. Press.

FITCH, F.B. [1980a]. A consistent combinatory logic with an inverse to equality. JSL 45, pp. 529-43. (See also Math. Reviews 1982, rev. 82i: 03023).

FITCH, F.B. [1980b]. An extension of a system of combinatory logic. *In* Hindley and Seldin 1980, pp. 125-40.

FORTUNE, S., LEIVANT, D., O'DONNELL, M. [1983]. The expressiveness of simple and second-order type structures. J. ACM 30, pp. 151-85.

FRIEDMAN, H. [1971]. Axiomatic recursive function theory. *In* Logic Colloq.'69, ed. R.O. Gandy et al. N-H, pp. 113-37.

FRIEDMAN, H. [1973]. Equality between functionals. SLNM 453, pp. 22-37.

GANDY, R.O. [1977]. The simple theory of types. *In* Logic Colloq.'76, ed. R.O. Gandy et al. N-H, pp. 173-81.

GANDY, R.O. [1980a]. An early proof of normalization by A.M. Turing. *In* Hindley and Seldin 1980, pp. 453-55.

GANDY, R.O. [1980b]. Proofs of strong normalization. *In* Hindley and Seldin 1980, pp. 457-78.

GIERZ, G., HOFMANN, K., KEIMEL, K., LAWSON, J., MISLOVE, M., SCOTT, D. [1980]. A compendium of Continuous Lattices. Springer.

GIRARD, J-Y. [1971]. Une extension de l'interprétation de Gödel à l'analyse. *In* Proc. 2nd Scandinavian Logic Symposium, ed. J.E. Fenstad. N-H, pp.63-92.

GIRARD, J-Y. [1972]. 'Interprétation fonctionnelle et élimination des coupures de l'arithmétique d'ordre supérieur'. Ph.D. thesis. University of Paris VII, 2, Place Jussieu, Paris.

GÖDEL, K. [1958]. Über eine bisher noch nicht benützte Erweiterung des finiten Standpunktes. Dialectica 12, pp. 280-87. (Engl. trans., J. Philos. Logic 9 (1980), pp. 133-42.)

GOODMAN, N.D. [1970]. A theory of constructions equivalent to arithmetic. *In* Intuitionism and Proof Theory, ed. A. Kino et al. N-H, pp. 101-20.

GOODMAN, N.D. [1972]. A simplification of combinatory logic. JSL 37, pp. 225-46.

GORDON, M., MILNER, R., WADSWORTH, C. [1979]. Edinburgh LCF, a mechanical logic of computation. SLNCS 78.

GRZEGORCZYK, A. [1964]. Recursive objects in all finite types. Fundamenta Math. 54, pp. 73-93.

HENKIN, L. [1950]. Completeness in the theory of types. JSL 15, pp. 81-91.

HENKIN, L. [1963]. A theory of propositional types. Fund. Math. 52, pp. 323-44.

HENNESSY, M., PLOTKIN, G.D. [1979]. Full abstraction for a simple parallel programming language. SLNCS 74, pp. 108-20.

HINDLEY, J.R. [1967]. Axioms for strong reduction in combinatory logic. JSL 32, pp. 224-36.

HINDLEY, J.R. [1969a]. The principal type-scheme of an object in combinatory logic. Trans. American Math. Soc. 146, pp. 29-60.

HINDLEY, J.R. [1969b]. An abstract form of the Church-Rosser theorem, I. JSL 34, pp. 545-60.

HINDLEY, J.R. [1974]. An abstract Church-Rosser theorem, II: applications. JSL 39, pp. 1-21.

HINDLEY, J.R. [1977]. Combinatory reductions and lambda reductions compared. ZML 23, pp. 169-80.

HINDLEY, J.R. [1978a]. Reductions of residuals are finite. Trans. American Math. Soc. 240, pp. 345-61.

HINDLEY, J.R. [1978b]. Standard and normal reductions. Trans. American Math. Soc. 241, pp. 253-71.

HINDLEY, J.R. [1979]. The discrimination theorem holds for combinatory weak reduction. TCS 8, pp. 393-94.

HINDLEY, J.R. [1982]. The simple semantics for Coppo-Dezani-Sallé types. SLNCS 137, pp. 212-26.

HINDLEY, J.R. [1983]. The completeness theorem for typing λ-terms. TCS 22, pp. 1-17 (cf. also TCS 22, pp. 127-33).

HINDLEY, J.R., LERCHER, B. [1970]. A short proof of Curry's normal form theorem. Proc. American Math. Soc. 24, pp. 808-10.

HINDLEY, J.R., LERCHER, B., SELDIN, J.P. [1972]. Introduction to Combinatory Logic. Cambridge University Press (London Math. Soc. Lecture-notes No. 7).

HINDLEY, J.R., LERCHER, B., SELDIN, J.P. [1975]. Introduzione alla Logica Combinatoria. Boringhieri, Turin (Italian ed. of above, revised, corrected).

HINDLEY, J.R., LONGO, G. [1980]. Lambda-calculus models and extensionality. ZML 26, pp. 289-310.

HINDLEY, J.R., SELDIN, J.P. (eds.) [1980]. To H.B. Curry, Essays on Combinatory Logic, Lambda Calculus and Formalism. Academic Press.

HOWARD, W.A. [1980]. The formulae-as-types notion of construction. *In* Hindley and Seldin 1980, pp. 479-90. (MS written in 1969).

HUET, G. [1973]. The undecidability of unification in third order logic. Information and Control 22, pp. 257-67.

HUET, G. [1975]. An unification algorithm for typed λ-calculus. TCS 2, pp. 27-57.

HUET, G. [1980]. Confluent reductions. J. ACM 27, pp. 797-821.

HUET, G. [1981]. A complete proof of correctness of the Knuth-Bendix completeness algorithm. J. Comput. Sys. Sci. 23, pp. 11-21.

HUET, G., OPPEN, D. [1980]. Equations and rewrite rules, a survey. *In* Formal Languages: Perspectives and Open Problems, ed. R.V. Book. Academic Press.

HYLAND, J.M.E. [1976]. A syntactic characterization of the equality in some models for the lambda calculus. J. London Math. Soc. (2) 12, pp. 361-70.

KAHN, G., MACQUEEN, D.B., PLOTKIN, G.D. (eds.) [1984]. Semantics of Data Types. SLNCS 173.

KELLEY, J.L. [1955]. General Topology. Van Nostrand Co.

KLEENE, S.C. [1936]. λ-definability and recursiveness. Duke Math. J. 2, pp. 340-53.

KLEENE, S.C. [1952]. Introduction to Metamathematics. Van Nostrand Co.

KLEENE, S.C. [1981]. Origins of Recursive function theory. Annals of the History of Computing 3, pp. 52-67.

KLEENE, S.C., ROSSER, J.B. [1935]. The inconsistency of certain formal logics. Annals Maths. (2) 36, pp. 630-36.

KLOP, J.W. [1980]. Combinatory Reduction Systems. Thesis, Univ. Utrecht; publ. Math. Centre, Kruislaan 413, Amsterdam, Holland.

KLOP, J.W. [1982]. Extending partial combinatory algebras. Bull. Europ. Ass. Theor. Computer Sci. 16, pp. 30-34.

KNUTH, D., BENDIX, P. [1967]. Simple word problems in universal algebras. *In* Computational Problems in Abstract Algebra, ed. J. Leech. Pergamon Press.

KOYMANS, C.P.J. [1982]. Models of the lambda calculus; Information and Control 52, pp. 306-32.

KOYMANS, C.P.J. [1984]. Models of the Lambda Calculus. Thesis, Math. Instituut, de Uithof, Utrecht, Netherlands.

KREISEL, G. [1959]. Interpretation of analysis by means of constructive functionals of finite type. *In* Constructivity in Mathematics, ed. A. Heyting. N-H, pp. 101-28.

LAMBEK, J. [1958-72]. Deductive systems and categories. Part I: Math. Systems Theory 2 (1958), pp. 287-318. II: SLNM 86 (1969), pp. 76-122. III: SLNM 274 (1972), pp. 57-82.

LAMBEK, J. [1980]. From λ-calculus to cartesian closed categories. *In* Hindley and Seldin 1980, pp. 375-402.

LAMBEK, J., SCOTT, P.J. [198-]. Introduction to Higher Order Categorical Logic. Cambridge University Press, forthcoming.

LANDIN, P.J. [1965]. A correspondence between ALGOL-60 and Church's lambda notation. Comm. ACM 8, pp. 89-101, 158-65.

LANDIN, P.J. [1966]. A lambda-calculus approach. *In* Advances in Programming and Non-numerical Computation, ed. L. Fox. Pergamon Press, pp. 97-141.

LÄUCHLI, H. [1965]. Intuitionistic propositional calculus and definably non-empty terms (Abstract). JSL 30, p. 263.

LÄUCHLI, H. [1970]. An abstract notion of realizability for which intuitionistic predicate calculus is complete. *In* Intuitionism and Proof Theory, ed. A. Kino et al. N-H, pp. 227-34.

LEIVANT, D. [1983]. Polymorphic type inference. 10th ACM Symposium on Principles of Prog. Langs., publ. ACM, pp. 88-98 (cf. also pp. 155-66).

LEIVANT, D. [198-]. Typing and computational properties of lambda expressions. TCS, to appear.

LERCHER, B. [1967a]. Strong reduction and normal form in combinatory logic. JSL 32, pp. 213-23.

LERCHER, B. [1967b]. The decidability of Hindley's axioms for strong reduction. JSL 32, pp. 237-39.

LERCHER, B. [1976]. Lambda-calculus terms that reduce to themselves. NDJ 17, pp. 291-92.

LÉVY, J-J. [1976]. An algebraic interpretation of λβK-calculus, and an application of a labelled λ-calculus. TCS 2, pp. 97-114.

LÉVY, J-J. [1980]. Optimal reductions in the lambda-calculus. *In* Hindley and Seldin 1980, pp. 159-91.

LONGO, G. [1983]. Set-theoretical models of λ-calculi: theories, expansions, isomorphisms. Ann. Pure and Applied Logic (formerly Ann. Math. Logic) 24, pp. 153-88.

LONGO, G., MARTINI, S. [1984]. Computability in higher types and the universal domain Pω. SLNCS 166, pp. 186-97.

LUCKHARDT, H. [1973]. Extensional Gödel functional interpretation. SLNM 306.

MAASS, W. [1974]. Church-Rosser Theorem für λ-Kalküle mit unendlich langen Termen. SLNM 500, pp. 257-63.

McCRACKEN, N.J. [1979]. An investigation of a programming language with a polymorphic type structure. Ph.D. Dissertation, Syracuse University, N.Y. 13210, U.S.A.

MACLANE, S. [1971]. Categories for the Working Mathematician. Springer.

MACQUEEN, D., PLOTKIN, G.D., SETHI, R. [1984]. An ideal model for recursive polymorphic types. 11th ACM Symposium on Principles of Prog. Langs., publ. ACM, pp. 165-74.

MACQUEEN, D.B., SETHI, R. [1982]. A semantic model of types for applicative languages. ACM Symposium on Lisp and Functional Programming, Pittsburgh, publ. ACM, pp. 243-52.

MAEHARA, S. [1969]. A system of simple type theory with type variables. Ann. Japan Assoc. Philos. Sci. 3, pp. 131-37.

MARTIN-LÖF, P. [1972]. Infinite terms and a system of natural deduction. Compositio Mathematica 24, pp. 93-103.

MARTIN-LÖF, P. [1973]. Hauptsatz for intuitionistic simple type theory. *In* Logic, Methodology and Philosophy of Science IV, ed. P. Suppes et al. N-H, pp. 279-90.

MARTIN-LÖF, P. [1975a]. An intuitionistic theory of types: predicative part. *In* Logic Colloq. '73, ed. H.E. Rose et al. N-H 1975, pp. 73-118.

MARTIN-LÖF, P. [1975b]. About models for intuitionistic type theories and the notion of definitional equality. *In* Proc. 3rd Scandinavian Logic Symposium, ed. S. Kanger. N-H, pp. 81-109.

MARTIN-LÖF, P. [1982]. Constructive logic and computer programming. *In* Logic, Methodology and Philosophy of Science VI, ed. L.J. Cohen et al. N-H, pp. 153-75.

MENDELSON, E. [1964]. Introduction to Mathematical Logic. Van Nostrand Co. (Refs. are to 2nd ed., 1979).

MEYER, A. [1982]. What is a model of the lambda calculus?; Information and Control 52, pp. 87-122.

MEYER, A.R. [198-]. Type theory and data types in programming; selected bibliography. Lab. for Computer Sci., M.I.T., Cambridge, Mass. 02139, U.S.A.

MEZGHICHE, M. [1984]. Une nouvelle β-réduction dans la logique combinatoire. TCS 31, pp. 151-64.

MILNER, R. [1977]. Fully abstract models of typed λ-calculi. TCS 4, pp. 1-22.

MILNER, R. [1978]. A theory of type polymorphism in programming. J. Comput. System Sci. 17, pp. 348-75.

MILNER, R., MORRIS, L., NEWEY, M. [1975]. A logic for computable functions with reflexive and polymorphic types. L.C.F. Report No. 1, Comp. Sci. Dept., King's Bldgs., Mayfield Road, Edinburgh, Scotland.

MITCHELL, J. [1984]. Type inference and type containment. SLNCS 173, pp. 257-77.

MITSCHKE, G. [1979]. The standardization theorem for the λ-calculus. ZML 25, pp. 29-31.

MITSCHKE, G. [1980]. Infinite terms and infinite reductions. *In* Hindley and Seldin 1980, pp. 243-58.

NEDERPELT, R.P. [1973]. Strong normalization in a typed lambda calculus with lambda structured types. Thesis, Tech. Hogeschool Eindhoven, Netherlands.

NEDERPELT, R.P. [1980]. An approach to theorem proving on the basis of a typed lambda calculus. SLNCS 87, pp. 182-94.

PLOTKIN, G.D. [1972]. A set-theoretical definition of application. Memo MIP-R-95, School of Artificial Intelligence, Univ. of Edinburgh, Scotland.

PLOTKIN, G.D. [1974]. The λ-calculus is ω-incomplete. JSL 39, pp. 313-17.

PLOTKIN, G.D. [1977]. LCF as a programming language. TCS 5, pp. 223-57.

PLOTKIN, G.D. [1978a]. T^{ω} as a universal domain. J. Comput. System Sci. 17, pp. 209-36.

PLOTKIN, G.D. [1978b]. The Category of Complete Partial Orders: a Tool for Making Meanings; lectures, summer school, Dipartimento di Informatica, Corso Italia 40, Pisa, Italy.

POTTINGER, G. [1978]. Proofs of the normalization and Church-Rosser theorems for the typed λ-calculus. NDJ 19, pp. 445-51.

PRAWITZ, D. [1965]. Natural deduction. Almqvist and Wiksell, Stockholm.

PRAWITZ, D. [1968]. Hauptsatz for higher order logic. JSL 33, pp. 452-57.

PRAWITZ, D. [1971]. Ideas and results in proof theory. *In* Proc. 2nd Scandinavian Logic Symposium, ed. J.E. Fenstad. N-H, pp. 235-307.

REYNOLDS, J.C. [1974]. Towards a theory of type structure. SLNCS 19, pp. 408-25.

REYNOLDS, J.C. [1984]. Polymorphism is not set-theoretic. SLNCS 173, pp. 145-56.

REYNOLDS, J.C. [1985]. Three approaches to type structure. SLNCS 185, pp. 97-138.

REZUS, A. [1982]. A Bibliography of Lambda-Calculi, Combinatory Logics and Related Topics. Publ. Math. Centre, Kruislaan 413, Amsterdam.

[illegible] Archiv. Math. Logik 20, pp. 65-74.

ROBINET, B. [1979]. Types et fonctionnalité. Actes de la Sixieme École de Printemps d'Informatique Théorique, ed. B. Robinet. Publ. L.I.T.P, Univ. Paris VII, 3, Place Jussieu, Paris, pp. 303-25.

RONCHI, S. [1981]. Discriminability of infinite sets of terms in the D_∞-models of the λ-calculus. SLNCS 112, pp. 350-64.

RONCHI, S. [198-]. Characterization theorems for a filter lambda model. Information and Control, to appear.

RONCHI, S., VENNERI, B. [1984]. Principal type schemes for an extended type theory. TCS 28, pp. 151-71.

ROSEN, B.K. [1973]. Tree manipulation systems and Church-Rosser theorems. J. ACM 20, pp. 160-87.

ROSENBLOOM, P. [1950]. The Elements of Mathematical Logic. Dover Publs. Inc., New York.

ROSSER, J.B. [1935]. A mathematical logic without variables; part I, Annals of Maths. (2) 36, pp. 127-50; II, Duke Math. J. 1, pp. 328-55.

ROSSER, J.B. [1984]. Highlights of the history of the lambda calculus. Annals Hist. Computing 6, pp. 337-49.

ROUSSOU, A. [1983]. Two Scott Models of the λ-calculus. M.Sc. thesis, Math. Dept., Concordia Univ., Montreal, Canada.

SALLÉ, P. [1978]. Une extension de la théorie des types en λ-calcul. SLNCS 62, pp. 398-410.

SANCHIS, L.E. [1964]. Types in combinatory logic. NDJ 5, pp. 161-80.

SANCHIS, L.E. [1967]. Functionals defined by recursion. NDJ 8, pp. 161-74.

SANCHIS, L.E. [1979]. Reducibilities in two models for combinatory logic. JSL 44, pp. 221-34.

SANCHIS, L.E. [1980]. Reflexive domains. *In* Hindley and Seldin 1980, pp. 339-61.

SCHÖNFINKEL, M. [1924]. Über die Bausteine der mathematischen Logik. Math. Annalen 92, pp. 305-16. (Engl. trans. in From Frege to Gödel, ed. J. van Heijenoort, Harvard Univ. Press 1967, pp. 355-66).

SCHÜTTE, K. [1960]. Syntactical and semantical properties of simple type theory. JSL 25, pp. 305-26.

SCHÜTTE, K. [1968]. On simple type theory with extensionality. *In* Logic, Methodology and Philos. Sci. III, ed. B. van Rootselaar et al., N-H, pp. 179-84.

SCHÜTTE, K. [1977]. Proof Theory. Springer Verlag.

SCHWICHTENBERG, H. [1975]. Elimination of higher type levels in definitions of primitive recursive functions by means of transfinite recursion. *In* Logic Colloq. '73, ed. H.E. Rose et al. N-H, pp. 279-303.

SCHWICHTENBERG, H. [1976]. Definierbare Funktionen im λ-Kalkül mit Typen. Archiv Math. Logik 17, pp. 113-14.

SCOTT, D.S. [1970a]. Outline of a Mathematical Theory of Computation. Tech. Monograph PRG-2, Programming Res. Group, 8-11 Keble Road, Oxford, England.

SCOTT, D.S. [1970b]. Constructive validity. SLNM 125, pp. 237-75.

SCOTT, D.S. [1972]. Continuous lattices. SLNM 274, pp. 97-136.

SCOTT, D.S. [1973]. Models for various type-free calculi. *In* Logic, Methodology and Philos. Sci. IV, ed. P. Suppes et al. N-H, pp. 157-87.

SCOTT, D.S. [1975a]. Lambda calculus and recursion theory. Proc. 3rd Scandinavian Logic Symposium, ed. S. Kanger. N-H, pp. 154-93.

SCOTT, D.S. [1975b]. Combinators and classes. *In* Böhm 1975, pp. 1-26.

SCOTT, D.S. [1976]. Data types as lattices. S.I.A.M. J. Computing 5, pp. 522-87.

SCOTT, D.S. [1980a]. Lambda calculus: some models, some philosophy. *In* The Kleene Symposium, ed. J. Barwise et al. N-H, pp. 223-65.

SCOTT, D.S. [1980b]. Relating theories of the λ-calculus. *In* Hindley and Seldin 1980, pp. 403-50.

SCOTT, D.S. [1982]. Domains for denotational semantics. SLNCS 140, pp. 577-613.

SCOTT, D.S., STRACHEY, C. [1971]. Towards a Mathematical Semantics for Computer Languages. Tech. Monograph PRG-6, Programming Res. Group, 8-11 Keble Road, Oxford, England.

SEELY, R.A.G. [1984]. Locally cartesian closed categories and type theory. Math. Proc. Cambridge Phil. Soc. 95, pp. 33-48.

SELDIN, J.P. [1968]. Studies in Illative Combinatory Logic. Thesis, Univ. Amsterdam, Inst. voor Grondslagenonderzoek.

SELDIN, J.P. [1973]. Equality in $\mathcal{F}_{21}$. JSL 38, pp. 571-75.

SELDIN, J.P. [1975]. Equality in $\mathcal{F}_{22}$. *In* Logic Colloq. '73, ed. H.E. Rose et al. N-H, pp. 433-44.

SELDIN, J.P. [1976]. Recent advances in Curry's program. Rend. Sem. Mat. Univers. Politecn. Torino 35, pp. 77-88.

SELDIN, J.P. [1977a]. The Q-consistency of $\mathcal{F}_{22}$. NDJ 18, pp. 117-27.

SELDIN, J.P. [1977b]. A sequent calculus for type assignment. JSL 42, pp. 11-28.

SELDIN, J.P. [1978]. A sequent calculus formulation of type assignment with equality rules for the λβ-calculus. JSL 43, pp. 643-49.

SELDIN, J.P. [1979]. Progress report on generalized functionality. Ann. Math. Logic 17, pp. 29-59.

SELDIN, J.P. [1980]. Curry's program. *In* Hindley and Seldin 1980, pp.3-34.

SETHI, R. [1974]. Testing for the Church-Rosser property. J. ACM 21, pp. 671-79.

SHABUNIN, L.V. [1983]. On the interpretation of combinators with weak reduction. JSL 48, pp. 558-63.

SMULLYAN, R.M. [1985]. To mock a mocking-bird. Knopf, New York.

SMYTH, M.B., PLOTKIN, G.D. [1982]. The category-theoretic solution of recursive domain equations. S.I.A.M. J. Computing 11, pp. 761-83.

SPECTOR, C. [1962]. Provably recursive functionals of analysis. *In* Recursive Function Theory. Proc. Symposia Pure Maths. 5, Amer. Math. Soc., pp. 1-28.

STAPLES, J. [1975]. Church-Rosser theorems for replacement systems. SLNM 450, pp. 291-307.

STATMAN, R. [1979]. The typed λ-calculus is not elementary recursive. TCS 9, pp. 73-82.

STATMAN, R. [1980]. On the existence of closed terms in the typed λ-calculus. Part I, *in* Hindley and Seldin 1980, pp. 511-34; Part II, TCS 15 (1981), pp. 329-38.

STATMAN, R. [1982a]. Completeness, invariance and λ-definability. JSL 47, pp. 17-26.

STATMAN, R. [1982b]. λ-definable functionals and $\beta\eta$-conversion. Archiv Math. Logik 22, pp. 1-6.

STOUGHTON, A. [198-]. Substitution revisited. MS 1985, Computer Science Dept., King's Bldg., Mayfield Road, Edinburgh, Scotland.

STOY, J.E. [1977]. Denotational Semantics: the Scott-Strachey Approach to Programming Languages. M.I.T. Press, Cambridge, Mass., U.S.A.

STRONG, H.R. [1968]. Algebraically generalized recursive function theory. IBM J. Research and Development 12, pp. 465-75.

SZABO, M. [1969]. The Collected Papers of Gerhard Gentzen. N-H.

TAIT, W.W. [1965]. Infinitely long terms of transfinite type. *In* Formal Systems and Recursive Functions, ed. J. Crossley et al. N-H, pp. 176-85.

TAIT, W.W. [1967]. Intensional interpretations of functionals of finite type. JSL 32, pp. 198-212.

TAKAHASHI, M. [1967]. A proof of cut-elimination in simple type-theory. J. Math. Soc. Japan 19, pp. 399-410.

TROELSTRA, A.S. (ed.) [1973]. Metamathematical Investigations of Intuitionistic Arithmetic and Analysis. SLNM 344.

TURNER, D.A. [1976]. SASL Language Manual. University of St Andrews, Scotland.

TURNER, D.A. [1979]. A new implementation technique for applicative languages. Software - Practice and Experience 9, pp. 31-49.

VISSER, A. [1980]. Numerations, λ-calculus and arithmetic. *In* Hindley and Seldin 1980, pp. 259-84.

WADSWORTH, C.P. [1976]. The relation between computational and denotational properties for Scott's D_∞-models of the lambda-calculus. S.I.A.M. J. Computing 5, pp. 488-521.

WADSWORTH, C.P. [1978]. Approximate reduction and lambda calculus models. S.I.A.M. J. Computing 7, pp. 337-56.

WAGNER, E. [1969]. Uniformly reflexive structures. Trans. American Math. Soc. 144, pp. 1-41.

WEGNER, P. [1980]. Programming with ADA: An Introduction by Means of Graduated Examples. Prentice-Hall.

INDEX

Roman letters come first, then Greek letters, then other symbols. Names of rules, etc., are not indexed separately. Abbreviations:

CL for 'combinatory logic';
λ for 'λ-calculus'.